T0340328

QUANTITATIVE TRADING
Algorithms, Analytics, Data, Models, Optimization

QUANTITATIVE TRADING
Algorithms, Analytics, Data, Models, Optimization

Xin Guo
University of California, Berkeley, USA

Tze Leung Lai
Stanford University, California, USA

Howard Shek
Tower Research Capital, New York City, New York, USA

Samuel Po-Shing Wong
5Lattice Securities Limited, Hong Kong, China

CRC Press
Taylor & Francis Group
Boca Raton London New York

CRC Press is an imprint of the
Taylor & Francis Group, an **informa** business
A CHAPMAN & HALL BOOK

Figures 1.1, 8.3, 8.4, and 8.6 are reprinted with permission of the CME Group, 2016.

CRC Press
Taylor & Francis Group
6000 Broken Sound Parkway NW, Suite 300
Boca Raton, FL 33487-2742

First issued in paperback 2019

© 2017 by Taylor & Francis Group, LLC
CRC Press is an imprint of Taylor & Francis Group, an Informa business

No claim to original U.S. Government works

ISBN-13: 978-1-4987-0648-3 (hbk)
ISBN-13: 978-0-367-87181-9 (pbk)

Library of Congress Cataloging-in-Publication Data

Names: Guo, Xin, 1969- author.
Title: Quantitative trading : algorithms, analytics, data, models,
optimization / Xin Guo, Tze Leung Lai, Howard Shek & Samuel Po-Shing Wong.
Description: Boca Raton, FL : CRC Press, [2017] | Includes bibliographical
references and index.
Identifiers: LCCN 2016026683 | ISBN 9781498706483 (hardback)
Subjects: LCSH: Investments--Mathematical models. | Speculation--Mathematical
models. | Investments--Data processing. | Electronic trading of securities.
Classification: LCC HG4515.2 .G87 2017 | DDC 332.64/50151--dc23
LC record available at https://lccn.loc.gov/2016026683

Visit the Taylor & Francis Web site at
http://www.taylorandfrancis.com

and the CRC Press Web site at
http://www.crcpress.com

To our loved ones

Xin: To my husband Ravi Kumar for your support and understanding when I was working intensively with my coauthors, and to my sons Raman and Kannan for bringing me joy, and in memory of my parents.

Tze: To my wife Letitia for your contributions to this book through your extensive practical experience in banking, investment, and analytics, and your innate love of data and models.

Howard: To my parents, and in memory of my grandparents, for your love and unwavering support over the years, and to my goddaughter Isabelle for bringing me endless joy every time I see you.

Sam: To my mother S.C. Lee, my wife Fanny, and my son Theodore, for your love and support.

Contents

Preface

After the tumultuous period marked by the 2007-2008 Financial Crisis and the Great Recession of 2009, the financial industry has entered a new era. Quantitative strategies, together with statistical models and methods, knowledge representation and data analytics, and algorithms and informatics for their development and implementation, are of increasing importance in this new era. The onset of this era is marked by two "revolutions" that have transformed modern life and business. One is technological, dubbed "the FinTech revolution" for financial services by the May 9, 2015, issue of *The Economist* which says: "In the years since the crash of 2007-08, policymakers have concentrated on making finance safer. . . . Away from the regulator spotlight, another revolution is under way. . . . From payments to wealth management, from peer-to-peer lending to crowdfunding, a new generation of startups is taking aim at the heart of the industry – and a pot of revenues that Goldman Sachs estimates is worth \$4.7 trillion. Like other disrupters from Silicon Valley, fintech firms are growing fast." The other is called "big data revolution". In August 2014, the UN Secretary General commissioned an Independent Advisory Group to make recommendations on "bringing about a data revolution" in sustainable development. The October 2012 issue of *Harvard Business Review* features an article on "Big Data: The Management Revolution". On August 20, 2015, the Premier of the People's Republic of China asked different government departments to share their data and implement a big data action plan. Soon afterward, on September 5, 2015, the country's State Council issued an action plan to develop and promote big data applications in economic planning, finance, homeland security, transportation, agriculture, environment, and health care.

To respond to the opportunities and challenges of this new era and the big data and FinTech revolutions that have fascinated their students, the two academics (Guo and Lai) on the author team, who happen to be teaching students in the greater Silicon Valley, developed and taught new courses in the Financial Engineering/Mathematics Curriculum at Berkeley and Stanford in the past three years and exchanged their course material. They also invited practitioners from industry, in particular the other two co-authors (Shek and Wong), to give guest lectures and seminars for these courses. This informal collaboration quickly blossomed into an intense concerted effort to write up the material into the present book that can be used not only to teach these courses more effectively but also to give short courses and training programs elsewhere, as we have done at Shanghai Advanced Institute of Finance, Fudan University Tsinghua University, Chinese University of Hong Kong, Hong Kong University of Science & Technology, National University of Singapore, National Taiwan University, and Seoul National University. A prerequisite or co-requisite of these courses is a course at the level of STATS 240 (Statistical Methods in Finance) at Stanford, which covers the first six chapters of Lai and Xing (2008). We will therefore make ample references to the relevant sections of these six chapters, summarizing their main results without repeating the details.

The website for this book can be found at `http://lait.web.stanford.edu/quantstratbook/`. The datasets for the exercises and examples can be downloaded from the website. We want to highlight in the book an interdisciplinary approach to the development and implementation of algorithmic trading and quantitative strategies. The interdisciplinary approach, which involves computer science and engineering, finance and economics, mathematics and statistics, law and regulation, is reflected not only in the research activities of the recently established Financial and Risk Modeling Institute (FARM) at Stanford, but also in the course offerings of Berkeley's Financial Engineering and Stanford's Financial Mathematics that has currently been transformed to the broader Mathematical and Computational Finance program to reflect the greater emphasis on data science, statistical modeling, advanced programming and high performance computing. Besides the interdisciplinary approach, another distinctive feature of the book is the effort to bridge the gap between academic research/education and the financial industry, which is also one of the missions of FARM. Different parts of the book can be used in short thematic courses for practitioners, which are currently being developed at FARM.

Acknowledgments

We want to express our gratitude to Cindy Kirby for her excellent editing and timely help in preparing the final manuscript. The first two authors thank their current and former Ph.D. students: Joon Seok Lee and Renyuan Xu at Berkeley, and Pengfei Gao, Yuming Kuang, Ka Wai Tsang, Milan Shen, Nan Bai, Vibhav Bukkapatanam, Abhay Subramanian, Zhen Wei, Zehao Chen, Viktor Spivakovsky, and Tiong-Wee Lim at Stanford for their research and teaching assistance, as well as students of IEOR 222 from 2011 to 2016 and IEOR 230X in Spring 2015 at UC Berkeley and Keith Sollers from UC Davis. They also acknowledge grant support by the National Science Foundation, under DMS 1008795 at Berkeley and DMS 1407828 at Stanford, for research projects related to the book. In addition, the first author would like to thank her collaborators Adrien de Larrard, Isaac Mao, Zhao Ruan and Lingjiong Zhu in research on algorithmic trading, funding support from the endowment of the Coleman Fung Chair Professorship, and the NASDAQ OMX education group for generous data and financial support. She also wants to thank her colleague Prof. Terry Hendershott who co-taught with her a high-frequency finance course at the Haas Business School. The last author wants to thank Prof. Myron Scholes for his valuable help and advice and Ted Givens for the excellent book cover design, while the second author wants to thank his colleague Prof. Joseph Grundfest of Stanford Law School for insightful discussions on regulatory issues in high-frequency trading.

Department of Industrial Engineering and Operations Research,
University of California at Berkeley   *Xin Guo*

Department of Statistics, Stanford University   *Tze Leung Lai*

Tower Research Capital, LLC   *Howard Shek*

5Lattice Securities Limited   *Samuel Po-Shing Wong*

List of Figures

List of Tables

1

Introduction

This chapter gives an overview and some historical background of quantitative trading, the title of this book. It begins with a historical review of the evolution of the trading infrastructure, from verbal communication and hand signaling in an exchange to electronic platforms, in Section 1.1. It then gives in Section 1.2 an introduction to quantitative trading strategies and in particular, the time-scales associated with different classes of strategies. In this connection, we give in Section 1.3 a brief historical account of the paradigm shift from the "efficient market hypothesis" (EMH) to arbitrage opportunities via quantitative trading. In Section 1.4, we describe "quant funds" that use these quantitative trading strategies, and also the closely related mutual funds and hedge funds. An overview of the algorithms, analytics, data, models, and optimization methods — the subtitle of this book — used in quantitative trading is given in Section 1.5. Section 1.6 discusses the interdisciplinary background of the book and the anticipated diversity of its target audience. It also provides suggestions on how the book can be used by different groups of readers. Supplements and problems are given in Section 1.7.

1.1 Evolution of trading infrastructure

From the perspective of economics, a stock exchange is basically a double auction system. Bids and offers are made in a stock exchange giving all participants a chance to compete for the order with the best price, analogous to an auction that results in "efficient price discovery". Before the advent of electronic trading platforms, trading floors (or "trading pits") were the venues where buyers and sellers of stocks and bonds (or futures and options) gathered at an exchange to trade. *Open outcry* was a method of communication, involving shouting and use of hand signals to transfer information about buy and sell orders. The hand signals used to communicate information in an open outcry environment consist of palm facing out and hands away from the body to gesture wishes to sell, and palms facing in and hands holding up to gesture wishes to buy; see Figure 1.1. Open outcry differs from a typical auction system that has a clear price-time priority because it allows multiple participants to respond to a bid or offer and they may each get a piece of the resulting trade.

FIGURE 1.1: Hand signals for trading in an open outcry system. (Used with permission of CME.)

Major advances in information technology in the 1980s led to electronic trading platforms that made buy and sell transactions easier to complete, monitor, clear and settle than the open outcry system. The NASDAQ, founded by the National Association of Securities Dealers (NASD) in 1971, is the world's first electronic stock market although it was initially only a computer bulletin board (with "automatic quotations" or AQ) and did not actually connect buyers and sellers. Until 1987, most trading of NASDAQ stocks proceeded via the telephone. On the other hand, Instinet was the first electronic communication network (ECN) that started electronic trading among institutional clients in 1969. The platform grew rapidly in the mid-1980s and became the dominant ECN by the time the US Securities and Exchange Commission (SEC) introduced the Order Handling Rules and ATS (alternative trading systems) regulation in the late 1990s. The London Stock Exchange moved to electronic trading in 1986. In 1992, the Chicago Mercantile Exchange (CME) launched its electronic trading platform Globex, which was conceived in 1987, allowing access to a variety of financial markets including treasuries, foreign exchanges and commodities. The Chicago Board of Trade (CBOT) developed a rival system based on an electronic platform branded "E Open Outcry" allowing electronic trading to take place alongside that in the CBOT pits. The Toronto Stock Exchange adopted electronic trading in 1997. The New York Mercantile Exchange (NYMEX) switched to electronic trading by using the Globex platform in 2006, after losing market share in oil futures trading to the electronic-based Intercontinental Exchange (ICE). The New York Stock Exchange (NYSE) moved to electronic trading in 2006.

Before the wide adoption of electronic exchange trading platforms, transacted prices, rather than bid and offer prices, were the most accurate data because of their recording for settlement. The real-time transactions data were delivered from the trading pits to the "public" by stock tickers: machines which printed abbreviated company names as alphabetic symbols followed by numeric stock transaction price and volume information. As noted by Powers (1966) about the "old world of trading floors", traders shouted out bids or offers and acceptances of a bid or offer. They recorded details of these trades on special trading cards, and this information would then be keypunched into the exchange's clearing system. "If any one of these trade descriptors didn't match, the system rejected them as an out trade, a trade that did not match and did not stand until the two parties resolved the differences." Out trades were considered a cost of trading, and "some exchanges would impose penalties on clearing firms if the percent of their trades that resulted in out trades exceeded a certain level."

FIGURE 1.2: Stock ticker manufactured by Western Union Telegraph Company in the 1870s and now an exhibit at the Computer History Museum in Mountain View, California. Originally, only transacted prices and abbreviated stock symbols were printed on the ticker tape; after the 1930s, traded volume was also printed. (Photo credit: Wikimedia Commons/Don DeBold.)

Electronic trading (sometimes called "etrading") platforms are typically proprietary software running on COTS (commercial off the shelf) operating systems, often using common communications protocols for the Internet and similar networks such as TCP/IP. Exchanges typically develop their own systems and offer two methods to access their systems: (a) an exchanged-provided graphic user interface (GUI) allowing traders to connect and run on their own desktops, and (b) an application programming interface (API) allowing dealers to connect their in-house systems directly to the exchange. Most exchanges provide gateways that sit on a company's network to connect to the exchange's system. Investment firms on both the buy side and the sell side have considerably more complex technology requirements since they have to deal with multiple exchanges and counterparties, and have been increasing their spending on technology for electronic trading while removing floor traders and executing traders in stocks, bonds, foreign exchanges and financial derivatives. They have increasingly relied on algorithms to analyze markets and then execute their orders electronically.

One important advantage of electronic trading is the reduced cost of transactions. Transaction costs are an important consideration in trading strategies although they are often assumed to be negligible in classical financial theories of investment and option pricing; see Lai and Xing (2008, p. 183, 300). Incorporating transaction costs introduces much greater complexity into these theories. Some recent developments in this area will be described in Chapter 3. Through "straight-through processing" (STP) of trades in electronic platforms, transaction costs can be brought down to the extent that increased trading volumes do not result in significantly larger transaction costs, thereby increasing the returns for the trades. Other improvements provided by electronic platforms are increased liquidity due to more buyers and sellers in a financial market and increased transparency of the market because the information of the bid and ask prices of securities flows around the world electronically. Since one can trade on a computer at the click of a button and does not need to go through a broker or pass orders to a trader on the exchange floor, electronic trading has removed some major barriers to global investing. Another important impact of electronic platforms is tighter spreads for market makers. The "spread" on a trade is the difference between the best buy and sell prices being quoted and represents the profit made by the market makers.

The move to electronic trading continues to increase not only in major exchanges around the world but also in business-to-business (B2B) and business-to-consumers (B2C) trading. In B2C trading, retail and institutional clients buy and sell from brokers (dealers) who act as intermediaries between the clients and the B2B markets. Institutional clients include hedge funds, fund managers and insurance companies trading large amounts of securities, while retail clients are typically individuals buying and selling relatively small amounts. Many retail brokers (e.g., Charles Schwab, E-Trade) went online in the late 1990s and the majority of retail trading in the US has moved from over the phone to electronic transactions. Large institutional clients usually place electronic orders via proprietary electronic trading platforms such as Bloomberg Terminal, Reuters 300Xtra, Thomson TradeWeb, or their broker's proprietary software. In B2B trading, large investment banks and brokers trade directly with one another, transacting large amounts of securities. Connecting counterparties through electronic trading is supported by the FIX (Financial Information eXchange) Protocol, which is the industry standard for pre-trade messaging and trade execution. The protocol was initially developed for trading stocks, and subsequently enhanced to trade commodities, currencies, derivatives, and fixed income securities.

1.2   Quantitative strategies and time-scales

Quantitative trading strategies are investment strategies based on quantitative analysis of financial markets and prediction of future performance. The strategy and associated predictions depend on the time-scale of the investment, as exemplified by the following classes of quantitative strategies.

- Fundamentally motivated quant (FMQ): The popular strategy of buying a stock if it is undervalued with respect to fundamentals is an FMQ. Since its associated fundamentals are based on quarterly earning reports and forecasts, the FMQ has a quarterly time-scale. Some common themes in the analysis of fundamentals are earning quality, value of the firm, and investor sentiment.

- Systematic macro: Macro strategies use macroeconomic analysis of market events and trends to identify opportunities for investment. They can be divided into discretionary and systematic strategies, with the discretionary ones carried out by investment managers who identify and select the investments. Systematic macro strategies are model-based and executed by software with limited human involvement. An example is to buy US dollars if the macroeconomic analysis suggests a rising trend for the dollar. Since the Federal Reserve Bank has to hold collaterals mainly in US Treasury debts and since the shortest maturity of Treasury bills is 28 days (about a month), the above systematic macro has a monthly time scale.

- Convergence or relative value trades and other statistical arbitrage (StatArb) strategies: Convergence trades refer to trading in similar assets that are expected to converge in value. For example, the most recently issued (called "on the run") 30-year US Treasury bond generally trades at a liquidity premium relative to the $29\frac{1}{2}$-year bond because it is more liquid. After a few months when a new 30-year bond is issued, the old bonds are now both off the run, resulting in decrease of the liquidity premium. Hence by selling the 30-year bond short and buying the $29\frac{1}{2}$-year bond, one can potentially benefit from a change from the liquidity premium by waiting a few months. Relative value strategies likewise take simultaneously a short position in an overvalued asset and a long position in an undervalued asset, with the expectation that their spread will decrease over time from the current spread. The assets include stocks, bonds and derivatives, and relative value strategies aim at consistent performance in various market conditions of these assets. These strategies are examples of StatArb strategies which have time scales ranging from minutes to months. An overview of statistical arbitrage is given in the next section.

- High-frequency trading (HFT): For reasons that will be explained in Section 1.5, the time scale of HFT is in milliseconds and the holding period of the traded securities is usually less than one second.

1.3   Statistical arbitrage and debates about EMH

The efficient market hypothesis was introduced by Eugene Fama in the 1960s as an outgrowth of his Ph.D. dissertation at the University of Chicago. Its theoretical background was the Brownian motion model (or random walk in discrete time) introduced by Louise Bachelier in his 1900 Ph.D. thesis *The Theory of Speculation* at the University of Paris (Sorbonne). Empirical studies by Cowles (1933, 1944) suggested that professional investors were usually unable to outperform the market, lending support for the random walk model. EMH became a prominent theory in financial economics in the 1970s after the publication of Fama's dissertation, Samuelson's (1965) version of the EMH, and Fama's (1970) review of the theory and evidence of EMH. Besides utility-maximizing agents, EMH assumes the agents to have rational expectations and to update their expectations whenever new information appears. It assumes that on average the agents are right even though individual agents can be wrong about the market so that the net effect of their reactions is random and cannot be reliably exploited to make an abnormal profit, especially in the presence of transaction costs. Empirical evidence has been mixed and the interpretation of the empirical results in testing EMH has led to inconclusive debates.

Behavioral finance offers an alternative theory to EMH. It points out departures from efficiency in financial markets due to cognitive biases such as overconfidence and over-reaction, information bias and other human errors in reasoning and processing information; see Shiller (2015) who shared the 2013 Nobel Prize in Economics with Fama. On the other hand, defenders of EMH argue that the biases are in individuals and not in the efficient market; see Malkiel (2003, Chap. 11). An attempt to reconcile the economic theories, based on EMH and behavioral finance, respectively, is the *adaptive market hypothesis* introduced by Lo (2004) who regards it as a modification of EMH derived from evolutionary principles. Earlier Lo and MacKinlay (1999) argued against random walk models assumed by EMH. Under the evolutionary principles in an adaptive market, prices reflect as much information as dictated by the combination of economic conditions and the number and nature of different groups of market participants: pension fund managers, hedge fund managers, market makers, retail investors, etc. If multiple members of a single group are competing for scarce resources in a single market, then the market is likely to exhibit high efficiency (e.g., the Treasury notes market). On the other hand, if a small number of members are competing for abundant resources, the market is likely to be inefficient. The degree of market efficiency can change with time and is related to "environmental" factors characterizing "market ecology", such as the number of competitors in the market, profit opportunities, and how the market participants adapt to changing environments. Under the adaptive market hypothesis, much of what behavioral finance cites as counterexamples to economic rationality is consistent with the evolutionary principles of financial interactions: competition, adaptation and natural selection; see Lo (2004).

EMH postulates that prices in an efficient market fully reflect all available information, hence the impossibility of profiting in such markets by trading on the basis of an information set. The taxonomy of information has led to three forms of market efficiency: (i) the weak form, in which the information set includes only the history of prices or returns, (ii) the semi-strong form, in which the information set includes all publicly available information known to all market participants, and (iii) the strong form, in which the information set includes all private information known to all market participants. Contrary to EMH, statistical arbitrage believes that there are statistical arbitrage opportunities (SAOs) such as probable mispricings of risky assets in financial markets. As opposed to deterministic arbitrage that yields a sure profit (which does not exist in efficient markets) by taking long and short positions in different securities, StatArb attempts to take advantage of SAOs to generate trading strategies with positive expected payoffs; a StatArb strategy can have negative payoffs with positive probability. Bondarenko (2003) has shown that ruling out SAOs imposes a martingale-type restriction on the dynamics of security prices.

StatArb is often used to refer to a particular class of hedge funds that use trading strategies to take advantage of SAOs. These funds typically consider a large number of securities and use advanced statistical modeling and data analytic methods together with high-performance computing and information technology. Portfolio construction often consists of two phases. In the first phase, each security is assigned a score that reflects its desirability, with high scores suggesting candidates for picking the long positions and low scores suggesting those for shorting. In the second phase, the stocks are combined into a portfolio to reduce market and factor risk. Whereas the scoring method is highly proprietary, commercially available risk models are available for the risk reduction phase, including those by MSCI/Barra and Northfield Information Systems.

1.4 Quantitative funds, mutual funds, hedge funds, and fund performance

A *quantitative* (or quant) *fund* is an investment fund that uses models and computing machinery instead of human judgement to make investment decisions. The investment process of the fund has three key components: (a) an input system that provides all necessary inputs including rules and data, (b) a forecasting engine that generates estimates of model parameters to forecast prices and returns and evaluate risk, and (c) a portfolio construction engine that uses optimization and heuristics to generate recommended portfolios.

In recent years, quantitative investment decisions have become a popular approach adopted by newly launched mutual funds. A *mutual fund* is a professionally managed collective investment fund that pools money from many investors to purchase securities. The fund is usually regulated, registered with the SEC in the US, overseen by a board of directors (or trustees if organized as an investment trust instead of a corporation or partnership), managed by a registered investment adviser, and sold to the general public so that investors can buy or sell shares of the fund at any time. Mutual funds play an important role in household finances, most notably in retirement planning. They are generally classified by their principal investments, with money market funds, bonds or fixed income funds, stock or equity funds, and hybrid funds being the main classes of funds. Investors in a mutual fund pay the fund's expenses, which include management fees, transaction and other fees and thereby reduce the investors' returns. A mutual fund can be open-ended, in which case it must be willing to buy shares from investors every business day, or closedended. Closely related to mutual funds are *exchange-traded funds* (ETFs). An ETF holds assets and trades close to the net asset values over the course of a trading day. Large brokers-dealers that have entered into agreements with the ETF can buy or sell shares of ETF directly from and to the ETF. Individuals can use a retail broker to trade ETF shares on the secondary market.

A hedge fund is a collective investment fund, often structured as a privately owned limited partnership that invests private capital speculatively. It is often open-ended and allows additions or withdrawals by its investors. Its value increases (or decreases) with the fund's return from investment strategies (or expenses), and is reflected in the amount an investor can withdraw from the fund. Hedge fund managers often invest their own money in the fund they manage, thereby aligning their interests with those of the fund's investors. A hedge fund is not sold to the general public or retail investors, and typically pays its investment managers an annual management fee, which is a percentage of the fund's net asset value, and a performance fee if the fund's net asset value increases during the year. Hedge funds use a wide range of trading strategies that can be broadly classified into global macro, directional, event-driven, and relative value arbitrage. Strategies can be "discretionary/qualitative", meaning that they are solicited by managers, or chosen by a computerized system, in which case they are called "quantitative" or "systematic". Global macro strategies use macroeconomic trends and events to identify investment opportunities in the global market. Directional strategies use market movements or inconsistencies to pick investments across a variety of markets, including technology, healthcare, pharmaceuticals, energy, basic materials, and emerging markets. Event-driven strategies identify investment opportunities in corporate transactional events such as consolidation, mergers and acquisitions, recapitalizations, bankruptcies, liquidation, and lawsuits. Relative value arbitrage strategies take advantage of relative discrepancies in prices between securities.

Several performance indices have been introduced to evaluate the performance of a fund or a portfolio manager. Assuming that the market has a risk-free asset with interest rate r_f, the capital asset pricing model (CAPM) has led to the *Sharpe ratio* SR $= (\mu - r_f)/\sigma$ as a performance measure of a portfolio whose log return r has mean μ and variance σ^2; see Lai and Xing (2008, Sect. 3.3). The *Treynor ratio* $(\mu - r_f)/\beta$ is another performance measure inspired by CAPM, where $\beta = \text{Cov}(r, r_M)/\sigma_M^2$, in which M denotes the market portfolio and σ_M^2 is its variance. Instead of the interest rate r_f in SR of the risk-free asset in CAPM, a more commonly used performance index to evaluate fund performance is the *information ratio* IR that replaces r_f by the log return r_B of a benchmark portfolio B and σ by the standard deviation of $r - r_B$; specifically, IR $= (\mu - \mu_B)/\sigma_{r-r_B}$. Some hedge funds use IR to calculate the performance fee. The Jensen index is related to the generalization of CAPM to the regression model $r - r_f = \alpha + \beta(\mu_M - r_f) + \epsilon$, in which ϵ represents random noise, and α (which is 0 under CAPM) is the *Jensen index*. An investment with positive α is considered to perform better than the market. The *Sortino ratio* is another modification of the Sharpe ratio and is defined by $(\mu - \tau)/\sigma_{\tau-}$, where τ is the target or required minimum rate of return and $\sigma_{\tau-}^2 = E\{(r - \tau)^2 | r \leq \tau\}$ is the downside semi-variance. Another refinement of the Sharpe ratio is the *risk-adjusted performance* (RAP) measure to rank portfolios, proposed by Franco Modigliani, who won the Nobel prize in economics in 1985, and his granddaughter Leah. RAP, also called M2 because of the last names of its proponents, is defined as $r_f + (\sigma_M/\sigma)(\mu - r_f)$; see Modigliani and Modigliani (1997).

The preceding performance measures involve only the first two moments of the returns of the portfolio and of the benchmark or market portfolio. If these returns are normally distributed, the first two moments are sufficient statistics. For non-Gaussian returns, other risk-adjusted performance measures have been developed, and a brief review is given in Section 2.9. The time horizon in which these performance measures are calculated depends on the schedule for reports to a fund's investors or for comparing the performance of different funds by companies such as Morningstar which has 1-month, year-to-date, 1-year, 3-year, and 5-year time scales (http://www.morningstar.com). The returns and their standard deviations are annualized for ease of comparison across funds. In the Morningstar Rating^{TM} for funds, funds are rated for up to three periods, and the ratings are based on an enhanced Morningstar Risk-Adjusted Return Score, which also accounts for all sales charges, or redemption fees. Funds are ranked by their scores within their categories, with the lowest 10% receive a single star, the next 22.5% two stars, the next 35% three stars, the next 22.5% four stars, and the top 10% five stars.

1.5   Data, analytics, models, optimization, algorithms

Among the five alphabetically listed topics in the subtitle of the book, "data" is arguably the most important to quantitative trading, irrespective of whether one adopts the arbitrage seeking or efficient market viewpoint. Analytics, the second topic in the book's subtitle, goes far beyond analysis of the data. There are many ways to analyze a set of data, and which way to choose often depends on how the analysis would be used for making good/optimal decisions based on the data. The decisions in the present context are trading strategies, and therefore analytics refers to both analysis of the data and development of data-driven strategies. Models provide the connection between the data and the chosen decisions and are one of the basic ingredients of statistical decision theory. Data, analytics, and models feature prominently in various quantitative strategies for different trading applications in Chapters 2–7, which also study basic optimization problems associated with the data analytics. Optimization is an important theme in the subtitle and is closely connected to data-driven trading decisions.

The front cover of the book shows the five topics in the order of this section's title, portraying their flow in quantitative trading. The data cloud is first processed by an analytics machine. Then the information extracted is used by the models and optimization procedures for the development of quantitative trading strategies. Algorithms are step-by-step procedures for computing the solutions of not only optimization but also other mathematical and data analysis problems. *Algorithmic trading* (also called *algo trading* or *automated trading*), however, refers to using algorithms to enter trade orders on electronic platforms. These algorithms execute trading strategies by using programs whose variables may include timing, price, and quantity of order. They are the output of the linear flow in the schematic diagram of quantitative trading on the book's front cover. Algo trading is widely used by mutual funds, hedge funds and other buy-side (investor-driven) institutional traders. Sell-side financial institutions such as market makers, investment banks and some hedge funds that provide liquidity to the market also mostly generate and execute orders electronically. A special class of algo trading is *high-frequency trading* (HFT), which is the topic of Chapters 6 and 7 that also describe some of the algorithms and strategies. Electronic trading platforms described in Section 1.1 paved the way for the growth of algorithmic trading. In 2006, a third of all EU and US stock trades were driven by algorithms, according to the Aite Group (`http://www.aitegroup.com`). The proportion of algo trading rose to about 50% of the US equity trading volume in 2012, according to a December 20, 2012, article on high-frequency trading in the New York Times. Algo trading has also permeated into foreign exchange and options markets. It has switched jobs once done by human traders to computers. The speed of computer communications, measured in milliseconds and even microseconds, has become very important. In HFT, a trader on the buy side needs to have the trading system, often called an "order management system", to understand a rapidly proliferating flow of new algorithmic order types. Traders on the sell side need to express algo orders in a form that is compatible with the buy-side traders' order management systems; see Shetty and Jayaswal (2006) for details. Chapter 8 describes these order management systems and other details concerning the informatics and computational support for HFT.

1.6 Interdisciplinary nature of the subject and how the book can be used

As the preceding sections have shown, quantitative trading is a multifaceted, interdisciplinary subject that traverses academia and industry, and our book tries to capture this character. The authors, coming from both academia and industry, combine their backgrounds in the different disciplines involved and their diverse professional experiences to put together what they believe to be the core of this subject for the target audience, consisting not only of students taking related courses or training programs such as those mentioned in the Preface, but also of traders, quantitative strategists, and regulators. Because of the diversity of backgrounds and interests of the target audience, we have organized the book in such a way that different groups of readers can focus on different self-contained parts of the book.

We first consider the target audience of students. As mentioned in the Preface, we have given different courses and training programs for master's-level students and advanced undergraduates at different universities. A one-semester course giving an introduction to high-frequency/algorithmic trading has been taught in the first semester at Berkeley out of Chapter 1, Sections 4.1, 4.2, 5.1, 5.2, 7.1, 7.2, 8.1 and 8.2 which form the first part of the course, the second part of which covers Section 5.5 and Chapter 6. This course typically begins with a review of some basic material from Appendices A, B and D. For a two-semester course, the second semester following the above course could begin with machine learning techniques covered in Supplement 9 of Section 2.9, Supplement 7 of Section 3.4 and Supplement 6 of Section 7.8, and would also cover Sections 5.3, 5.4, 7.3, 7.4, 7.5, 7.7, 8.3 and 8.5. A higher-level one-semester course has been given at Stanford to students who satisfy the pre-requisite of having taken an introductory course at the level of Lai and Xing (2008, Chapters 1–6). The course is divided into three parts, the first of which covers portfolio management, the second electronic trading and high-frequency econometrics, and the third algorithmic trading and market making. It therefore covers all chapters of the book, albeit focusing only on certain sections for the "big picture" and letting students choose topics for further reading and their term projects, depending on their home departments. For example, students from computer science or electrical engineering typically choose Chapters 7 and 8, while students from economics and statistics tend to choose Chapters 3, 4 and 5. Students are allowed to form interdisciplinary teams to do team projects so that they can learn from each other and share the workload. Another course on asset management and financial econometrics based on Chapters 1–4 is under development. A summer program on data analytics and quantitative trading has also been given, with parts of Chapters 2, 4, and 5 as the reading material and team projects. For all these courses, the book attempts to strike the delicate balance between methodological development and practical insights, to integrate the main ideas and contributions from different disciplines seamlessly, and to help students understand cutting-edge methods and how they work in practice and relate to data, particularly in the Supplements and Problems section of each chapter.

We next consider junior traders and quantitative strategists who have graduate degrees in some quantitative discipline listed in the Preface and who want to learn more about cutting-edge methodologies and modern trading practices. The book can become part of their arsenal of reference materials to solidify understanding of the theoretical foundation of some of the modeling techniques and to inspire fresh thinking on data analytics, optimal execution and limit order book modeling, and on exploring informatics related to exchange message protocol and matching engine. While individual quants might be experts on a specific area, the broader scope of the book presented with methodological rigor will help consolidate their understanding of the whole end-to-end trading process. For execution traders, our self-contained treatment will help them gain a deeper understanding of the models behind each trading decision, and of the effect of their tweaking the parameters of these models. Furthermore, a basic grasp of the strengths and weaknesses of the latest statistical models will help traders identify microstructure patterns and market behaviors that can be modeled. Moreover, the book discusses many regulatory issues, such as pre-trade and post-trade risk management, circuit breakers, and so on, that are pivotal to all quantitative traders. Since the implementation of these technological and regulatory infrastructures may change quite substantially even in the near future, we recommend that readers focus on the key ideas rather than the details. Thus many topics, such as the mechanism of matching engines in the exchanges, are only introduced in a broad-brush, conceptual manner.

1.7 Supplements and problems

1. **Economic theory of stock trading infrastructure: double auction markets** Unlike the conventional auction market with one auctioneer and multiple buyers, a double auction market consists of multiple sellers and multiple buyers, and has many items for which sellers offer for sale and buyers make bids. The transacted price of an item is the highest price a buyer is willing to pay and the lowest price a seller is willing to accept. A fundamental problem in the economic theory of double auction markets is to find a *competitive equilibrium*: an equilibrium condition in which the quantity supplied is equal to the quantity demanded in a competitive market where freely determined prices would give rise to an *equilibrium price* in this equilibrium condition. The competitive equilibrium is also called *Walrasian equilibrium*, named after Léon Walras who introduced this concept in the context of a single auctioneer and multiple agents calculating their demands for (or supplies of) a good at every possible price and submitting their bids to the auctioneer. The auctioneer presumably would match supply and demand in a market of perfect competition, with complete information and no transaction costs. Walras proposed to achieve the equilibrium and find the clearing (equilibrium) price by using an iterative process called *tâtonnement* (French for "trial and error"). Prices are announced by the auctioneer and agents state how much of the good they would like to buy (demand) or to offer (supply). Above-equilibrium prices when there is excess supply will be lowered and excess demand will cause prices to be raised in this iterative process. Many variations and mechanisms have been developed to implement this basic Walrasian idea and apply it in different settings; see, e.g., Vickrey (1961), Demange et al. (1986), Roth and Sotomayor (1990), Richter and Wong (1999), Krishna (2009), Milgrom (2004), Varian (2007), Edelman et al. (2007). Extensions of the Walrasian equilibrium theory to double auction markets has been an ongoing research area in economics since the 1980s. Because of the complexity of the problem, it was natural to start with a single buyer and a single seller, for which the double auction market reduces to a bilateral trade scenario. In this scenario, if the seller sets the price S (at least to cover the cost of production) and the buyer prices the good at B, then there is a transaction if $B \geq S$ with equilibrium price being any number $p \in [S, B]$. In practice, the buyer and seller only know their own valu-

ations. Suppose a prior uniform distribution over an interval is assumed for each counterparty. Myerson and Satterthwaite (1983) have shown how the Bayesian game of the buyer and seller setting their bid and ask prices can have a Nash equilibrium in which both can maximize their respective expected utilities by reporting their true valuations.

For general double auction markets with multiple buyers and sellers, there is an additional complication due to "heterogeneously informed" market participants. The seminal works of Glosten and Milgrom (1985) and Kyle (1985) show how the economic concept of adverse selection due to information asymmetry works in double auction markets that have market makers. Under the assumption of sufficient adverse selection in the market, Back and Baruch (2013) have established the existence and the optimizing characteristic of the Nash equilibrium in a limit order market. Another direction of research in double auction markets with multiple buyers and sellers is market/mechanism design. McAfee (1992) has considered a double auction mechanism that provides dominant strategies[1] for buyers and sellers. Using data on order submissions and execution and cancellation histories, Hollifield et al. (2006) have developed a method to estimate the current gains from the trade, the gains from trade in a perfectly liquid market, and the gains from trade with a monopoly liquidity supplier, and have applied this method to the data, providing empirical evidence that the current limit-order market is a good market design.

2. **Mechanism design** The references Vickrey (1961), Myerson and Satterthwaite (1983) and McAfee (1992) cited in Supplement 1 belong to the important subfield of economics called "mechanism design", which uses an engineering approach to designing economic mechanisms and incentives. The design problem is often cast in a game-theoretic setting in which the players (agents) act rationally and the mechanism $y(\cdot)$ mapping the "type space" of the agents into a space of outcomes is determined by a principal. An agent's type profile θ refers to the actual information (or preference) the agent has on the quality of a good to sell (or buy). The agent reports to the principal, possibly dishonestly, a type profile $\hat{\theta}$, and receives outcome $y(\hat{\theta})$ after the mechanism is executed. As a benchmark for mechanism design, the principal would choose a *social choice function* f to optimize a certain payoff function assuming full information θ. Instead of using $f(\theta)$ for the allocation of goods received or rendered by the agent with true type θ, a mechanism maps $\hat{\theta}$ to outcome $y(\hat{\theta})$. If the principal still hopes to implement the social choice function, then an "incentive compatibility" constraint is needed in the game so that agents find it optimal to report the type profile truthfully, i.e., $\hat{\theta} = \theta$. Alternatively, the principal may aim at maximizing another payoff function, taking into consideration that agents may report profiles different from the actual ones. In particular,

---

[1]In game theory, strategic dominance refers to one strategy having a better outcome than the other for a player, irrespective of how the player's opponents may play.

Myerson and Satterthwaite's 1983 paper that Supplement 1 has referred to is in fact about designing mechanisms that are "Bayesian incentive compatible[2] and individually rational", and assumes that there is a broker (principal) to whom the buyer and seller report their prices. The 2007 Nobel Prize in Economic Sciences was awarded to Leonid Hurwicz, Eric Maskin and Roger Myerson "for having laid the foundations of mechanism design theory". Vickrey (1961) was also recognized as a foundational paper that won him a share of the 1996 Nobel Prize in Economic Sciences.

3. **Adverse selection** arises in markets where there is information asymmetry between buyers and sellers. A classic example in economics is Akerlof's used car market, in which sellers of used cars know the quality of their cars (either "lemons" for defective cars or good ones that have been carefully maintained), but buyers do not have this information before their purchase and therefore are only willing to pay a price for the average quality of cars in the market.[3]

Exercise Read Glosten and Milgrom (1985) and explain how adverse selection leads to positive bid-ask spreads in an exchange platform with informed traders ("insiders") who know the value of a firm before its earnings announcement, uninformed traders using a distribution that represents the consensus guess of the value given all public information, and a specialist (market maker) whose expected profit from providing liquidity to the market is 0 and who can update the information every time a trade occurs.

4. *Exercise on McAfee mechanism* Read McAfee (1992) and explain why the McAfee mechanism "satisfies the $1/n$ convergence to efficiency of the buyer's bid" and "always produces full information first best prices; the inefficiency arises because the least valuable profitable trade may be prohibited by the mechanism", as mentioned in his abstract. The McAfee mechanism, presented in the second section of the paper, can be described as follows:

   - Order the buyers' bids $b_1 \geq b_2 \geq \cdots$ and the sellers' prices $s_1 \leq s_2 \leq \cdots$.
   - Find the breakeven index k such that $b_k \leq s_k$ and $b_{k+1} > s_{k+1}$.
   - Let $p = (b_{k+1} + s_{k+1})/2$.
   - If $s_k \leq p \leq b_k$, then the k buyers with bids at or above b_k and the k sellers with prices at or below s_k trade at price p.

---

[2] A mechanism is said to be Bayesian incentive compatible if honest reporting yields a Nash equilibrium.

[3] Another well-known example of adverse selection is the economics of health insurance pioneered by Joseph Stiglitz, who shared the 2001 Nobel Prize in Economic Sciences with George Akerlof and Michael Spence for their analyses of markets with asymmetric information.

- If $p \notin [s_k, b_k]$, then the $k - 1$ buyers with bids at or above b_{k-1} pay at price b_k, and the $k - 1$ sellers with asks at or below s_{k-1} receive price s_k from the market maker.

The mechanism requires the presence of a market maker who keeps the excess revenue $(k - 1)(b_k - s_k)$ generated by the mechanism in case $p \notin [s_k, b_k]$.

5. Mechanism design and double auction theory are not only basic research areas in game theory and microeconomics (see, e.g., Fudenberg et al., 2007) but have also attracted much interest in computer science. For example, Nisan and Ronen (2007) have shown that essentially all known heuristic approximations for combinatorial auctions yield non-truthful VCG-based[4] mechanisms. They have also developed a modification of the VCG-based mechanisms to circumvent this problem. Another example is the recent work of Dütting et al. (2014) that views double auctions as being composed of ranking algorithms for each side of the market and a composition rule. The paper applies this framework to a wide range of feasibility-constrained environments, and translates approximation guarantees for greedy algorithms into welfare guarantees for the double auction.

6. **Efficient capital markets and martingales**   This is the title of a paper by Le Roy (1989) who elucidated the important role of martingales, saying that the independent increments assumption in the traditional random walk model is "too restrictive to be generated" by a competitive equilibrium involving optimizing agents and that the weaker martingale model already "captures the flavor of the random walk arguments", as first noted by Samuelson (1965).[5] Observing that "much recent research has found evidence that equity returns can be predicted with some reliability", Lehmann (1990a) says that this is still compatible with an efficient market in which the time-varying returns have predictable mean reversions, as in the case of a martingale price process. He then uses this martingale structure under EMH to derive a statistical test of the hypothesis, and carries out an empirical study that shows inefficiency in the market for liquidity in common stocks (specifically NYSE), mainly around large price changes that can be attributed to investors' overreaction and cognitive biases.

---

[4]VCG stands for Vickrey, Clarke and Groves who have proposed mechanism designs for auctions in 1961, 1971 and 1973, respectively. VCG mechanisms involve computationally infeasible optimal outcomes, and VCG-based mechanisms refer to using suboptimal approximations to implement the VCG mechanisms.

[5]Le Roy credits "Samuelson's to be the most important paper because of its role in effecting this shift from the random walk to the martingale model."

2

Statistical Models and Methods for Quantitative Trading

We begin this chapter with data and their time-scales. Section 2.1 describes some stylized facts about stock price data and emphasizes the distinction between low-frequency time-scales (daily, weekly or monthly) and high frequency (tick-by-tick) transactions. Empirical studies of the EMH discussed in Sections 1.3 and 1.7 have used primarily low-frequency data. Although one of the stylized facts of high-frequency data (negative lag-1 autocorrelation in Section 2.1.2) has been used as evidence against EMH, it was subsequently refuted by incorporating market constraints (called *microstructure*) into the time series analysis of transaction prices. Whereas models of high-frequency transactions data will be deferred to Chapters 4 and 5, Sections 2.2 and 2.6 give an overview of statistical models of low-frequency data that have been developed over time to be consistent with EMH. These models and the associated future returns have been used to develop investment strategies under EMH. We review in Section 2.3 Modern Portfolio Theory, which has been described by Malkiel (2003) as a "walking shoe" to generate good returns (that are optimal in some sense) under the random walk model (one of the simplest statistical models consistent with EMH) "down Wall Street". Because the model contains unknown parameters, its implementation has spawned a large literature that is summarized in Section 2.4, in which we also describe associated statistical methods used for data analysis. A new approach that incorporates unknown parameters (reflecting model uncertainty) into the risk component of the portfolio is given in Section 2.5, which also illustrates the difference between "data analysis" using statistical methods and application-driven "data analytics" that may require new methods for the problem at hand. Besides an overview of the statistical models that have been developed to address the discrepancies between the random walk model for log prices and the stylized facts on low-frequency returns, Section 2.7 extends the approach in Section 2.5 to time series models that are consistent with EMH and also match the stylized facts. In Section 2.8 we review data-driven trading strategies that assume instead of EMH the existence of statistical arbitrage opportunities. We describe some statistical methods underlying these strategies. In addition, we also summarize strategies that are based on economic considerations and behavioral finance. Supplements and problems are given in Section 2.9.

2.1 Stylized facts on stock price data

Let P_t denote the price of a stock (or a more general asset) at time t. Suppose the asset does not have dividend over the period from time $t-1$ to time t. Then the *one-period return* on the asset is $R_t = (P_t - P_{t-1})/P_{t-1}$. The *logarithmic* or *continuously compounded return* is $r_t = \log(P_t/P_{t-1}) = p_t - p_{t-1}$, where $p_t = \log P_t$. Hence, the k-period log return $r_t(k) = \log(P_t/P_{t-k})$ is the sum of k single-period returns: $r_t(k) = r_t + \cdots + r_{t-k+1}$; see pages 64–65 of Lai and Xing (2008).

2.1.1 Time series of low-frequency returns

Section 6.1 of Lai and Xing (2008) summarizes various stylized facts from the empirical literature on the time series of daily and weekly log returns of stock data. These include (a) non-normality because of excess kurtosis, (b) small autocorrelations of the r_t series, (c) strong autocorrelations of the r_t^2 series, (d) asymmetry of magnitudes in upward and downward movements of stock returns, and (e) marked changes of volatility in response to exogenous (e.g., macroeconomic) variables and external events (such as scheduled earnings announcements). Since the r_t^2 series can be regarded as measuring the underlying time-varying volatilities (pages 139 and 145–146 of Lai and Xing, 2008), the stylized fact (c) is often related to clustering of large changes in returns, or "volatility clustering." The asymmetry in (d) refers to the "leverage effect" that the volatility response to a large positive return is considerably smaller than that of a negative return of the same magnitude.

2.1.2 Discrete price changes in high-frequency data

Transaction prices are quoted in discrete units or ticks. On the NYSE, the tick size was \$0.125 before June 24, 1997, and \$0.0625 afterward until January 29, 2001, when all NYSE and AMEX stocks started to trade in decimals. For futures contracts on the S&P 500 index traded on the CME, the tick size is 0.05 index points. Because of the discrete transaction prices, the observed price changes can only take a few distinct values (typically including 0). This also leads to "price clustering", which is the tendency for transaction prices to cluster around certain values.

Even though the tick-by-tick transaction price data were made available by stock tickers in the late 19th century and became popular in the early 20th century, most published empirical studies only analyzed weekly or monthly data until the 1950s when Osborne (1959) studied daily data published by *The Wall Street Journal*. Niederhoffer and Osborne (1966) pioneered the analysis of ticker tape data that were provided by Francis Emory Fitch Inc. The data set they used was a collection of tick-by-tick transactions for six stocks in DJIA

(Dow Jones Industrial Averages) on 22 trading days of October 1964, consisting of 10,536 observations. Besides the discreteness of transaction prices noted in the preceding paragraph, they also found intraday periodicity and seasonality of trading intensity, which subsequent authors (Andersen and Bollerslev, 1997; Hasbrouck, 1999) further elaborated. On stock exchanges, the seasonality is called U-shaped because transactions tend to be heaviest near the beginning and close of trading hours and lightest around the lunch hour. Consequently, the durations between transactions display a daily periodic pattern. The Eurodollar market exhibits a similar U-shaped daily pattern but with the first half of the trading day (when the European markets are still open) more active than the second half (when the European markets have already closed). In addition, Niederhoffer and Osborne (1966) found negative lag 1 autocorrelation in price changes from one transaction to the next.

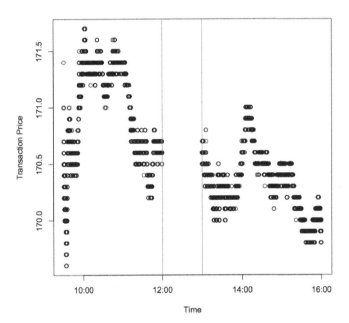

FIGURE 2.1: Time series plot of the tick-by-tick transaction prices P_t of Stock Code 388 on Oct 10, 2014.

Example 2.1 This example illustrates the above stylized facts by using the tick-by-tick transaction prices of The Stock Exchange of Hong Kong Limited collected on October 10, 2014. To avoid confusion of SEHK with its share price, we refer to its shares by its stock code 388. Figure 2.1 is the time series plot of the transaction prices. The vertical lines of Figure 2.1 indicate the lunch break between 12:00 to 13:00 Hong Kong Time. We do not connect the data points by lines to highlight the discrete price levels. There are 10,455

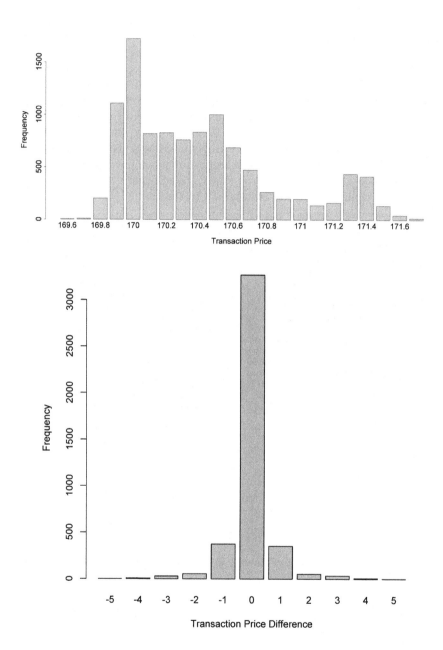

FIGURE 2.2: Distribution of tick-by-tick transaction prices (top panel) and of price differences in the morning session (bottom panel).

transactions that day and the histogram of the transaction prices is shown in Figure 2.2 (top panel). The highest and the lowest transaction prices are HK\$ 169.6 and 171.7, respectively. Since the tick size is 0.1, this is equivalent to a price range of 21 ticks. The mean and median of the transaction prices are 170.4 and 170.3, respectively, and the mode is 170.0, reflecting investors' preference of integers.

Let P_i denote the price of the ith transaction and $D_i = P_i - P_{i-1}$ be the corresponding price change. The histogram of price changes D_i (in the unit of tick size) in the bottom panel of Figure 2.2 shows a preponderance (3,268 out of 4,189 transactions in the morning session) of zeros. Figure 2.3, which plots the autocorrelation function of D_i in the morning session, shows significantly negative lag 1 autocorrelation and that the autocorrelations at other lags are insignificant or small albeit reaching significance.

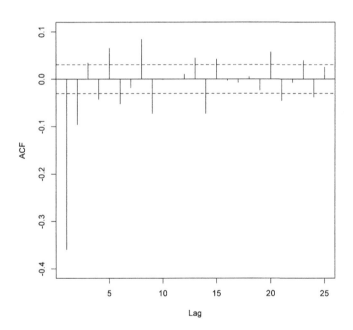

FIGURE 2.3: ACF of transaction price differences of 388 in the morning session of Oct 10, 2014. Dashed lines represent rejection boundaries of 5%-level tests of zero ACF at the indicated lag.

2.2   Brownian motion models for speculative prices

According to Jovanovic and Le Gall (2001) and Le Gall (2007), the earliest scientific analyses of stock price data were performed by Regnault (1863) and Bachelier (1900). Both of them employed quote data collected from the Paris Exchange to study the mean deviation of price changes. They found that such deviation is proportional to the square root of the time interval, i.e., $E[|P_{t+\Delta} - P_t|] \propto \sqrt{\Delta}$ where P_t is the price at time t. While Regnault (1863) gave a primarily empirical study that suggested such square-root law, Bachelier's 1900 Ph.D. thesis derived it from a heuristic version of the functional central limit theorem leading to Brownian motion; see Appendix A of Lai and Xing (2008) for an introduction to Brownian motion and a rigorous version of the functional central limit theorem for more general martingales. By an information fairness consideration for an equilibrium market, Bachelier (1900, 1938) showed that the price changes D_i could be modeled as independent symmetric random variables, with equal probability that the transaction price goes up, or down, s ticks. Restricting the continuous-time Brownian motion B_t to evenly spaced times yields a random walk with normal increments $N(0, \sigma^2)$. Samuelson (1973) reviewed empirical studies by Cowles (1933, 1944) and Kendall (1953) supporting this random walk model for "speculative prices" used in the title of Bachelier's thesis, saying that "as measured by the absence of significant serial correlation, 18 English common-stock-price series were found to look like random walks." Noting, however, that Bachelier had in fact applied the Brownian motion model to rational pricing of warrants or calls in the Paris Exchange, Samuelson critiqued that "seminal as the Bachelier model is, it leads to ridiculous results," and argued for a geometric Brownian motion for the stock price P_t and therefore an i.i.d. model of log returns for low-frequency data; see pages 66 and 328 of Lai and Xing (2008).

2.3   Modern Portfolio Theory (MPT) as a "walking shoe" down Wall Street under EMH

Still within the confines of EMH, Part 3 of Malkiel (2003) begins with a new walking shoe, which is the "new investment technology" of modern portfolio theory, introduced by Markowitz, that "tells investors how to combine stocks in their portfolios to give them the least risk possible, consistent with the return they seek." It then continues with the Capital Pricing Model (CAPM) which is a refinement of MPT due to Sharpe, Lintner and Black.

Chapter 3 of Lai and Xing (2008) gives details about Markowitz's portfolio optimization theory and CAPM. For a portfolio consisting of m assets (e.g., stocks) with expected returns μ_i, let w_i be the weight of the portfolio's value invested in asset i such that $\sum_{i=1}^{m} w_i = 1$, and let $\mathbf{w} = (w_1, \ldots, w_m)^\top$, $\boldsymbol{\mu} = (\mu_1, \ldots, \mu_m)^\top$, $\mathbf{1} = (1, \ldots, 1)^\top$. The portfolio return has mean $\mathbf{w}^\top \boldsymbol{\mu}$ and variance $\mathbf{w}^\top \boldsymbol{\Sigma} \mathbf{w}$, where $\boldsymbol{\Sigma}$ is the covariance matrix of the asset returns. Given a target value μ_* for the mean return of a portfolio, Markowitz characterizes an efficient portfolio by its weight vector $\mathbf{w}_{\text{eff}}$ that solves the optimization problem

$$\mathbf{w}_{\text{eff}} = \arg \min_{\mathbf{w}} \mathbf{w}^\top \boldsymbol{\Sigma} \mathbf{w} \qquad \text{subject to } \mathbf{w}^\top \boldsymbol{\mu} = \mu_*, \ \mathbf{w}^\top \mathbf{1} = 1, \ \mathbf{w} \geq \mathbf{0}. \quad (2.1)$$

When short selling is allowed, the constraint $\mathbf{w} \geq \mathbf{0}$ (i.e., $w_i \geq 0$ for all i) in (2.1) can be removed, yielding the following problem that has an explicit solution:

$$\mathbf{w}_{\text{eff}} = \arg \min_{\substack{\mathbf{w}:\mathbf{w}^\top \boldsymbol{\mu} = \mu_*, \\ \mathbf{w}^\top \mathbf{1} = 1}} \mathbf{w}^\top \boldsymbol{\Sigma} \mathbf{w}$$

$$= \left\{ B\boldsymbol{\Sigma}^{-1}\mathbf{1} - A\boldsymbol{\Sigma}^{-1}\boldsymbol{\mu} + \mu_* \left(C\boldsymbol{\Sigma}^{-1}\boldsymbol{\mu} - A\boldsymbol{\Sigma}^{-1}\mathbf{1} \right) \right\} / D,$$

where $A = \boldsymbol{\mu}^\top \boldsymbol{\Sigma}^{-1} \mathbf{1} = \mathbf{1}^\top \boldsymbol{\Sigma}^{-1} \boldsymbol{\mu}$, $B = \boldsymbol{\mu}^\top \boldsymbol{\Sigma}^{-1} \boldsymbol{\mu}$, $C = \mathbf{1}^\top \boldsymbol{\Sigma}^{-1} \mathbf{1}$, and $D = BC - A^2$. For $m \geq 3$, the set of points (σ, μ) that correspond to the volatility σ and mean μ of the returns of the portfolios of the m assets is called a feasible region, and is convex to the left in the sense that the line segment joining any two points in the region does not cross the left boundary of the region. The upper left boundary of the region is called the *efficient frontier*.

Extending Markowitz's mean-variance portfolio optimization to a market that has a risk-free asset with interest rate r_f besides the m risky assets, CAPM has an efficient frontier that is a straight line and is tangent to the original feasible region for the m risky assets at a point which corresponds to a single fund M of the risky assets. The "one-fund theorem" refers to that any efficient portfolio can be constructed as a linear combination of the fund M and the risk-free asset. Letting σ_M and μ_M denote the standard deviation (volatility) and mean of the return r_M of this fund, CAPM relates the excess return $r_i - r_f$ of asset i to that of the market portfolio via the regression model

$$r_i - r_f = \beta (r_M - r_f) + \epsilon_i, \quad (2.2)$$

in which $E(\epsilon_i) = 0$, $\text{Cov}(\epsilon_i, r_M) = 0$, and $\beta_i = \text{Cov}(r_i, r_M)/\sigma_M^2$. From (2.2) it follows that

$$\sigma_i^2 = \beta_i^2 \sigma_M^2 + \text{Var}(\epsilon_i), \quad (2.3)$$

which decomposes the variance σ_i^2 of the ith asset return as a sum of *systematic risk* $\beta_i^2 \sigma_M^2$ (associated with the market) and *idiosyncratic risk* that is unique to the asset and can be diversified in a portfolio. The *beta* of a portfolio with return r is similarly defined as $\beta = \text{Cov}(r, r_M)/\sigma_M^2$, which can be used as a measure of the sensitivity of the asset return to market movements. A widely used proxy for the market portfolio, which is a fund consisting of the weighted assets in a universe of investments, is S&P's 500 index, a leading index of the US equity market.

2.4 Statistical underpinnings of MPT

Since $\boldsymbol{\mu}$ and $\boldsymbol{\Sigma}$ in Markowitz's efficient frontier are actually unknown, a natural idea is to replace them by the sample mean vector $\widehat{\boldsymbol{\mu}}$ and covariance matrix $\widehat{\boldsymbol{\Sigma}}$ of a training sample of historical returns $\mathbf{r}_t = (r_{1t}, \ldots, r_{mt})^\top, 1 \le t \le n$. However, this plug-in frontier is no longer optimal because $\widehat{\boldsymbol{\mu}}$ and $\widehat{\boldsymbol{\Sigma}}$ actually differ from $\boldsymbol{\mu}$ and $\boldsymbol{\Sigma}$, and Frankfurter et al. (1976) and Jobson and Korkie (1980) have reported that portfolios associated with the plug-in frontier can perform worse than an equally weighted portfolio that is highly inefficient. Michaud (1989) comments that the minimum variance (MV) portfolio $\mathbf{w}_{\text{eff}}$ based on $\widehat{\boldsymbol{\mu}}$ and $\widehat{\boldsymbol{\Sigma}}$ has serious deficiencies, calling the MV optimizers "estimation-error maximizers". His argument is reinforced by subsequent studies, e.g., Best and Grauer (1991), Chopra et al. (1993), Canner et al. (1997), Simaan (1997), and Britten-Jones (1999). Three approaches have emerged to address the difficulty during the past two decades. The first approach uses multifactor models to reduce the dimension in estimating $\boldsymbol{\Sigma}$, and the second approach uses Bayes or other shrinkage estimates of $\boldsymbol{\Sigma}$. Both approaches use improved estimates of $\boldsymbol{\Sigma}$ for the plug-in efficient frontier. They have also been modified to provide better estimates of $\boldsymbol{\mu}$, for example, in the quasi-Bayesian approach of Black and Litterman (1991). The third approach uses bootstrapping to correct for the bias of $\widehat{\mathbf{w}}_{\text{eff}}$ as an estimate of $\mathbf{w}_{\text{eff}}$.

2.4.1 Multifactor pricing models

Multifactor pricing models relate the ith asset returns r_i to k factors $f_1, \ldots, f_k$ in a regression model of the form

$$r_i = \alpha_i + (f_1, \ldots, f_k)^\top \boldsymbol{\beta}_i + \epsilon_i, \tag{2.4}$$

in which α_i and $\boldsymbol{\beta}_i$ are unknown regression parameters and ϵ_i is an unobserved random disturbance that has mean 0 and is uncorrelated with $\mathbf{f} := (f_1, \ldots, f_k)^\top$. The case $k = 1$ is called a *single-factor* (or *single-index*) model. Under CAPM which assumes, besides known $\boldsymbol{\mu}$ and $\boldsymbol{\Sigma}$, that the market has a risk-free asset with return r_f (interest rate) and that all investors minimize the variances of their portfolios for their target mean returns, (2.4) holds with $k = 1$, $\alpha_i = r_f$ and $f = r_M - r_f$, where r_M is the return of a hypothetical market portfolio M which can be approximated in practice by an index fund such as S&P 500 Index. The arbitrage pricing theory (APT), introduced by Ross (1976), involves neither a market portfolio nor a risk-free asset and states that a multifactor model of the form (2.4) should hold approximately in the absence of arbitrage for sufficiently large m. The theory, however, does not specify the factors and their number. Methods for choosing factors in (2.4) can be broadly classified as economic and statistical, and commonly used statistical methods include factor analysis and principal component analysis, while the widely used Fama-French three-factor model is derived from economic considerations and empirical observations; see Section 3.4 of Lai and Xing (2008). Section 2.9 gives some recent developments in factor models.

2.4.2 Bayes, shrinkage, and Black-Litterman estimators

A popular conjugate family of prior distributions for estimation of covariance matrices from i.i.d. normal random vectors $\mathbf{r}_t$ with mean $\boldsymbol{\mu}$ and covariance matrix $\boldsymbol{\Sigma}$ is

$$\boldsymbol{\mu}|\boldsymbol{\Sigma} \sim N(\boldsymbol{\nu}, \boldsymbol{\Sigma}/\kappa), \qquad \boldsymbol{\Sigma} \sim IW_m(\boldsymbol{\Psi}, n_0), \tag{2.5}$$

where $IW_m(\boldsymbol{\Psi}, n_0)$ denotes the inverted Wishart distribution with n_0 degrees of freedom and mean $\boldsymbol{\Psi}/(n_0 - m - 1)$. The posterior distribution of $(\boldsymbol{\mu}, \boldsymbol{\Sigma})$ given $(\mathbf{r}_1, \ldots, \mathbf{r}_n)$ is also of the same form:

$$\boldsymbol{\mu}|\boldsymbol{\Sigma} \sim N(\widehat{\boldsymbol{\mu}}, \boldsymbol{\Sigma}/(n+\kappa)), \qquad \boldsymbol{\Sigma} \sim IW_m((n + n_0 - m - 1)\widehat{\boldsymbol{\Sigma}}, n + n_0),$$

where $\widehat{\boldsymbol{\mu}}$ and $\widehat{\boldsymbol{\Sigma}}$ are the Bayes estimators of $\boldsymbol{\mu}$ and $\boldsymbol{\Sigma}$ which shrink the mean vector $\bar{\mathbf{r}}$ and the sample covariance matrix $n^{-1}\sum_{i=1}^{n}(\mathbf{r}_i - \bar{\mathbf{r}})(\mathbf{r}_i - \bar{\mathbf{r}})^\top$ toward their respective prior means; see Section 4.3.3 of Lai and Xing (2008).

Simply using $\bar{\mathbf{r}}$ to estimate $\boldsymbol{\mu}$, Ledoit and Wolf (2003, 2004) propose to shrink the MLE of $\boldsymbol{\Sigma}$ towards a structured covariance matrix, instead of using directly this Bayes estimator which requires specification of the hyperparameters $\boldsymbol{\mu}$, κ, n_0 and $\boldsymbol{\Psi}$. Their rationale is that whereas the MLE $\mathbf{S} = \sum_{t=1}^{n}(\mathbf{r}_t - \bar{\mathbf{r}})(\mathbf{r}_t - \bar{\mathbf{r}})^\top/n$ has a large variance when $m(m+1)/2$ is comparable with n, a structured covariance matrix $\mathbf{F}$ has much fewer parameters that can be estimated with smaller variances. They propose to estimate $\boldsymbol{\Sigma}$ by a convex combination of $\widehat{\mathbf{F}}$ and $\mathbf{S}$:

$$\widehat{\boldsymbol{\Sigma}} = \widehat{\delta}\widehat{\mathbf{F}} + (1 - \widehat{\delta})\mathbf{S}, \tag{2.6}$$

where $\widehat{\delta}$ is an estimator of the *optimal shrinkage constant* δ used to shrink the MLE toward the estimated structured covariance matrix $\widehat{\mathbf{F}}$. Besides the covariance matrix $\mathbf{F}$ associated with a single-factor model, they also propose to use a constant correlation model for $\mathbf{F}$ in which all pairwise correlations are identical, and have found that it gives comparable performance in simulation and empirical studies. They advocate using this shrinkage estimate in lieu of $\mathbf{S}$ in implementing Markowitz's efficient frontier.

The difficulty of estimating $\boldsymbol{\mu}$ well enough for the plug-in portfolio to have reliable performance has been noted by Jorion (1986) who proposes to use a shrinkage estimator similar to the Bayes estimator $\widehat{\boldsymbol{\mu}}$, and by Black and Litterman (1991) who propose a pragmatic Bayesian approach that basically amounts to shrinking an investor's subjective estimate of $\boldsymbol{\mu}$ to the market's estimate implied by an "equilibrium portfolio," which we discuss below.

The investor's subjective guess of $\boldsymbol{\mu}$ is described in terms of "views" on linear combinations of asset returns, which can be based on past observations and the investor's personal/expert opinions. These views are represented by $\mathbf{P}\boldsymbol{\mu} \sim N(\mathbf{q}, \boldsymbol{\Omega})$, where $\mathbf{P}$ is a $p \times m$ matrix of the investor's "picks" of the assets to express the guesses, and $\boldsymbol{\Omega}$ is a diagonal matrix that expresses the investor's uncertainties in the views via their variances. The "equilibrium portfolio", denoted by $\widetilde{\mathbf{w}}$, is based on a normative theory of an equilibrium market, in which $\widetilde{\mathbf{w}}$ is assumed to solve the mean-variance optimization problem $\max_{\mathbf{w}}(\mathbf{w}^\top \boldsymbol{\pi} - \lambda \mathbf{w}^\top \boldsymbol{\Sigma} \mathbf{w})$, with λ being the average risk-aversion level of the market and $\boldsymbol{\pi}$ representing the market's view of $\boldsymbol{\mu}$. This theory yields the relation $\boldsymbol{\pi} = 2\lambda \boldsymbol{\Sigma} \widetilde{\mathbf{w}}$, which can be used to infer $\boldsymbol{\pi}$ from the market capitalization or benchmark portfolio as a surrogate of $\widetilde{\mathbf{w}}$. Incorporating uncertainty in the market's view of $\boldsymbol{\mu}$, Black and Litterman assume that $\boldsymbol{\pi} - \boldsymbol{\mu} \sim N(\mathbf{0}, \tau \boldsymbol{\Sigma})$, in which $\tau \in (0,1)$ is a small parameter, and also set exogenously $\lambda = 1.2$. Combining $\mathbf{P}\boldsymbol{\mu} \sim N(\mathbf{q}, \boldsymbol{\Omega})$ with $\boldsymbol{\pi} - \boldsymbol{\mu} \sim N(\mathbf{0}, \tau \boldsymbol{\Sigma})$ under a working independence assumption between the two multivariate normal distributions yields the Black-Litterman estimate of $\boldsymbol{\mu}$:

$$\widehat{\boldsymbol{\mu}}^{\mathrm{BL}} = \left[(\tau\boldsymbol{\Sigma})^{-1} + \mathbf{P}^\top\boldsymbol{\Omega}^{-1}\mathbf{P}\right]^{-1}\left[(\tau\boldsymbol{\Sigma})^{-1}\boldsymbol{\pi} + \mathbf{P}^\top\boldsymbol{\Omega}^{-1}\mathbf{q}\right], \qquad (2.7)$$

with covariance matrix $[(\tau\boldsymbol{\Sigma})^{-1} + \mathbf{P}^\top\boldsymbol{\Omega}^{-1}\mathbf{P}]^{-1}$. Note that (2.7) involves $\boldsymbol{\Sigma}$, which Black and Litterman estimated by using the sample covariance matrix of historical data, and that their focus was to address the estimation of $\boldsymbol{\mu}$ for the plug-in portfolio. Clearly Bayes or shrinkage estimates of $\boldsymbol{\Sigma}$ can be used instead. Various modifications and extensions of (2.7) have been proposed; see Meucci (2005, 2010) and Bertsimas et al. (2012).

2.4.3 Bootstrapping and the resampled frontier

To adjust for the bias of $\widehat{\mathbf{w}}_{\mathrm{eff}}$ as an estimate of $\mathbf{w}_{\mathrm{eff}}$, Michaud (1989) uses the average of the bootstrap weight vectors:

$$\bar{\mathbf{w}} = B^{-1} \sum_{b=1}^{B} \widehat{\mathbf{w}}_b^*, \qquad (2.8)$$

where $\widehat{\mathbf{w}}_b^*$ is the estimated optimal portfolio weight vector based on the bth bootstrap sample $\{\mathbf{r}_{b1}^*, \ldots, \mathbf{r}_{bn}^*\}$ drawn with replacement from the observed sample $\{\mathbf{r}_1, \ldots, \mathbf{r}_n\}$. Specifically, the bth bootstrap sample has sample mean vector $\widehat{\boldsymbol{\mu}}_b^*$ and covariance matrix $\widehat{\boldsymbol{\Sigma}}_b^*$, which can be used to replace $\boldsymbol{\mu}$ and $\boldsymbol{\Sigma}$ in (2.1), thereby yielding $\widehat{\mathbf{w}}_b^*$. Thus, the resampled efficient frontier corresponds to plotting $\bar{\mathbf{w}}^\top \widehat{\boldsymbol{\mu}}$ versus $\sqrt{\bar{\mathbf{w}}^\top \widehat{\boldsymbol{\Sigma}} \bar{\mathbf{w}}}$ for a fine grid of μ_* values, where $\bar{\mathbf{w}}$ is defined by (2.8) in which $\widehat{\mathbf{w}}_b^*$ depends on the target level $\boldsymbol{\mu}_*$.

2.5 A new approach incorporating parameter uncertainty

A major difficulty with the "plug-in" efficient frontier (which uses $\mathbf{S}$ to estimate $\boldsymbol{\Sigma}$ and $\bar{\mathbf{r}}$ to estimate $\boldsymbol{\mu}$), or its variants that estimate $\boldsymbol{\Sigma}$ by (2.6) and $\boldsymbol{\mu}$ by the Black-Litterman method, or its "resampled" version, is that Markowitz's idea of using the variance of $\mathbf{w}^{\top}\mathbf{r}_{n+1}$ as a measure of the portfolio's risk cannot be captured simply by the plug-in estimates $\mathbf{w}^{\top}\widehat{\boldsymbol{\Sigma}}\mathbf{w}$ of $\mathrm{Var}(\mathbf{w}^{\top}\mathbf{r}_{n+1})$ and $\mathbf{w}^{\top}\widehat{\boldsymbol{\mu}}$ of $E(\mathbf{w}^{\top}\mathbf{r}_{n+1})$. This difficulty has been noted was recognized by Broadie (1993), who uses the terms *true frontier* and *estimated frontier* to refer to Markowitz's efficient frontier (with known $\boldsymbol{\mu}$ and $\boldsymbol{\Sigma}$) and the plug-in efficient frontier, respectively, and who also suggests considering the *actual* mean and variance of the return of an estimated frontier portfolio. Whereas the problem of minimizing $\mathrm{Var}(\mathbf{w}^{\top}\mathbf{r}_{n+1})$ subject to a given level μ_* of the mean return $E(\mathbf{w}^{\top}\mathbf{r}_{n+1})$ is meaningful in Markowitz's framework, in which both $E(\mathbf{r}_{n+1})$ and $\mathrm{Cov}(\mathbf{r}_{n+1})$ are known, the surrogate problem of minimizing $\mathbf{w}^{\top}\widehat{\boldsymbol{\Sigma}}\mathbf{w}$ under the constraint $\mathbf{w}^{\top}\widehat{\boldsymbol{\mu}} = \mu_*$ ignores the fact both $\widehat{\boldsymbol{\mu}}$ and $\widehat{\boldsymbol{\Sigma}}$ have inherent errors (risks) themselves. Lai, Xing, and Chen (2011b) (abbreviated by LXC in the remainder of this chapter) consider the more fundamental problem

$$\max\left\{E(\mathbf{w}^{\top}\mathbf{r}_{n+1}) - \lambda\mathrm{Var}(\mathbf{w}^{\top}\mathbf{r}_{n+1})\right\} \tag{2.9}$$

when $\boldsymbol{\mu}$ and $\boldsymbol{\Sigma}$ are unknown and treated as state variables whose uncertainties are specified by their posterior distributions given the observations $\mathbf{r}_1, \ldots, \mathbf{r}_n$ in a Bayesian framework. The weights $\mathbf{w}$ in (2.9) are random vectors that depend on $\mathbf{r}_1, \ldots, \mathbf{r}_n$. Note that if the prior distribution puts all its mass at $(\boldsymbol{\mu}_0, \boldsymbol{\Sigma}_0)$, then the minimization problem (2.9) reduces to Markowitz's portfolio optimization problem that assumes $\boldsymbol{\mu}_0$ and $\boldsymbol{\Sigma}_0$ are given. The Lagrange multiplier λ in (2.9) can be regarded as the investor's risk-aversion index when variance is used to measure risk.

2.5.1 Solution of the optimization problem

The problem (2.9) is not a standard stochastic optimization problem because of the term $[E(\mathbf{w}^{\top}\mathbf{r}_{n+1})]^2$ in $\mathrm{Var}(\mathbf{w}^{\top}\mathbf{r}_{n+1}) = E[(\mathbf{w}^{\top}\mathbf{r}_{n+1})^2] - [E(\mathbf{w}^{\top}\mathbf{r}_{n+1})]^2$. A standard stochastic optimization problem in the Bayesian setting is of the form $\max_{a\in\mathcal{A}} Eg(\mathbf{X}, \boldsymbol{\theta}, a)$, in which $g(\mathbf{X}, \boldsymbol{\theta}, a)$ is the reward when action a is taken, $\mathbf{X}$ is a random vector with distribution $F_{\boldsymbol{\theta}}$, $\boldsymbol{\theta}$ has a prior distribution and the maximization is over the action space $\mathcal{A}$. The key to its solution is the tower property of conditional expectations $Eg(\mathbf{X}, \boldsymbol{\theta}, a) = E\{E[g(\mathbf{X}, \boldsymbol{\theta}, a)|\mathbf{X}]\}$, which implies that the stochastic optimization problem can be solved by choosing a to maximize the posterior reward $E[g(\mathbf{X}, \boldsymbol{\theta}, a)|\mathbf{X}]\}$. LXC converts (2.9) to a standard stochastic control problem by using an additional parameter. Let $W = \mathbf{w}^{\top}\mathbf{r}_{n+1}$ and note that $E(W) - \lambda\mathrm{Var}(W) = h(EW, EW^2)$, where

$h(x, y) = x + \lambda x^2 - \lambda y$. Let $W_B = \mathbf{w}_B^\top \mathbf{r}_{n+1}$ and $\eta = 1 + 2\lambda E(W_B)$, where $\mathbf{w}_B$ is the Bayes weight vector. Then

$$
\begin{aligned}
0 &\geq h(EW, EW^2) - h(EW_B, EW_B^2) \\
&= E(W) - E(W_B) - \lambda\{E(W^2) - E(W_B^2)\} + \lambda\{(EW)^2 - (EW_B)^2\} \\
&= \eta\{E(W) - E(W_B)\} + \lambda\{E(W_B^2) - E(W^2)\} + \lambda\{E(W) - E(W_B)\}^2 \\
&\geq \{\lambda E(W_B^2) - \eta E(W_B)\} - \{\lambda E(W^2) - \eta E(W)\}.
\end{aligned}
$$

Moreover, the last inequality is strict unless $EW = EW_B$, in which case the first inequality is strict unless $EW^2 = EW_B^2$. This shows that the last term above is ≤ 0, or equivalently,

$$\lambda E(W^2) - \eta E(W) \geq \lambda E(W_B^2) - \eta E(W_B), \tag{2.10}$$

and that equality holds in (2.10) if and only if W has the same mean and variance as W_B. Hence the stochastic optimization problem (2.9) is equivalent to minimizing $\lambda E[(\mathbf{w}^\top \mathbf{r}_{n+1})^2] - \eta E(\mathbf{w}^\top \mathbf{r}_{n+1})$ over weight vectors $\mathbf{w}$ that can depend on $\mathbf{r}_1, \ldots, \mathbf{r}_n$. Since $\eta = 1 + 2\lambda E(W_B)$ is a linear function of the solution of (2.9), we cannot apply this equivalence directly to the unknown η. Instead LXC solves a family of standard stochastic optimization problems over η and then choose the η that maximizes the reward in (2.9).

To summarize, LXC solves (2.9) by rewriting it as the following maximization problem over η:

$$\max_\eta \left\{ E\left[\mathbf{w}^\top(\eta)\mathbf{r}_{n+1}\right] - \lambda \mathrm{Var}\left[\mathbf{w}^\top(\eta)\mathbf{r}_{n+1}\right] \right\}, \tag{2.11}$$

where $\mathbf{w}(\eta)$ is the solution of the stochastic optimization problem

$$\mathbf{w}(\eta) = \arg\min_{\mathbf{w}} \left\{ \lambda E\left[(\mathbf{w}^\top \mathbf{r}_{n+1})^2\right] - \eta E(\mathbf{w}^\top \mathbf{r}_{n+1}) \right\}. \tag{2.12}$$

2.5.2   Computation of the optimal weight vector

Let $\boldsymbol{\mu}_n$ and $\mathbf{V}_n$ be the posterior mean and second moment matrix given the set $\mathcal{R}_n$ of current and past returns $\mathbf{r}_1, \ldots, \mathbf{r}_n$. Since $\mathbf{w}$ is based on $\mathcal{R}_n$, it follows from $E(\mathbf{r}_{n+1}|\mathcal{R}_n) = \boldsymbol{\mu}_n$ and $E(\mathbf{r}_{n+1}\mathbf{r}_{n+1}^\top|\mathcal{R}_n) = \mathbf{V}_n$ that

$$E(\mathbf{w}^\top \mathbf{r}_{n+1}) = E(\mathbf{w}^\top \boldsymbol{\mu}_n), \qquad E\left[(\mathbf{w}^\top \mathbf{r}_{n+1})^2\right] = E(\mathbf{w}^\top \mathbf{V}_n \mathbf{w}). \tag{2.13}$$

Without short selling, the weight vector $\mathbf{w}(\eta)$ in (2.11) is given by the following posterior analog of (2.1):

$$\mathbf{w}(\eta) = \arg\min_{\substack{\mathbf{w}:\mathbf{w}^\top \mathbf{1}=1, \\ \mathbf{w}\geq 0}} \left\{ \lambda \mathbf{w}^\top \mathbf{V}_n \mathbf{w} - \eta \mathbf{w}^\top \boldsymbol{\mu}_n \right\}, \tag{2.14}$$

which can be computed by quadratic programming (e.g., by `quadprog` in MATLAB). When short selling is allowed but there are limits on short setting, the constraint $\mathbf{w} \geq \mathbf{0}$ can be replaced by $\mathbf{w} \geq \mathbf{w}_0$, where $\mathbf{w}_0$ is a vector of negative numbers. When there is no limit on short selling, the constraint $\mathbf{w} \geq \mathbf{0}$ in (2.14) can be removed and $\mathbf{w}(\eta)$ in (2.12) is given explicitly by

$$
\begin{aligned}
\mathbf{w}(\eta) &= \arg \min_{\mathbf{w}:\mathbf{w}^\top \mathbf{1}=1} \left\{ \lambda \mathbf{w}^\top \mathbf{V}_n \mathbf{w} - \eta \mathbf{w}^\top \boldsymbol{\mu}_n \right\} \\
&= \frac{1}{C_n} \mathbf{V}_n^{-1} \mathbf{1} + \frac{\eta}{2\lambda} \mathbf{V}_n^{-1} \left(\boldsymbol{\mu}_n - \frac{A_n}{C_n} \mathbf{1} \right),
\end{aligned}
\tag{2.15}
$$

where the second equality can be derived by using a Lagrange multiplier and

$$
A_n = \boldsymbol{\mu}_n^\top \mathbf{V}_n^{-1} \mathbf{1} = \mathbf{1}^\top \mathbf{V}_n^{-1} \boldsymbol{\mu}_n, \quad B_n = \boldsymbol{\mu}_n^\top \mathbf{V}_n^{-1} \boldsymbol{\mu}_n, \quad C_n = \mathbf{1}^\top \mathbf{V}_n^{-1} \mathbf{1}.
$$

Quadratic programming can be used to compute $\mathbf{w}(\eta)$ for more general linear and quadratic constraints than those in (2.14).

Let $\boldsymbol{\Sigma}_n$ denote the posterior covariance matrix given $\mathcal{R}_n$. Note that the tower property of conditional expectations, from which (2.13) follows, has the following analog for $\mathrm{Var}(W)$:

$$
\mathrm{Var}(W) = E\left[\mathrm{Var}(W|\mathcal{R}_n)\right] + \mathrm{Var}\left[E(W|\mathcal{R}_n)\right] = E(\mathbf{w}^\top \boldsymbol{\Sigma}_n \mathbf{w}) + \mathrm{Var}(\mathbf{w}^\top \boldsymbol{\mu}_n).
\tag{2.16}
$$

Using $\boldsymbol{\Sigma}_n$ to replace $\boldsymbol{\Sigma}$ in the optimal weight vector that assumes $\boldsymbol{\mu}$ and $\boldsymbol{\Sigma}$ to be known, therefore, ignores the variance of $\mathbf{w}^\top \boldsymbol{\mu}_n$ in (2.16), and this omission is an important root cause for the Markowitz optimization enigma related to "plug-in" efficient frontiers. For more flexible modeling, one can allow the prior distribution in the preceding Bayesian approach to include unspecified hyperparameters, which can be estimated from the training sample by maximum likelihood, or method of moments or other methods. For example, for the conjugate prior (2.5), we can assume $\boldsymbol{\nu}$ and $\boldsymbol{\Psi}$ to be functions of certain hyperparameters that are associated with a multifactor model of the type (2.4). This amounts to using an empirical Bayes model for $(\boldsymbol{\mu}, \boldsymbol{\Sigma})$ in the stochastic optimization problem (2.4). Besides a prior distribution for $(\boldsymbol{\mu}, \boldsymbol{\Sigma})$, (2.4) also requires specification of the common distribution of the i.i.d. returns to evaluate $E_{\boldsymbol{\mu}, \boldsymbol{\Sigma}}(\mathbf{w}^\top \mathbf{r}_{n+1})$ and $\mathrm{Var}_{\boldsymbol{\mu}, \boldsymbol{\Sigma}}(\mathbf{w}^\top \mathbf{r}_{n+1})$. The bootstrap provides a nonparametric method to evaluate these quantities, as described below.

2.5.3 Bootstrap estimate of performance and the NPEB rule

To begin with, note that we can evaluate the frequentist performance of asset allocation rules by making use of the bootstrap method. The bootstrap samples $\{\mathbf{r}_{b1}^*, \dots, \mathbf{r}_{bn}^*\}$, drawn with replacement from the observed sample $\{\mathbf{r}_1, \dots, \mathbf{r}_n\}$ for $1 \leq b \leq B$, can be used to estimate $E_{\boldsymbol{\mu}, \boldsymbol{\Sigma}}(\mathbf{w}_n^\top \mathbf{r}_{n+1}) = E_{\boldsymbol{\mu}, \boldsymbol{\Sigma}}(\mathbf{w}_n^\top \boldsymbol{\mu})$ and $\mathrm{Var}_{\boldsymbol{\mu}, \boldsymbol{\Sigma}}(\mathbf{w}_n^\top \mathbf{r}_{n+1}) = E_{\boldsymbol{\mu}, \boldsymbol{\Sigma}}(\mathbf{w}_n^\top \boldsymbol{\Sigma} \mathbf{w}_n) + \mathrm{Var}_{\boldsymbol{\mu}, \boldsymbol{\Sigma}}(\mathbf{w}_n^\top \boldsymbol{\mu})$ of various portfolios Π whose weight vectors $\mathbf{w}_n$ may depend on $\mathbf{r}_1, \dots, \mathbf{r}_n$ and a non-informative prior for $(\boldsymbol{\mu}, \boldsymbol{\Sigma})$. In particular, one can use Bayes or other estimators for $\boldsymbol{\mu}_n$ and $\mathbf{V}_n$ in (2.14) or (2.15) and then choose η to maximize the bootstrap estimate of $E_{\boldsymbol{\mu}, \boldsymbol{\Sigma}}(\mathbf{w}_n^\top \mathbf{r}_{n+1}) - \lambda \mathrm{Var}_{\boldsymbol{\mu}, \boldsymbol{\Sigma}}(\mathbf{w}_n^\top \mathbf{r}_{n+1})$. Using the empirical distribution of $\mathbf{r}_1, \dots, \mathbf{r}_n$ to be the common distribution of the returns and a non-informative prior for $(\boldsymbol{\mu}, \boldsymbol{\Sigma})$, LXC substitutes $\boldsymbol{\mu}_n$ by $\bar{\mathbf{r}}$ in (2.14) and $\mathbf{V}_n$ by the second moment matrix $n^{-1} \sum_{t=1}^{n} \mathbf{r}_t \mathbf{r}_t^\top$ in (2.15), thereby yielding a "non-parametric empirical Bayes" variant, which LXC abbreviates by NPEB, of the optimal rule in Section 2.5.1. LXC reports a simulation study that assumes i.i.d. annual returns (in %) of $m = 4$ assets whose mean vector and covariance matrix are generated from the normal and inverted Wishart prior distribution (2.5) with $\kappa = 5$, $n_0 = 10$, $\boldsymbol{\nu} = (2.48, 2.17, 1.61, 3.42)^\top$ and the hyperparameter $\boldsymbol{\Psi}$ given by $\boldsymbol{\Psi}_{11} = 3.37$, $\boldsymbol{\Psi}_{22} = 4.22$, $\boldsymbol{\Psi}_{33} = 2.75$, $\boldsymbol{\Psi}_{44} = 8.43$, $\boldsymbol{\Psi}_{12} = 2.04$, $\boldsymbol{\Psi}_{13} = 0.32$, $\boldsymbol{\Psi}_{14} = 1.59$, $\boldsymbol{\Psi}_{23} = -0.05$, $\boldsymbol{\Psi}_{24} = 3.02$, $\boldsymbol{\Psi}_{34} = 1.08$. The simulation study considers four scenarios for the case $n = 6$ without short selling. The first scenario assumes this prior distribution and studies the Bayesian reward for $\lambda = 1, 5$ and 10. The other scenarios consider the frequentist reward at three values of $(\boldsymbol{\mu}, \boldsymbol{\Sigma})$ generated from the prior distribution, denoted by Freq 1, Freq 2, Freq 3 in Table 2.1 which compares the Bayes rule that maximizes (2.9), called "Bayes" hereafter, with three other rules: (a) the "oracle" rule that assumes $\boldsymbol{\mu}$ and $\boldsymbol{\Sigma}$ to be known, (b) the plug-in rule that replaces $\boldsymbol{\mu}$ and $\boldsymbol{\Sigma}$ by the sample estimates of $\boldsymbol{\mu}$ and $\boldsymbol{\Sigma}$, and (c) the NPEB rule. Note that although both (b) and (c) use the sample mean vector and sample covariance (or second moment) matrix, (b) simply plugs the sample estimates into the oracle rule while (c) uses the empirical distribution to replace the common distribution of the returns in the Bayes rule. For the plug-in rule, the quadratic programming procedure may have numerical difficulties if the sample covariance matrix is nearly singular. If it should happen, the default option of adding $0.005\mathbf{I}$ to the sample covariance matrix is used. Table 2.1 summarizes LXC's simulation results, each of which is based on 500 simulations. In each scenario, the reward of the NPEB rule is close to that of the Bayes rule and somewhat smaller than that of the oracle rule. The plug-in rule has substantially smaller rewards, especially for larger values of λ.

2.6   From random walks to martingales that match stylized facts

We have assumed that the return vectors $\mathbf{r}_1, \ldots, \mathbf{r}_{n+1}$ are i.i.d. in the stochastic optimization problem (2.9) with solution (2.14)–(2.15) in Sections 2.5.2 and 2.5.3 so that the results can be compared with other procedures in the literature that are based on this assumption. Moreover, we implicitly assume a non-informative prior distribution for $(\boldsymbol{\mu}, \boldsymbol{\Sigma})$ in NPEB so that its performance can be compared with those of other procedures that do not involve prior distributions. However, as pointed out by LXC, one should use an improved model when diagnostic checks of the "working i.i.d. model" reveal its inadequacy. In this section we continue the review in Section 2.2 on models of asset prices and then describe how LXC applies (2.14) and (2.15) to time

TABLE 2.1: Rewards of four portfolios formed from $m = 4$ assets

		Bayes	Plug-in	Oracle	NPEB
$\lambda=1$	Bayes	0.0324	0.0317	0.0328	0.0324
	Freq 1	0.0332	0.0324	0.0332	0.0332
	Freq 2	0.0293	0.0282	0.0298	0.0293
	Freq 3	0.0267	0.0257	0.0268	0.0267
$\lambda=5$	Bayes	0.0262	0.0189	0.0267	0.0262
	Freq 1	0.0272	0.0182	0.0273	0.0272
	Freq 2	0.0233	0.0183	0.0240	0.0234
	Freq 3	0.0235	0.0159	0.0237	0.0235
$\lambda=10$	Bayes	0.0184	0.0067	0.0190	0.0183
	Freq 1	0.0197	0.0063	0.0199	0.0198
	Freq 2	0.0157	0.0072	0.0168	0.0159
	Freq 3	0.0195	0.0083	0.0198	0.0196

series model of asset returns that match the stylized facts of returns data in Section 2.1.2. The brief chronological survey is intended to reflect the interplay of models and data (two keywords in the subtitle of the book), while the description of LXC's use of the second-generation models under EMH also aims at illustrating "analytics" and "optimization" in the book's subtitle.

2.6.1 From Gaussian to Paretian random walks

In Section 2.2, we mentioned the empirical justification, due to Kendall (1953) and others, of the random walk model for log prices of stocks. Fama (1963) summarized subsequent developments in modeling the distribution of the log returns, which are the i.i.d. increments of the random walk, focusing on the "stable Paretian hypothesis" due to Mandelbrot (1963). Paretian refers to the distribution, introduced by Pareto, that has probability $cx^{-\alpha}$ of exceeding x as $x \to \infty$. The Pareto distribution has infinite mean for $\alpha \leq 1$, and infinite variance for $1 < \alpha \leq 2$. Mandelbrot (1963) studied the daily differences of logarithmic closing cotton prices from 1900 to 1905, and the monthly differences from 1880 to 1940. The data, obtained from the US Department of Agriculture, exhibited much fatter tails than normal distributions, which are special cases of stable distributions, denoted by Stable$(\alpha, \beta, \gamma, \delta)$. In particular, the case $\alpha = 2$ and $\beta = 0$ corresponds to $Y \sim N(\delta, \gamma^2)$, while the case $\alpha = 1$ and $\beta = \delta = 0$ corresponds to $Y \sim \text{Cauchy}(\gamma)$ with density function $f(y) = \{\pi\gamma(1 + (y/\gamma)^2)\}^{-1}, -\infty < y < \infty$. An important property of stable distributions is that they are the "attractors" (i.e., all possible limiting distributions) of suitably centered and scaled sums of i.i.d. random variables (a.k.a. random walks). Another important property is self-decomposability: Let $Y_1, \ldots, Y_n$ be i.i.d. Stable$(\alpha, \beta, \gamma, \delta)$. Then there exist

$c_n > 0$ and d_n such that $Y_1 + \cdots + Y_n$ has the same distribution as $c_n Y + d_n$, where $Y \sim \text{Stable}(\alpha, \beta, \gamma, \delta)$. In particular, for the case $\beta = \delta = 0$ associated with symmetric stable distributions, we can choose $c_n = n^{1/\alpha}$ and $d_n = 0$. Mandelbrot calls non-normal stable distributions "stable Paretian distributions" because they are the limiting distributions of normalized random walks with Paretian (a.k.a. power law) increments. Because the density function of $\text{Stable}(\alpha, \beta, \gamma, \delta)$ does not have explicit formulas except for special cases such as normal, Cauchy, and the inverse Gaussian distribution (i.e., Lévy's distribution of the first passage time of Brownian motion), estimation of its parameters is far from being routine and has evolved during the past 50 years after Mandelbrot's seminal paper.

2.6.2   Random walks with optional sampling times

Besides replacing Brownian motion by more general stable processes (with stable distributions for its increments), Mandelbrot also introduced another wider class of models by using a *subordinator*, which is a "random clock" for the original process. The subordinated process model of Mandelbrot and Taylor (1967) uses $B_{T(t)}$ to model the price P_t, where $\{B_s, s \geq 0\}$ is Brownian motion and the subordinator $T(t)$ is a stable process with positive increments and is independent of $\{B_s, s \geq 0\}$. Another well-known subordinated process model in finance is the *variance-gamma process*, introduced by Madan and Seneta (1990), which has the form of $\sigma B_{\tau(t)} + \theta \tau(t)$, where $\tau(t)$ has independent increments such that $\tau(t)$ has the gamma distribution with mean $\gamma t/\lambda$ and variance $\gamma t/\lambda^2$.

Mandelbrot and Taylor's process $B_{T(t)}$ has stationary independent increments and the increments have infinite variance because $T(t)$ is a stable process with positive increments. Clark (1973) removed the stable process assumption on $T(t)$. Instead he assumed $v_t = T(t) - T(t-1)$ to be i.i.d. positive random variables with mean μ and variance $\sigma^2 > 0$. Then conditional on v_t, $\Delta_t = P_t - P_{t-1}$ is $N(0, \gamma^2 v_t)$. Therefore, Δ_t has mean 0 and $\text{Var}(\Delta_t) = E[\text{Var}(\Delta_t|v_t)] = \gamma^2 E(V_t) = \gamma^2 \mu$. Since $E(\Delta_t^4) = E[E(\Delta_t^4|v_t)] = 3\gamma^4 E(v_t^2) = 3\gamma^4(\mu^2 + \sigma^2)$, the kurtosis of Δ_t is equal to

$$\frac{E(\Delta_t^4)}{[\text{Var}(\Delta_t)]^2} = 3 \times \frac{\mu^2 + \sigma^2}{\mu^2} > 3,$$

showing that Δ_t is leptokurtic while still having finite variance, which matches the constraint that prices should be bounded over a finite time interval. Clark (1973) also relaxed the assumption of stationarity for the independent increments of $T(t)$. He provided the important insight that $v_t = T(t) - T(t-1)$ might vary with the trading activity on day t. In particular, he proposed a parametric model of the trading volume V_t on day t as a proxy for v_t. To be specific, he proposed to approximate $\gamma^2 v_t$ by AV_t^a or $\exp(B + bV_t)$.

Example 2.6   Because Clark's proxy involves nonlinear parameters if we

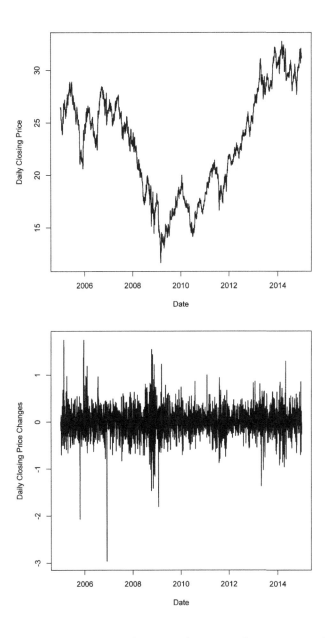

FIGURE 2.4: Time series of P_t (top panel) and Δ_t (bottom panel) for Pfizer closing prices from January 1, 2005, to December 31, 2014.

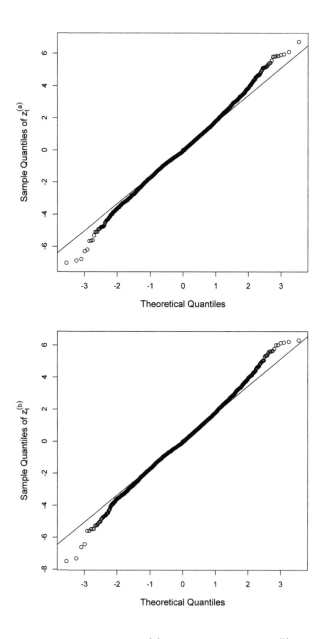

FIGURE 2.5: Normal QQ-plots of $z_t^{(a)}$ (top panel) and of $z_t^{(b)}$ (bottom panel) with $z_t^{(a)}$ and $z_t^{(b)}$ defined in (2.18).

regress Δ_t on V_t, it is more convenient to regress $\log \Delta_t^2$, which has the distribution of the logarithm of $\gamma^2 v_t \chi_1^2$, using one of the linear models

$$\log \Delta_t^2 = \log A + a \log V_t + \varepsilon_t \quad \text{or} \quad \log \Delta_t^2 = B + bV_t + \varepsilon_t. \tag{2.17}$$

We use the Pfizer daily closing prices between January 1, 2005 and December 31, 2014 together with the corresponding trading volumes to fit both regression models, after removing 43 cases that have $\Delta_t = 0$. Figure 2.4 plots the time series of P_t in the top panel, and of the price changes $\Delta_t = P_t - P_{t-1}$ in the bottom panel. The mean of V_t is of the order of 10^7, and the estimates are $\log(\hat{A}) = -25.2$, $\hat{a} = 1.23$; $\hat{B} = -4.83$, $\hat{b} = 2.47 \times 10^{-8}$. Since $\Delta_t / \sqrt{v_t} \sim N(0, \gamma^2)$ under Clark's model, we can use normal QQ-plots of

$$z_t^{(a)} = \frac{\Delta_t}{\sqrt{\hat{A} \times V_t^{\hat{a}}}}, \qquad z_t^{(b)} = \frac{\Delta_t}{\sqrt{\exp\left\{\hat{B} + \hat{b}V_t\right\}}} \tag{2.18}$$

to check the adequacy of the regression models in (2.17).[1] Figure 2.5 shows a much better fit for Δ_t than that of the normal model presented in the top panel of Figure 2.6 . Moreover, the kurtosis of $z_t^{(a)}$ and $z_t^{(b)}$ are 3.50 and 3.56, respectively, which are much closer to 3 than the kurtosis 9.90 of Δ_t.

2.6.3   From random walks to ARIMA, GARCH and general martingale regression models for time series data

Box and Jenkins (1970) used a transformation (taking logarithm) to convert daily closing prices of IBM stock to a time series whose first-order differences (hence "I" standing for "integrated" in ARIMA) can be approximated by an ARMA (autoregressive, moving average) model based on diagnostic checks with the autocorrelation and partial autocorrelation functions. Subsequently, Box and Pierce (1970) and Ljung and Box (1978) introduced statistical tests, based on the sample autocorrelations, of the null hypothesis of independence for time series data. Chapter 5 of Lai and Xing (2008) provides a summary of the Box-Jenkins methodology and its Wold decomposition background for weakly stationary models together with software implementation. Another major direction in econometric time series modeling after the Box-Jenkins methodology took center stage was to address the volatility clustering patterns found in stock returns and other econometric time series, culminating in the ARCH model introduced by Engle (1982) and its generalization GARCH by Bollerslev, which are basically extensions of the ARMA model with martingale difference (instead of i.i.d.) innovations, to the squared returns series r_t^2. Chapter 6 of Lai and Xing (2008)gives an introduction to GARCH models and its variants such as EGARCH (exponential GARCH) to address the leverage effect mentioned in Section 2.1.1. It also describes how an ARMA model

---

[1]Although $\hat{b}$ is of the order 10^{-8}, it cannot be ignored in $z_t^{(b)}$ because the mean of V_t is of the order 10^7.

for the level of r_t can be combined with the GARCH or EGARCH model for
its volatility, leading to ARMA-GARCH and ARMA-EGARCH models and
the software packages for model fitting and model-based forecasting. A recent
update is the R package `rugarch`, in which u stands for "univariate" and r
stands for "realized".

Hence, by 1982 when Engle's seminal paper[2] was published, time series
models matching the stylized facts of low-frequency log returns of stock data
were already available to replace the random walk model for prices, or their
logarithms, giving a much more definitive fix of the random walk model than
those by Mandelbrot, Taylor and Clark in the previous two decades, which
have been reviewed in the preceding two sections. The year 1982 also marks
the beginning of an even more general approach that includes ARMA and
GARCH models as special cases, and which we shall call "martingale regression" modeling based on time series data. This approach was initially
introduced by Lai and Wei (1982) under the name "stochastic regression"
to analyze the behavior of stochastic input-output systems for which the inputs are generated by a feedback control system so that the current input
vector $\mathbf{x}_t$ depends on the past outputs and inputs and produces the output
$y_t = \boldsymbol{\beta}^\top \mathbf{x}_t + \varepsilon_t$, in which ε_t represents random noise that is assumed to form
a martingale difference sequence. Specifically, letting $\mathcal{F}_t$ be the σ-field (infor-
mation set) generated by $(\mathbf{x}_s, y_s)$, $s \leq t$, it is assumed in stochastic regression
that $E(\varepsilon_t|\mathcal{F}_{t-1}) = 0$ and $\mathbf{x}_t$ is $\mathcal{F}_{t-1}$-measurable. The AR(p) model is a special
case with $\mathbf{x}_t = (y_{t-1}, \dots, y_{t-p})^\top$. The ARMA($p, q$) model is more complicated
because the random disturbances come in form $\varepsilon_t + c_1 \varepsilon_{t-1} + \cdots + c_q \varepsilon_{t-q}$. On
the other hand, if $\varepsilon_{t-1}, \dots, \varepsilon_{t-q}$ were observable, then this would reduce to a
stochastic regression model with $\mathbf{x}_t = (y_{t-1}, \dots, y_{t-p}, \varepsilon_{t-1}, \dots, \varepsilon_{t-q})^\top$. This is
basically the idea of Lai and Wei (1986) in their analysis of the extended least
squares estimator of the ARMA parameters. If the parameters were known,
then one could indeed retrieve $\varepsilon_{t-1}, \dots, \varepsilon_{t-q}$ from $(\mathbf{x}_s, y_s)$, $1 \leq s \leq t$, by
assuming $\mathbf{x}_s = \mathbf{0}$ and $\varepsilon_s = 0$ for $s \leq 0$, which is the basis of computing
the likelihood function. Lai and Ying (1991b) subsequently applied martin-
gale theory similar to that for extended least squares to analyze the recursive
maximum likelihood estimator.

The assumption of martingale difference (instead of i.i.d. zero-mean) ran-
dom errors ε_t also allows time series modeling of ε_t to incorporate dynamic changes in volatility. In fact, the GARCH(h, k) model $\varepsilon_t = \sigma_t \zeta_t$,
$\sigma_t^2 = \omega + \sum_{i=1}^h \beta_i \sigma_{t-i}^2 + \sum_{j-1}^k \alpha_j \varepsilon_{t-j}^2$ in which ζ_t are i.i.d. with mean 0 and
variance 1, can be written as an ARMA model for ε_t^2:

$$\varepsilon_t^2 = \omega + \sum_{j=1}^{\max(h,k)} (\alpha_j + \beta_j)\varepsilon_{t-j}^2 + \eta_t - \sum_{i=1}^h \beta_i \eta_{t-i},$$

in which $\eta_t = \varepsilon_t^2 - \sigma_t^2$ is a martingale difference sequence and $\alpha_j = 0$ for $j > k$,

---

[2]Robert Engle was awarded the 2003 Nobel Prize in Economic Sciences for this ground-
breaking work that has been very influential in econometric time series modeling.

$\beta_i = 0$ for $i > h$; see p. 147 of Lai and Xing (2008). We therefore use the more general term "martingale regression" to allow dynamic volatility modeling of the martingale difference noise ε_t in the stochastic regression model $y_t = \boldsymbol{\beta}^\top \mathbf{x}_t + \varepsilon_t$. This volatility modeling also involves a martingale structure with martingale difference innovations η_t. These martingale regression models are particularly useful for replacing the random walk model for log prices in MPT in Sections 2.4 and 2.5 by more general time series models that reflect the factor structure discussed in Section 2.4.1 and match the stylized facts of asset returns in Section 2.1.1. In particular, Lai et al. (abbreviated by LXC in Section 2.5) proposed to replace the "working i.i.d. model" for the returns r_{it} in Section 2.5 by a martingale regression model of the form

$$r_{it} = \boldsymbol{\beta}_i^\top \mathbf{x}_{i,t-1} + \epsilon_{it}, \tag{2.19}$$

where the components of $\mathbf{x}_{i,t-1}$ include 1, factor variables such as the return of a market portfolio like S&P 500 at time $t - 1$, and lagged variables $r_{i,t-1}, r_{i,t-2}, \ldots$. The basic idea underlying (2.19) is to introduce covariates (including lagged variables to account for time series effects) so that the errors ϵ_{it} can be regarded as i.i.d., as in the working i.i.d. model. The regression parameter $\boldsymbol{\beta}_i$ can be estimated by the method of moments, which is equivalent to least squares. We can also include heteroskedasticity by assuming that $\epsilon_{it} = s_{i,t-1}(\boldsymbol{\gamma}_i)z_{it}$, where z_{it} are i.i.d. with mean 0 and variance 1, $\boldsymbol{\gamma}_i$ is a parameter vector which can be estimated by maximum likelihood or generalized method of moments, and $s_{i,t-1}$ is a given function that depends on $r_{i,t-1}, r_{i,t-2}, \ldots$. A well-known example is the GARCH(1, 1) model

$$\epsilon_{it} = s_{i,t-1}z_{it}, \qquad s_{i,t-1}^2 = \omega_i + a_i s_{i,t-2}^2 + b_i r_{i,t-1}^2, \tag{2.20}$$

for which $\boldsymbol{\gamma}_i = (\omega_i, a_i, b_i)$. These martingale regression models show how much more versatile martingales can be used to capture information in the semi-strong form of EMH than random walks, reinforcing LeRoy's and Lehmann's points in Supplement 6 of Section 1.7.

2.7   Neo-MPT involving martingale regression models

We now return to LXC's new approach, described in Section 2.5, to MPT in ignorance of the mean vector and covariance matrix of the vector of asset returns $\mathbf{r}_{n+1}$ in the next period, after observing $\mathbf{r}_1, \ldots, \mathbf{r}_n$ up to now. The solution and computation of the optimal weight vector $\mathbf{w}(\mathbf{r}_1, \ldots, \mathbf{r}_n)$ in Sections 2.5.1 and 2.5.2 allows models which fit the data better than the working i.i.d. model, assumed in Section 2.5.3 mainly to compare with the existing methods that assume i.i.d. returns. In particular, LXC proposes to use the martingale regression model (2.20), which is consistent with EMH, instead of the i.i.d. model that does not match the stylized facts of asset returns. Another practical issue in the implementation of their new approach to MPT, which we shall simply call neo-MPT (or *Neomodern Portfolio Theory*), is the choice of λ in (2.9). We address this issue in Section 2.7.2 after describing how LXC extends the NPEB procedure to the martingale regression model (2.20). Finally, Section 2.7.3 describes LXC's empirical study of this neo-MPT.

2.7.1 Incorporating time series effects in the NPEB procedure

As noted in Section 2.5.2, a key ingredient in the optimal weight vector that solves the optimization problem (2.9) is $(\boldsymbol{\mu}_n, \mathbf{V}_n)$, where $\boldsymbol{\mu}_n = E(\mathbf{r}_{n+1}|\mathcal{R}_n)$ and $\mathbf{V}_n = E(\mathbf{r}_{n+1}\mathbf{r}_{n+1}^\top|\mathcal{R}_n)$. Instead of the classical model of i.i.d. returns, one can combine domain knowledge of the m assets with time series modeling to obtain better predictors of future returns via $\boldsymbol{\mu}_n$ and $\mathbf{V}_n$. The regressors $\mathbf{x}_{i,t-1}$ in (2.19)–(2.20) can be chosen to build a combined substantive-empirical model for prediction; see Section 7.5 of Lai and Xing (2008). Since the model (2.19)–(2.20) is intended to produce i.i.d. $\mathbf{z}_t = (z_{1t}, \ldots, z_{mt})^\top$, we can still use the NPEB approach to determine the optimal weight vector, bootstrapping from the estimated common distribution of $\mathbf{z}_t$. Note that (2.19)–(2.20) models the asset returns separately, instead of jointly in a multivariate regression or multivariate GARCH model which has too many parameters to estimate. While the vectors $\mathbf{z}_t$ are assumed to be i.i.d., (2.20) does not assume their components to be uncorrelated since it treats the components separately rather than jointly. The conditional cross-sectional covariance between the returns of assets i and j given $\mathcal{R}_n$ is given by

$$\text{Cov}(r_{i,n+1}, r_{j,n+1}|\mathcal{R}_n) = s_{i,n}(\boldsymbol{\gamma}_i)s_{j,n}(\boldsymbol{\gamma}_j)\text{Cov}(z_{i,n+1}, z_{j,n+1}|\mathcal{R}_n), \quad (2.21)$$

for the model (2.19)–(2.20). Note that (2.20) determines $s_{i,n}^2$ recursively from $\mathcal{R}_n$, and that $\mathbf{z}_{n+1}$ is independent of $\mathcal{R}_n$ and therefore its covariance matrix can be consistently estimated from the residuals $\hat{\mathbf{z}}_t$. The NPEB approach uses the following formulas for $\boldsymbol{\mu}_n$ and $\mathbf{V}_n$ in (2.14):

$$\boldsymbol{\mu}_n = (\widehat{\boldsymbol{\beta}}_1^\top \mathbf{x}_{1,n}, \ldots, \widehat{\boldsymbol{\beta}}_m^\top \mathbf{x}_{m,n})^\top, \quad \mathbf{V}_n = \boldsymbol{\mu}_n\boldsymbol{\mu}_n^\top + (\widehat{s}_{i,n}\widehat{s}_{j,n}\widehat{\sigma}_{ij})_{1 \leq i,j \leq n},$$

in which $\widehat{\boldsymbol{\beta}}_i$ is the least squares estimate of $\boldsymbol{\beta}_i$, and $\widehat{s}_{l,n}$ and $\widehat{\sigma}_{ij}$ are the usual estimates of $s_{l,n}$ and $\text{Cov}(z_{i,1}, z_{j,1})$ based on $\mathcal{R}_n$.

2.7.2 Optimizing information ratios along efficient frontier

As pointed out in Section 2.3, the λ in Section 2.5 is related to how risk-averse one is when one tries to maximize the expected utility of a portfolio. It represents a penalty on the risk that is measured by the variance of the portfolio's return. In practice, it may be difficult to specify an investor's risk aversion parameter λ that is needed in the theory in Section 2.5.1. A commonly used measure of a portfolio's performance, especially in connection with the evaluation of a fund's performance relative to those of other funds, is the information ratio $(\mu - \mu_0)/\sigma_e$, which is the excess return per unit of risk; the excess is measured by $\mu - \mu_0$, where $\mu_0 = E(r_0)$, r_0 is the return of the benchmark investment and σ_e^2 is the variance of the excess return. We can regard λ as a tuning parameter, and choose it to maximize the information ratio by modifying the NPEB procedure in Section 2.5.2, where the bootstrap estimate of $E_{\boldsymbol{\mu},\boldsymbol{\Sigma}}[\mathbf{w}^{\top}(\eta)\mathbf{r}] - \lambda \mathrm{Var}_{\boldsymbol{\mu},\boldsymbol{\Sigma}}[\mathbf{w}^{\top}(\eta)\mathbf{r}]$ is used to find the portfolio weight $\mathbf{w}_\lambda$ that solves the optimization problem (2.9). Specifically, we use the bootstrap estimate of the information ratio

$$E_{\boldsymbol{\mu},\boldsymbol{\Sigma}}(\mathbf{w}_\lambda \mathbf{r} - r_0)\Big/ \sqrt{\mathrm{Var}_{\boldsymbol{\mu},\boldsymbol{\Sigma}}(\mathbf{w}_\lambda^{\top} \mathbf{r} - r_0)} \qquad (2.22)$$

of $\mathbf{w}_\lambda$, and maximize the estimated information ratios over λ in a grid that will be illustrated in Section 2.7.3.

2.7.3   An empirical study of neo-MPT

We now describe LXC's empirical study of the out-of-sample performance of the proposed approach and other methods for mean-variance portfolio optimization when the means and covariances of the underlying asset returns are unknown. The study uses monthly stock market data from January 1985 to December 2009, which are obtained from the Center for Research in Security Prices (CRSP) database,[3] and evaluates out-of-sample performance of different portfolios of these stocks for each month after the first ten years (120 months) of this period to accumulate training data. Following Ledoit and Wolf (2004), at the beginning of month t, with t varying from January 1995 to December 2009, LXC selects $m = 50$ stocks with the largest market values among those that have no missing monthly prices in the previous 120 months, which are used as the training sample. The portfolios for month t to be considered are formed from these m stocks.

Note that this period contains highly volatile times in the stock market, such as around "Black Monday" in 1987, the Internet bubble burst and the September 11 terrorist attacks in 2001, and the "Great Recession" that began in 2007 with the default and other difficulties of subprime mortgage loans. LXC uses sliding windows of $n = 120$ months of training data to construct portfolios of the stocks for the subsequent month. In contrast to the Black-Litterman approach described in Section 2.4.2, the portfolio construction is based solely on these data and uses no other information about the stocks and their associated firms, since the purpose of the empirical study is to illustrate the basic statistical aspects of the proposed method and to compare

---

[3]CRSP stands for Center for Research in Security Prices, University of Chicago Booth School of Business. The CRSP database can be accessed through the Wharton Research Data Services at the University of Pennsylvania (http://wrds.wharton.upenn.edu).

it with other statistical methods for implementing Markowitz's mean-variance portfolio optimization theory. Moreover, for a fair comparison, LXC does not assume any prior distribution as in the Bayes approach, and only uses NPEB in this study.

Performance of a portfolio is measured by the excess returns e_t over a benchmark portfolio. As t varies over the monthly test periods from January 1995 to December 2009, LXC (i) adds up the realized excess returns to give the *cumulative realized excess return* $\sum_{l=1}^{t} e_l$ up to time t, and (ii) uses the average realized excess return and the standard deviation to evaluate the *realized information ratio* $\sqrt{12}\bar{e}/s_e$, where $\bar{e}$ is the sample average of the monthly excess returns and s_e is the corresponding sample standard deviation, using $\sqrt{12}$ to annualize the ratio as in Ledoit and Wolf (2004). Noting that the realized information ratio is a summary statistic of the monthly excess returns in the 180 test periods, LXC supplements this commonly used measure of investment performance with the time series plot of cumulative realized excess returns, from which the realized excess returns e_t can be retrieved by differencing.

Using the S&P 500 Index as the benchmark portfolio u_t, LXC considers the excess returns e_{it} over u_t, and fits the model (2.19)–(2.20) to the training sample to predict the mean and volatility of e_{it} for the test period. Short selling is allowed, with the constraint $w_i \geq -0.05$ for all i. Note that since $\sum_{i=1}^{m} w_i = 1$,

$$\sum_{i=1}^{m} w_i r_{it} - u_t = \sum_{i=1}^{m} w_i(r_{it} - u_t) = \sum_{i=1}^{m} w_i e_{it}, \qquad (2.23)$$

whereas active portfolio optimization considers weights $\widetilde{w}_i = w_i - w_{i,B}$ that satisfy the constraint $\sum_{i=1}^{m} \widetilde{w}_i = 0$. In view of (2.23), when the objective is to maximize the mean return of the portfolio subject to a constraint on the volatility of the excess return over the benchmark (which is related to achieving an optimal information ratio), we can replace the returns r_{it} by the excess returns e_{it} in the portfolio optimization problem (2.1) or (2.9) and model e_{it} by simpler stationary time series models than r_{it}. Besides the model (2.19)–(2.20) with $e_{i,t} = (1, e_{i,t-1}, u_{t-1})\beta_i + \epsilon_{i,t}$ that yields the NPEB$_{\text{SRG}}$ portfolio for the test sample (in which SRG denotes stochastic regression with GARCH(1,1) errors), LXC also fits the simpler AR(1) model $e_{it} = \alpha_i + \gamma_i e_{i,t-1} + \epsilon_{it}$ to compare with the plug-in, covariance-shrinkage and resampled portfolios which assume the i.i.d. working model and which use the annualized target return $\mu_* = 0.015$.

For the NPEB$_{\text{AR}}$ and NPEB$_{\text{SRG}}$ portfolios, LXC uses the training sample as in Section 2.7.2 to choose λ by maximizing the information ratio over the grid $\lambda \in \{2^i : i = -3, -2, \ldots, 6\}$. Figure 4 of LXC plots the time series of cumulative realized excess returns over the S&P 500 Index during the test period of 180 months, for NPEB$_{\text{SRG}}$, NPEB$_{\text{AR}}$, the plug-in portfolio described in the first paragraph of Section 2.4, the Ledoit-Wolf portfolio (2.6), and Michaud's bootstrap portfolio (2.8), using $\mu_* = 0.015$ as the target return. It shows NPEB_{SRG} to perform markedly better than NPEB_{AR}, which already greatly outperforms the other three procedures that perform similarly to each other. Table 2.2 gives the annualized realized information ratios of the five portfolios, with the S&P 500 Index as the benchmark portfolio. In addition to $\mu_* = 0.015$, the table also considers $\mu_* = 0.02, 0.03$ for which problem (2.1) can be solved for the Plug-in (P), Shrink (S), Bootstrap (B) portfolios for all 180 test periods under the weight constraint.

TABLE 2.2: Realized information ratios and average realized excess returns (in square brackets) with respect to S&P 500

$\mu_* = 0.015$			$\mu_* = 0.02$			$\mu_* = 0.03$			NPEB	
P	S	B	P	S	B	P	S	B	AR	SRG
0.527	0.352	0.618	0.532	0.353	0.629	0.538	0.354	0.625	0.370	1.169
[0.078]	[0.052]	[0.077]	[0.078]	[0.051]	[0.078]	[0.076]	[0.050]	[0.077]	[0.283]	[0.915]

2.8 Statistical arbitrage and strategies beyond EMH

Sections 1.3 and 1.7 have reviewed empirical tests and debates about EMH. In this section, we put off EMH and consider a wide variety of quantitative trading strategies that attempt to generate expected returns exceeding a certain level by searching for and capitalizing on statistical arbitrage opportunities (SAOs). We elaborate here on these strategies and the SAOs that have been described briefly in Section 1.3, and discuss the statistical or economic principles underlying them.

2.8.1 Technical rules and the statistical background in non-parametric regression and change-point modeling

Technical rules are based on the premise that past market prices contain information about profitable patterns and trends that will recur in the future. They use "technical analysis" to identify these patterns from historical data and thereby to develop a quantitative strategy for buying and selling securities. Because technical analysis usually uses data visualization "charts", it is often called "charting" and has received derogatory comments from proponents of EMH such as Malkiel (2003) who said: "Obviously, I am biased against the chartist.... Technical analysis is anathema to the academic world...the method is patently false." Presumably the reason for his strong dislike of technical analysis is that it is strictly data-driven and lacks economic justification. On the other hand, Brock et al. (1992, p. 1732) point out that despite the negative attitude of academics such as Malkiel towards technical analysis, it "has been enjoying a renaissance on Wall Street" and that "all major brokerage firms publish technical commentary on the market and individual securities... based on technical analysis." Brock et al. (1992), Lo et al. (2000) and Section 11.1.1 of Lai and Xing (2008) have given some details on nonparametric regression and change-point detection methods as the statistical background of a number of technical rules. Here we provide further details and additional references.

Filter rule and equivalent CUSUM change-point detection rule. A popular trading rule is the filter rule, which involves the *drawdown* $D_t = \max_{0 \le k \le t} P_k - P_t$, and the *drawup* $U_t = P_t - \min_{0 \le k \le t} P_k$ at time t. A filter rule gives a buy signal if $D_t/(\max_{0 \le k \le t} P_k) \ge c$ (suggesting a $100c\%$ drop from the running maximum price), and likewise gives a sell signal if $U_t/(\min_{0 \le k \le t} P_k) \ge c$. Letting $r_k = \log(P_k/P_{k-1})$ and $S_t = \log P_0 + \sum_{k=1}^{t} r_k$, note that

$$U_t \Big/ \left(\min_{0 \le k \le t} P_k \right) \ge c \iff S_t - \min_{0 \le k \le t} S_k \ge \log(1 + c) \qquad (2.24)$$

Hence the sell signal of the filter rule corresponds to the CUSUM rule in quality control: take corrective action as soon as $S_t - \min_{0 \le k \le t} S_k \ge b$; see Lai and Xing (2008, p. 279) and Page (1954). Alexander (1961) and Fama and Blume (1966) showed empirically that after taking transaction costs into account, the filter rule did not outperform the buy-and-hold strategy that simply buys the stock and holds it throughout the time period under consideration. Lam and Yam (1997) consider more general CUSUM rules and convert them to the filter rule form that involves drawups and drawdowns, and provide empirical evidence of the improvement of these rules over classical filter rules; see Section 2.9.

Moving average rules and window-limited GLR fault detection schemes. Lai and Xing (2008, p. 277), Brock et al. (1992), and Sullivan et al. (1999) have described some moving average rules that have been popular in technical analysis because of their simplicity and intuitive appeal. Besides fixed-length moving averages, Brock et al. also consider variable-length moving averages. These variable-length moving averages are closely related to the window-limited GLR fault detection schemes introduced by Willsky and Jones (1976) and Lai (1995), details of which are given in Section 2.9.

Directional trading using neural networks. Gencay (1998) describes a class of directional trading strategies whose cumulative log-return for an investment horizon of n days is $S_n = \sum_{t=1}^{n} \hat{y}_t r_t$, where $\hat{y}_t = 1$ (or -1) for a long (or short) position of a stock/security/currency at time t. The $\hat{y}_t$ chosen is an estimate of the Bernoulli variable $y_t = 2 \times (I_{\{r_t > 0\}} - 1)$ based on observations up to time $t-1$. Gencay (1998) uses a single-layer neural network estimate and cross-validation to determine the number of hidden units in the neural network; see Sections 7.3.2 and 7.4.4 of Lai and Xing (2008) for an introduction to cross-validation and the details of neural networks as a nonparametric regression method. He also reports an empirical study that shows markedly better performance than the buy-and-hold strategy. Note that transaction costs and the prediction errors $y_t - \hat{y}_t$ are left out in his development and evaluation of this trading strategy. Incorporating these considerations suggests a modified directional trading rule

$$
\hat{y}_t = \begin{cases} 1 & \text{if } \hat{\pi}_t^1 \geq q \\ -1 & \text{if } \hat{\pi}_t^2 \geq q \\ 0 & \text{otherwise,} \end{cases}
$$

where $\hat{\pi}_t^1$, or $\hat{\pi}_t^2$, is the estimate of the conditional probability that $r_t \geq c_1$, or $r_t \leq c_2$, respectively, given the information set $\mathcal{F}_{t-1}$ to time $t-1$, and $2q < 1$. The conditional probabilities can be estimated by logistic regression; see Section 4.1 of Lai and Xing (2008).

2.8.2   Time series, momentum, and pairs trading strategies

As noted by Lai and Xing (2008, p. 278), time series modeling and forecasting have been used, either explicitly or implicitly, in the development of technical rules: "Whereas the (nonparametric) regression approach focuses on patterns of past data with the hope that such patterns can be extrapolated to the future, a suitably chosen time series model can address prediction more directly." They go on to describe momentum strategies and pairs trading strategies as examples of the usefulness of time series as a framework for technical analysis and analyzing technical rules; see in particular their Section 11.1.3. In the remainder of this section, we elaborate on cross-sectional and time-series momentum which they have not addressed.

Cross-sectional momentum strategies rely on continuation into the future of the relative performance of a group of stock securities, thereby buying recent "winners" that outperform others and selling recent "losers" that underperform. Jegadeesh and Titman (1993) gave empirical evidence, showing that firms with relatively high returns over the past 3 to 12 months continue to outperform firms with relatively lower returns over the same sample period in their study. About a decade later, Lewellen (2002)explored some time series aspects of the momentum puzzle which "suggests that prices are not even weak-form efficient", and examined autocorrelations of the returns of portfolios formed from NYSE, AMEX, and NASDAQ common stocks, and also the autocorrelations of market-adjusted returns, using the CRSP database. Moskowitz et al. (2012) introduced time-series momentum as an alternative framework for momentum strategies, and gave empirical evidence that time-series momentum strategies (which are based on each security's past absolute performance, not relative to those of other securities) performed well relative to cross-sectional momentum strategies. In particular, they found substantial abnormal returns in a diversified portfolio of time series momentum strategies across all asset classes that performed best during extreme markets. Moreover, examination of the trading activities of speculators and hedgers led to their conclusion that "speculators profit from time series momentum at the expense of hedgers".

2.8.3   Contrarian strategies, behavioral finance, and investors' cognitive biases

Contrarian strategies monitor markets and investor sentiments to detect cognitive biases prevailing in the market. Dreman (1979, 2012), a widely acclaimed leader in contrarian investing because of his books and his contrarian investment firm Dreman Value Management, LLC, recognizes the important role that the psychological and interpretational aspects of human investment decisions play in driving stock prices, often pushing prices away from their intrinsic values. Psychology, and in particular the tendency to gravitate toward herd behavior, is part of human nature. Another human limitation is the interpretational difficulty in estimating future company value from annual reports, news and commentaries. While the price of a stock should ultimately move toward its intrinsic value, poor estimates of the value, due to the interpretational limitation and emotions prevalent in the market, may lead to periods of undervaluation or overvaluation. Contrarian strategies invest against (contrary to) the market, buying the out-of-favor stocks or shorting the preferred ones, and waiting for price reversals when the market rediscovers value in the out-of-favor stocks or shuns the high-fliers.

Contrarian strategies are therefore closely related to behavioral finance, a relatively new subfield of Finance and of Economics that brings psychological theory and human behavior into financial modeling, predictions and decisions, and economic analysis and policy. As noted in Section 1.3, behavioral finance provides an alternative view of the market to EMH under which prices should reflect their intrinsic values. In fact, it changes modeling risky ("speculative") prices of stocks and other securities in fundamental ways. It uses the *prospect theory* of Kahneman and Tversky (1979) for decision making under uncertainty; see Section 2.9 for further details and discussion.

2.8.4   From value investing to global macro strategies

The "intrinsic value" of a listed company also plays a major role in *value investing*, which dates back to the classic text by Graham and Dodd (1934)[4] and is also discussed in Section 11.1.5 of Lai and Xing (2008), where its connection to contrarian strategies and herd behavior is also discussed. Fundamental analysis of the company and its business sector plays an important role in de-

---

[4]Both Graham and Dodd were professors at Columbia Business School and their most famous student was Warren Buffett, who ran successful value investing funds before taking over Berkshire Hathaway.

termining the company's intrinsic value, particularly for value investing that also involves predicting future cash flows of the company over an investment horizon. These predictions may also involve analysis of macroeconomic trends.

The analysis and prediction of global macroeconomic developments feature prominently in *global macro* strategies that invest on a large scale around the world based on these predictions and geopolitical developments including government policies and inter-government relations. The analytics component of systematic global macro strategies used by hedge funds involves monitoring interest rate trends, business cycles, the global network of flow of funds, global imbalance patterns, and changing growth models of emerging economies. As noted in Section 1.4, hedge fund stategies can be broadly classified as "discretionary", relying on the skill and experience of the fund managers, and "systematic", relying on quantitative analysis of data and computer models and algorithms to implement the trading positions. Many global macro funds trade in the commodities and futures markets. George Soros, the legendary head of the Soros Fund Management LLC, has achieved outstanding returns in his Global Macro Fund and is best known for netting $1 billion profit by taking a short position of the pound sterling in 1992 when he correctly predicted that the British government would devalue the pound sterling at the time of the European Rate Mechanism debacle.

2.8.5 In-sample and out-of-sample evaluation of investment strategies and statistical issues of multiple testing

Section 11.1.4 of Lai and Xing (2008) discusses the issues with data snooping in the evaluation of the profitability of trading strategies, and reviews the bootstrap methods of White (2000) and Hansen (2005), and the approach of Romano and Wolf (2005) that uses stepdown bootstrap tests to control the family-wise error rate in multiple testing. We can use these methods for valid in-sample testing of trading strategies during their development. Clearly out-of-sample (ex-post) evaluation of their performance is more definitive but may be too late because they have already been used and resulted in losses. An in-sample (ex-ante) surrogate of an ex-post performance measure can be implemented by k-fold cross-validation; see Hastie et al. (2009) and Lai and Tsang (2016). Although the aforementioned bootstrap methods can maintain the family-wise error rate and avoid the pitfalls of data snooping in empirical testing of the profitability of trading strategies, there is a lack of systematic simulation study and theoretical development of the power of these tests of a large number of hypotheses. Lai and Tsang (2016) have recently filled the gap by developing a comprehensive theory and methodology for efficient postselection testing.

How to evaluate the performance of investment managers/strategies has actually been a long-standing problem in finance, as noted by Henriksson and Merton (1981) who proposed statistical methods to test for the forecasting skills of market-timers. A subsequent paper of Pesaran and Timmermann

(1992) developed a nonparametric test of the forecasting skill in predicting the direction of change of an economic variable under consideration, and applied it in two empirical studies in the British manufacturing sector. It should be noted, however, that these tests are based on the assumption of i.i.d. pairs of forecasts and outcomes. As we have seen in Section 2.6.3, time series effects should be incorporated in asset returns and therefore the test statistics should account for these effects. In addition, more sophisticated forecasts would incorporate uncertainties in the forecasts by giving, for example, the probability of price increase rather than whether the price will increase. Such probability forecasts are in fact implicit in the directional trading strategies at the end of Section 2.8.1. Testing the skills in probability forecasts when there are also time series effects is considerably more difficult, but Lai, Gross, and Shen (2011a) have recently made use of the martingale structure implicit in forecasting to resolve these difficulties.

2.9   Supplements and problems

1. *The Stable$(\alpha, \beta, \gamma, \delta))$ distribution* The distribution is characterized by its characteristic function:

$$E\left\{\exp(iuY)\right\} = \begin{cases} \exp\left\{-\gamma^\alpha |u|^\alpha \left[1 - i\beta \left(\tan \frac{\pi\alpha}{2}\right) \text{sign}(u)\right] + i\delta u\right\} & \alpha \neq 1 \\ \exp\left\{-\gamma |u| \left[1 + i\beta \frac{2}{\pi} \text{sign}(u) \log |u|\right] + i\delta u\right\} & \alpha = 1 \end{cases}$$

with $\gamma > 0$ and $i = \sqrt{-1}$. An early contribution to parameter estimation for stable distributions was the approximate quantile method of Fama and Roll (1971) for symmetric stable distributions; $E \exp(iuY)$ is real if Y is symmetric (i.e., Y and $-Y$ have the same distribution). McCulloch (1986) subsequently developed a general quantile approach, which can be implemented by McCullochParametersEstim in the R package StableEstim, version 2.0. This package also includes the R functions MLParametersEstim and GMMParametersEstim which correspond to the maximum likelihood estimation method and the generalized method of moments, respectively. Another estimation approach in StableEstim is KoutParametersEstim which uses an iterative Koutrouvelis regression method with different grids where the empirical characteristic functions are computed. To illustrate, we use McCullochParametersEstim to fit a symmetric stable distribution to the 2,517 daily closing price changes Δ_t of Pfizer Inc. shown in the bottom panel of Figure 2.4. The estimated parameters are $\hat{\alpha} = 1.652$, $\hat{\gamma} = 0.171$, $\hat{\beta} = \hat{\delta} = 0$. Figure 2.6 compares the QQ-plot of Δ_t of the fitted symmetric distribution with that of a fitted normal distribution. It shows that the data have thicker tails than the normal model but thinner tails than the stable model.

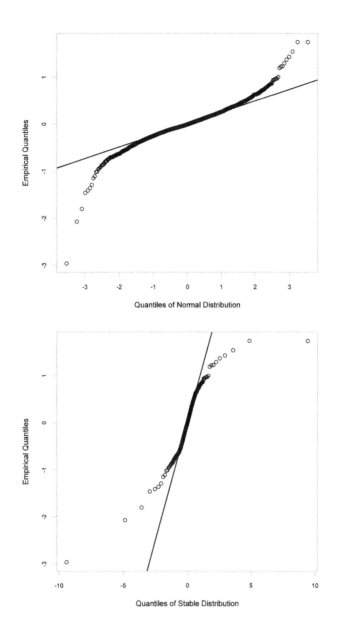

FIGURE 2.6: QQ-plots of Δ_t of normal model (top panel) and symmetric stable distribution (bottom panel).

2. *Risk-adjusted performance of a portfolio.* In Section 1.4, we have introduced the information ratio, the Sortino ratio, and the Modigliani RAP
   (risk-adjusted performance) measure to evaluate a portfolio's return adjusted for its risk profile. These performance measures, which are often
   used to rank investment funds, involve only the first two moments, or alternatively the mean and semi-variance, which are appropriate for Gaussian
   returns. For non-Gaussian returns, Gatfaoui (2010) reports a simulation
   study showing the impact of deviations from normality on the performance
   measures and rankings inferred from the sample estimates of the Sharpe
   ratio. Under the assumption of i.i.d. returns, the bootstrap method can
   be used to obtain bias-corrected estimates of these performance measures;
   see Section 3.5.2 of Lai and Xing (2008).

   Following Mandelbrot (1963), Rachev and Mittnik (2000) and Martin et al.
   (2005) propose to replace the common normal distribution of the asset
   returns by a stable distribution with finite mean, i.e., $1 < \alpha < 2$ in Section
   2.6.1. In place of σ which is infinite, they propose to use the conditional
   mean of the excess return $r - r_f$ given that it exceeds some upper quantile
   of the excess return distribution. This is called the expected shortfall in risk
   management; see Section 12.1.3 of Lai and Xing (2008). Note, however,
   that in practice $E(r - r_f)$ is replaced by its sample version $\bar{r} - r_f$ and
   that one uses the sample standard deviation s to replace the unknown σ
   (possibly infinite). It turns out that the self-normalized quantity $(\bar{r} - r_f)/s$
   has a bounded moment generating function in some neighborhood of the
   origin even when the returns have a stable distribution; see de la Peña
   et al. (2009). Although one may argue that the population version of the
   Sharpe ratio $E(r - r_f)/\sigma$ is not informative because it is 0 when $\sigma = \infty$ as
   in the case of stable distributions, what really counts is the sample Sharpe
   ratio which has a nonzero mean by the aforementioned property of self-
   normalized sums, and the bootstrap method for inference on the expected
   Sharpe ratio still works well in these settings.

3. *Black-Litterman asset allocation* Black and Litterman (1991) start with
   a normal assumption on the asset return $\mathbf{r}_t$ at time t with expected return
   $\boldsymbol{\mu} : \mathbf{r}_t|(\boldsymbol{\mu}, \boldsymbol{\Sigma}) \sim N(\boldsymbol{\mu}, \boldsymbol{\Sigma})$. To simplify the problem, they assume that $\boldsymbol{\Sigma}$
   is known (and substitute it by the sample covariance matrix based on
   historical data). They adopt a Bayesian approach with $\boldsymbol{\mu} \sim N(\boldsymbol{\pi}, \boldsymbol{\Sigma}_{\mu})$, or
   equivalently,

   $$\boldsymbol{\mu} = \boldsymbol{\pi} + \boldsymbol{\epsilon}^{(e)}, \qquad \boldsymbol{\epsilon}^{(e)} \sim N(\mathbf{0}, \tau\boldsymbol{\Sigma}), \tag{2.25}$$

   in which $\boldsymbol{\pi}$ corresponds to the "equilibrium risk premium" of the risky
   assets. Similar to the "elicitation" of the prior distribution in the Bayesian
   approach, they use "reverse optimization" to arrive at the specification of
   the prior mean in the form

   $$\boldsymbol{\pi} = \gamma\boldsymbol{\Sigma}\mathbf{w}^*, \tag{2.26}$$

   where γ is the risk aversion parameter of the market and $\mathbf{w}^*$ contains the market capitalization weights of the assets. They then make another simplifying assumption that $\boldsymbol{\Sigma}_\mu$ is proportional to the covariance matrix $\boldsymbol{\Sigma}$:

$$\boldsymbol{\Sigma}_\mu = \tau\boldsymbol{\Sigma}, \tag{2.27}$$

where τ is a scalar reflecting the uncertainty in the prior distribution of $\boldsymbol{\mu}$, for which a simple rule of thumb is the choice $\tau = \frac{1}{T}$, where T is the length of the sample period, thus setting $\tau = 0.017$ for 5 years of monthly data. Instead of considering the posterior distribution of $\boldsymbol{\mu}$ given the information set $\mathcal{F}_t$ of asset returns, $1 \le t \le n$, as in the first paragraph of Section 2.4.2, Black and Litterman allow investors to come up with their ways of processing this (and perhaps also other private) information to express their *views* via

$$\mathbf{P}\boldsymbol{\mu} = \mathbf{q} + \boldsymbol{\epsilon}^{(v)}, \qquad \boldsymbol{\epsilon}^{(v)} \sim N(\mathbf{0}, \boldsymbol{\Omega}), \tag{2.28}$$

in which $\mathbf{P}$ is a $p{\times}m$ matrix of the investors' individual picks of their assets, $\mathbf{q}$ is the "view vector" and $\boldsymbol{\Omega}$ is a diagonal covariance matrix representing the uncertainty of the views. Assuming that $\boldsymbol{\epsilon}^{(e)}$ and $\boldsymbol{\epsilon}^{(v)}$ are independent, they combine the investor's view (2.28) with the prior distribution (2.25)–(2.27) to derive the posterior distribution of $\boldsymbol{\mu}$ given the investor's view. This is the essence of the Black-Litterman approach.

Exercise   Use Bayes theorem to show that the posterior distribution is normal with mean (2.7) and covariance matrix in the line below (2.7).

4. The neo-MPT approach in Section 2.7.3 may perform better if the investor can combine domain knowledge with statistical modeling, although LXC has not done so in order to compare neo-MPT with other procedures that do not make use of domain knowledge. Indeed, using a purely empirical analysis of the past returns of these stocks to build the prediction model (2.19) does not fully utilize the power and versatility of the proposed approach, which is developed in Section 2.5 in a general Bayesian framework, allowing the skillful investor to make use of prior beliefs on the future return vector $\mathbf{r}_{n+1}$ and statistical models for predicting $\mathbf{r}_{n+1}$ from past market data. The prior beliefs can involve both the investor's and the market's "views", as in the Black-Litterman approach described in Section 2.4.2, for which the market's view is implied by the equilibrium portfolio. Note that Black and Litterman incorporate potential errors of these views in their normal priors whose covariance matrices reflect the uncertainties. Our Bayesian approach goes one step further to incorporate these uncertainties into the *actual* means and variances of the portfolio's return in the optimization problem (2.9), instead of simply using the *estimated* means and variances in the plug-in approach.

5. *Multi-objective optimization and efficient frontier*   Let $\mathbf{f} : \mathbb{R}^n \to \mathbb{R}^m$ be a vector-valued function of n decision variables such that $f_i(\mathbf{x})$ represents the ith objective function in a multi-objective optimization problem. There

may also be inequality and equality constraints on the decision variables so that the optimization problem can be represented by maximization of $\mathbf{f}(\mathbf{x})$ over $\mathbf{x} \in S$, where S is the constrained set and $\mathbf{x} \in S$ dominates $\mathbf{x}' \in S$ if $f_i(\mathbf{x}) \geq f_i(\mathbf{x}')$ for every $i = 1, \cdots, m$, with strict inequality for some i. A vector of decision variables is said to be *Pareto optimal* if there does not exist $\mathbf{x} \in S$ that dominates it. The set of Pareto optimal elements of S is called the *Pareto boundary*. If $\mathbf{x}$ is a random variable, then the f_i are expected functionals of $\mathbf{x}$. In particular, Markowitz's mean-variance portfolio optimization problem is a two-objective stochastic optimization problem, with $f_1(\mathbf{w}) = E(\mathbf{w}^\top \mathbf{r})$ and $f_2(\mathbf{w}) = -\mathrm{Var}(\mathbf{w}^\top \mathbf{r})$. Since one would like the mean of the portfolio return to be high and its variance to be low, the Pareto boundary in this case corresponds to the efficient frontier. A portfolio on Markowitz's efficient frontier can be interpreted as a minimum-variance portfolio achieving a target mean return, or a maximum-mean portfolio at a given volatility (i.e., standard derivation of returns). Portfolio managers prefer the former interpretation as target returns are appealing to investors, but this can cause problems for the plug-in approach, which will be discussed in Section 3.1.2.

6. *Fama-French 3-factor and 5-factor models*   The three factors in the Fama and French (1993) three-factor model are (a) the excess return $r_M - r_f$ on a market portfolio M as in CAPM, (b) the difference $r_S - r_L$ between the return on a portfolio S of small stocks and the return on a portfolio L of large stocks, and (c) the difference $r_h - r_l$ between the return r_h on a portfolio of high book-to-market stocks and the return r_l on a portfolio of low book-to-market stocks. Recently, Fama and French (2015) added two more factors in their five-factor model: (d) the difference between the returns on diversified portfolios of stocks with robust and weak profitability, and (e) the difference between the returns on diversified portfolios of low and high investment stocks. These factor models are used to develop and refine investment strategies for "financial advisors and institutional clients to engineer investment solutions," according to Kenneth French's website at Dartmouth's Tuck School of Business. The website also features a Data Library that gives current monthly returns for the factors and some associated portfolios of stocks in the CRSP database.

7. *Commentaries on statistical models*   George Box, to whom Section 2.6.3 has referred concerning time series models of stock returns, has the following words of wisdom on mathematical/statistical models and on model building: "Essentially, all models are wrong, but some are useful." This is particularly relevant not only to those in this chapter, but also to those that will be introduced in the subsequent chapters. This statement appeared in Box and Draper (1987, p. 424), and similar statements reflecting Box's philosophy that parsimonious models, though not strictly correct, can provide useful approximations appeared as early as Box (1976). Lai and Wong (2006) followed up on Box'es philosophy and combined it with

the subsequent commentaries of Cox (1990) and Lehmann (1990b) to come up with a substantive-empirical approach to time series modeling that combines domain knowledge (what Cox called "substantive") with empirical time series models (such as those in Section 2.6.3). In connection with the models in Section 5.3 on the limit order book, we should also mention Breiman (2001) who introduces *algorithmic modeling* to cope with the emergence of big data for statistical analysis.

8. *Factor models and high-dimensional covariance matrices* The multifactor model (2.4) associated with APT actually involves an economy with a large number of stocks. Chamberlain and Rothschild (1983) point out that Ross (1976) has only presented a heuristic justification of APT asserting that "asset prices are approximately linear functions of factor loadings" in an arbitrage-free market with a very large number of assets. They use a Hilbert space with the mean-square inner product to formulate (a) the prices of assets and their portfolios and (b) conditions for lack of arbitrage opportunities. In this framework they define a strict r-factor structure for the returns of the first N assets by $\mathbf{\Sigma}_N = \mathbf{B}_N\mathbf{B}_N^T + \mathbf{D}_N$, where $\mathbf{\Sigma}_N$ is the covariance matrix of these returns, $\mathbf{B}_N$ is $N \times r$ and $\mathbf{D}_N$ is a diagonal matrix whose elements are uniformly bounded by some constant. They also define an approximate r-factor structure if $\mathbf{D}_N$ is replaced by a nonnegative definite matrix $\mathbf{R}_N$ such that $\sup_N \lambda_{\max}(\mathbf{R}_N) < \infty$. They prove that an approximate r-factor structure is sufficient for Ross' theory. More importantly, they also show that if $\mathbf{\Sigma}_N$ has only r unbounded eigenvalues, then an approximate r-factor structure exists and is unique, and that the corresponding r unit-length eigenvectors converge and play the role of factor loadings. Sections 2.2 and 3.4.3 of Lai and Xing (2008) give an introduction to principal component analysis (PCA) and its application to estimating r and the factors in an approximate r-factor model. Section 3.4.1 of that book gives an introduction to factor analysis and its application to estimating r and the factors in an exact r-factor model. Under the classical assumption of i.i.d. returns, factor models as elucidated by Chamberlain and Rothschild have played a major role in dimension reduction and estimation of the covariance matrix $\mathbf{\Sigma}_N$.

Closely related to the developments in high-dimensional covariance matrix estimation and random matrix theory in statistical/machine learning for big data, there have been corresponding developments in factor models of asset returns in the presence of a large number of assets which we now summarize. The asymptotic distribution of the eigenvalues of a $p \times p$ real symmetric or complex Hermitian matrix is a basic theme in random matrix theory (RMT). In particular, Wigner (1955) used the empirical measure of the eigenvalues of a symmetric $p \times p$ matrix such that the entries above the diagonal are i.i.d. and are independent of those on the diagonal that are also i.i.d., with all entries being zero-mean random variables, to model the spectrum of a heavy atom, and established weak convergence of the empirical measure to a semi-circle law as $p \to \infty$. In 1967, Marčenko and Pastur (1967) published a foundational paper in RMT, in which they generalized random Hermitian matrices to random self-adjoint linear operators and introduced a powerful method involving Stieltjes transform to derive the limiting distribution of the eigenvalues of the operator. For the special case of sample covariance matrices, this technique was further developed by Bai (1999), Bai and Silverstein (1998) and others. El Karoui (2008) and Ledoit and Wolf (2012) made use of these refinements, together with the original Marčenko-Pastur equation that links the Stieltjes transform of the empirical spectral distribution (i.e., the probability distribution which puts equal mass on each eigenvalue of a $p \times p$ covariance matrix) to an integral with respect to the limiting spectral distribution as $t \to \infty$, to derive estimators of large-dimensional covariance matrices and to show that they have certain asymptotically optimal properties. Returning to factor models, Johnstone and Lu (2009) and Onatski (2010, 2012) developed an asymptotic theory for "weakly influential factors".

Johnstone and Lu assume i.i.d. ε_t with mean $\mathbf{0}$ in the factor model

$$\mathbf{x}_t = \mathbf{C}\mathbf{f}_t + \varepsilon_t, \quad t = 1, \ldots, T, \tag{2.29}$$

in which $\mathbf{f}_t = (f_{t1}, \ldots, f_{tr})^T$ is the vector of factors. Onatski allows cross-sectional and temporal correlations in ε_t of the form

$$\varepsilon_t = \mathbf{A} \left(b_1(L)u_{1t}, \ldots, b_p(L)u_{pt} \right)^\top, \tag{2.30}$$

where u_{ij} are i.i.d. with mean 0 and variance σ^2, $\mathbf{A}$ is a $p \times p$ coefficient matrix representing the cross-sectional correlations, and $b_i(L)$ is a polynomial of the lag operator L such that $b_i(L)u_{it} = u_{it} + b_{i,1}u_{i,t-1} + \ldots + b_{i,\nu_i}u_{i,t-\nu_i}$. Letting $\mathbf{F} = (\mathbf{f}_1, \ldots, \mathbf{f}_T)$ and normalizing the factors and loadings such that $T^{-1}\mathbf{F}^T\mathbf{F} = \mathbf{I}$ and $\mathbf{C}^T\mathbf{C}$ is diagonal, the assumption of weakly influential factors can be stated as

$$\mathbf{C}^T\mathbf{C} \to \mathrm{diag}(d_1, \ldots, d_r) \text{ with } \min_{1 \leq i \leq r} d_i > 0, \text{ as } T \to \infty \text{ and } p/T \to \rho, \tag{2.31}$$

for some constant ρ. Johnstone and Lu (2009) and Onatski (2010) show that if $\rho > 0$, then the factors determined by using PCA are inconsistent. They also show how the principal components estimators can be modified to obtain consistent estimates of "identifiable" factors and eigenvalues.

In contrast to assumption (2.31) on weakly influential factors, Fan et al. (2013) assume the "pervasiveness condition"

$$\liminf_{p \to \infty} p^{-1} \lambda_{\min}(\mathbf{C}^T\mathbf{C}) > 0 \tag{2.32}$$

and the "approximately sparse condition" that for some $0 \le \delta \le 1$,

$$\max_{i \le p} \sum_{j=1}^{p} |E(\varepsilon_i \varepsilon_j)|^{\delta} = O(1). \tag{2.33}$$

In this case, there exists $1 \le \nu \le r$ such that the largest ν eigenvalues of the $p \times p$ covariance matrix Σ of $\mathbf{x}_t$ are of order p and Fan et al. (2013) show how these eigenvalues and their corresponding unit-length eigenvectors can be estimated consistently by suitable thresholding of the entries of the residual matrix $\mathbf{R}_p$ in the Chamberlain-Rothschild approximate factor decomposition of Σ, yielding their estimator POET (principal orthogonal complement estimator by thresholding) using the sample covariance matrix in lieu of Σ. They also propose a consistent estimate of ν.

9. *High-dimensional regression: Sparsity and pathwise variable selection*
   The linear regression model

$$y_t = \alpha + \beta_1 x_{t1} + \cdots + \beta_p x_{tp} + \epsilon_t, \quad t = 1, \ldots, n, \tag{2.34}$$

in which the ϵ_t represent random, unobserved disturbances with $E(\epsilon_t) = 0$, is an extensively studied and widely used statistical model. By choosing the x_{ti} to be basis functions of observed input variables, (2.34) can be used for nonparametric regression; moreover, choosing some x_{ti} to be y_{t-i} and other x_{ti} as exogenous covariates and their lagged variables yields a rich class of time series models; Sections 7.1-7.4, 9.4 and 10.3 of Lai and Xing (2008). Despite the vast literature on its theory and applications, (2.34) became an active area of research again in the last decade, in response to the availability of big data that results in p (number of input variables) considerably larger than n (sample size). The $p \gg n$ problem appears hopeless at first sight since it involves many more parameters than the sample size and therefore the parameters cannot be well estimated, resulting in unacceptably large variances of the estimates. On the other hand, the regression function $f(x_1, \ldots, x_p) = \alpha + \beta_1 x_1 + \cdots + \beta_p x_p$ may still be estimable if the regression model is "sparse", and many applications indeed involve sparse regression models. There are two major issues with estimating $\boldsymbol{\beta} = (\beta_1, \cdots, \beta_p)^{\top}$ when $p \gg n$. The first is singularity of $\mathbf{X}^{\top}\mathbf{X}$, noting that the n values $\boldsymbol{\beta}^{\top}\mathbf{x}_1, \ldots, \boldsymbol{\beta}^{\top}\mathbf{x}_n$ cannot determine the p-dimensional vector $\boldsymbol{\beta}$ uniquely for $p > n$, where $\mathbf{X} = (x_{tj})_{1 \le t \le n, 1 \le j \le p}$ and $\mathbf{x}_i^{\top}$ is the ith row vector of $\mathbf{X}$. Assuming the ϵ_t to be i.i.d. normal and using a normal prior $\boldsymbol{\beta} \sim N(\mathbf{0}, \lambda \mathbf{I})$ can remove such singularity since the posterior mean of $\boldsymbol{\beta}$ is the ridge regression estimator $\hat{\boldsymbol{\beta}}^{\text{ridge}} = (\mathbf{X}^{\top}\mathbf{X} + \lambda \mathbf{I})^{-1}\mathbf{X}^{\top}\mathbf{Y}$, where $\mathbf{Y} = (y_1, \cdots, y_n)^{\top}$ (Lai and Xing, 2008, p.107). The posterior mean minimizes the penalized residual sum of squares $||\mathbf{Y} - \mathbf{X}\boldsymbol{\beta}||^2 + \lambda ||\boldsymbol{\beta}||^2$, with the L_2-penalty $||\boldsymbol{\beta}||^2 = \sum_{j=1}^{p} \beta_j^2$ and regularization parameter λ. The second issue with estimating $\boldsymbol{\beta}$ when $p \gg n$ is sparsity. Although the number of parameters is much larger than the sample size, one expects for the problem at hand that most of them are small and can be shrunk to 0. While the L_2-penalty does not lead to a sparse estimator $\hat{\beta}^{\text{ridge}}$, the L_1-penalty $\sum_{j=1}^{p} |\beta_j|$ does. The minimizer $\hat{\beta}^{\text{lasso}}$ of $||\mathbf{Y} - \mathbf{X}\beta||^2 + \lambda \sum_{j=1}^{p} |\beta_j|$ introduced by Tibshirani (1996), is called lasso (*least absolute shrinkage and selection operator*) because it sets some coefficients to be 0 and shrinks the others toward 0, thus performing both subset section and shrinkage. Unlike ridge regression, $\hat{\beta}^{\text{lasso}}$ does not have an explicit solution unless $\mathbf{X}$ has orthogonal columns, but can be computed by convex optimization algorithms. Zou and Hastie (2005) introduced the elastic net estimator $\hat{\beta}^{\text{enet}}$ that minimizes a linear combination of L_1- and L_2-penalties:

$$\hat{\beta}^{\text{enet}} = (1 + \lambda_2) \arg \min_{\beta} \{ ||\mathbf{Y} - \mathbf{X}\beta||^2 + \lambda_1 ||\beta||_1 + \lambda_2 ||\beta||^2 \}, \quad (2.35)$$

where $||\beta||_1 = \sum_{j=1}^{p} |\beta_j|$. The factor $(1+\lambda_2)$ in (2.35) is used to correct the "double shrinkage" effect of the naive elastic net estimator, which is Bayes with respect to the prior density proportional to $\exp\{-\lambda_2 ||\beta||^2 - \lambda_1 ||\beta||_1\}$, a compromise between the Gaussian prior (for ridge regression) and the double exponential prior (for lasso). Note that (2.35) is still a convex optimization problem. The choice of the regularization parameters λ_1 and λ_2 in (2.35), and λ in ridge regression or lasso, is carried out by k-fold cross-validation.[5] The R package glmnet can be used to compute $\hat{\beta}^{\text{lasso}}$ and $\hat{\beta}^{\text{enet}}$ and perform k-fold cross-validation.

Since the non-convex optimization problem of minimizing

$$\{ ||\mathbf{Y} - \mathbf{X}\beta||^2 + \lambda \sum_{j=1}^{p} I_{\{\beta_j \neq 0\}}, \quad (2.36)$$

which corresponds to the L_0-penalty, is infeasible for large p, lasso is sometimes regarded as an approximation of (2.36) by a convex optimization problem. Ing and Lai (2011) recently introduced a fast stepwise regression method, called the orthogonal greedy algorithm (OGA), following the terminology introduced by Temlyakov (2000), and used it in conjunction with a *high-dimensional information criterion* (HDIC) for variable selection along the OGA path. The method, which provides an approximate solution of the L_0-regularization problem, has three components. The first is the forward selection of input variables in a greedy manner so that the selected variables at each step minimize the residual sum of squares after ordinary least squares (OLS) is performed on it together

---

[5]Sections 7.3.2 of Lai and Xing (2008) explains the leave-one-out cross-validation method, in which observation i is removed from the sample as the test set and the remaining observations are used as the training set, $i = 1, \ldots, n$. A more flexible approach that uses a larger test sample is to split the data into k roughly equal-sized parts (of size n/k if n is divisible by k) and to use one part as the test sample and the remaining $k - 1$ parts as the training sample. The cross-validation estimate of the squared prediction error is the average of the squared prediction errors, over the k test samples.

with previously selected variables. This is carried out by OGA that orthogonalizes the included input variables sequentially so that OLS can be computed by component-wise linear regression, thereby circumventing matrix inversion; OGA is also called orthogonal matching pursuit in signal processing (Tropp and Gilbert, 2007). The second component of the procedure is a stopping rule to terminate forward inclusion after K_n variables are included. The choice $K_n = O((n/\log p_n)^{1/2})$ is based on a convergence rate result reflecting the bias-variance trade-off in the OGA iterations. The third component of the procedure is variable selection along the OGA path according to

$$HDIC(J) = n \log \hat{\sigma}_J^2 + \#(J) w_n \log p, \qquad (2.37)$$

where $J \subset \{1, \dots, p\}$ represents a set of selected variables, $\hat{\sigma}_J^2 = n^{-1} \sum_{t=1}^{n} (y_t - \hat{y}_{t;J})^2$ in which $\hat{y}_{t;J}$ denotes the fitted value of y_t when $\mathbf{Y}$ is projected into the linear space spanned by the column vectors $\mathbf{X}_j$ of $\mathbf{X}$, with $j \in J$, and w_n characterizes the information criterion used, e.g., $w_n = \log n$ for HDBIC, and $w_n = c$ for some constant c that does not change with n for HDAIC. Letting $\hat{J}_k = \{\hat{j}_1, \dots, \hat{j}_k\}$ denote the set of selected variables up to the kth step of OGA iterations, Ing and Lai (2011) choose $\hat{k}_n = \arg\min_{1 \le k \le K_n} HDIC(\hat{J}_k)$ and eliminate irrelevant variables $\hat{j}_l$ along the OGA path if $HDIC(\hat{J}_{\hat{k}_n} - \{\hat{j}_l\}) > HDIC(\hat{J}_{\hat{k}_n})$, $1 \le l \le \hat{k}_n$. The procedure is denoted by the acronym OGA+HDIC+Trim, which they showed to have the oracle property of being equivalent to OLS on an asymptotically minimal set of relevant regressors under a strong sparsity assumption.

10. *Dynamic factor models and multi-step ahead forecasts*   We now return to the factor model in Supplement 8 and introduce time series models on the factors. Barigozzi and Hallin (2016) have followed up on the approach of Fan et al. (2013) using the approximate sparsity assumption (2.33) and only ν "spiked" eigenvalues, for econometric analysis of high-dimensional covariance matrices based on time series data on the returns of a large number p stocks, with $T \ll p$. They point out, however, that the approach does not take into account the time series nature of the setting even though it allows ε_t to be strictly stationary (instead of i.i.d.) and strong mixing with sufficiently fast mixing rate. Motivated by the dynamic factor models in macroeconomic modeling that will be discussed below, they propose a general dynamic factor model (DFM) that has "level-common" and "level-idiosyncratic" components, in which "level" refers to the levels of the asset returns that are related to past returns via a vector autoregressive model. Moreover, they also propose a corresponding decomposition of a nonlinear transformation of the volatility of the level-idiosyncratic component. Their model is closely related to the martingale regression model of Lai et al. (LXC) in Section 2.6.3 that also has level-type components (2.19) and volatility-type components (2.20).

The DFMs considered by Barigozzi and Hallin (2016) have their origins in macroeconomic time series, particularly for multi-step ahead forecasts of key macroeconomic variables for the development and evaluation of monetary and other economic policies. Because multi-step ahead forecasts are involved in the analysis and prediction of the effect of a policy innovation on the economy, vector autoregression (VAR)[6] has been widely used to provide such forecasting models since the seminal works of Bernanke and Blinder (1992) and Sims (1992). However, Bernanke et al. (2005) have pointed out that "to conserve degrees of freedom, standard VARs rarely employ more than six to eight variables," leading to at least three problems with the forecasts thus constructed. The first is the "price puzzle" that predicts an increase (instead of the anticipated decrease) in price level following a contractionary monetary policy shock. The second is that "it requires taking a stand on specific observable measures" in the information sets in VAR analysis (e.g., industrial production or real GDP as the representative measure of economic activity). The third is that "impulse responses can be observed only for the included variables, which generally constitute only a small subset of variables that the researcher and policymakers care about" but does not consider how the effect of policy shocks on other variables may in turn impact on the included variables. They therefore introduce a factor-augmented vector autoregression (FAVAR) approach that "combines the standard VAR analysis with factor analysis." The factor or principal component analysis, which basically relates to the covariance matrix of the vector $\mathbf{x}_t$ of macroeconomic indicators, is used in the FAVAR approach to determine predictors that augment the vector $\mathbf{y}_t$ of basic economic variables in the VAR analysis; see Forni et al. (2005) and Stock and Watson (2002, 2011). Let $\mathbf{y}_t = (y_{t1}, \ldots, y_{tq})^T$ the vector of basic variables whose multi-step ahead forecasts are to be constructed. The FAVAR model is a VAR model for the factor-augmented time series $\mathbf{A}(L)^\top \mathbf{y}_t = \mathbf{G}(L)^\top \mathbf{f}_{t-1} + \boldsymbol{\eta}_t$, $\boldsymbol{\Phi}(L)^\top \mathbf{f}_t = \boldsymbol{\Psi}(L)^\top \mathbf{y}_{t-1} + \mathbf{w}_t$, where $\mathbf{f}_t$ are the factors in the factor model (2.29) for $\mathbf{x}_t$, and $\mathbf{A}(L) = \mathbf{I} - \sum_{s=1}^{d_1} \mathbf{A}_s L^s$, $\mathbf{G}(L) = \sum_{s=0}^{d_2} \mathbf{G}_s L^s$, $\boldsymbol{\Phi}(L) = \mathbf{I} - \sum_{s=1}^{d_3} \boldsymbol{\Phi}_s L^s$, and $\boldsymbol{\Psi}(L) = \sum_{s=0}^{d_4} \boldsymbol{\Psi}_s L^s$ are matrix polynomials of the lag operator L.

11. *Change-point detection and related technical rules* The connection between the filter rule in Section 2.8.1 and the CUSUM rule for change-point detection can be extended to construct other technical rules from corresponding change-point detection rules. Lai (1995) has provided a general method to convert a hypothesis test into a change-point detection rule. Lemma 1 of his paper says that if $X_1, X_2, \ldots$ are i.i.d. random variables and τ is a stopping time with respect to $X_1, X_2, \ldots$ such that $P(\tau < \infty) \leq \alpha$, and if N_k is the stopping time obtained by applying τ to $X_k, X_{k+1}, \ldots$, then $N = \min_{k \geq 1}(N_k + k - 1)$ is a stopping time and $EN \geq 1/\alpha$. Consider the one-sided sequential probability ratio test

---

[6]Section 9.4 of Lai and Xing (2008) gives an introduction to VAR models.

(SPRT) which stops sampling at stage

$$\tau = \inf\{t : \sum_{i=1}^{t} \log\left(\frac{f_1(X_i)}{f_0(X_i)}\right) \geq \log\gamma\}$$

and which rejects $H_0 : f = f_0$ on stopping in favor of the simple alternative $H_1 : f = f_1$. For this one-sided SPRT, $P_0(\tau < \infty) \leq \exp\{-\log\gamma\} = \gamma^{-1}$ and the lemma then yields that $E_0(N) \geq \gamma$, where P_0 and E_0 denote the probability measure and expectation under which $X_1, X_2, \ldots$, are i.i.d. with common density function f_0. Moreover, N reduces to the one-sided CUSUM rule

$$N = \inf\{n \geq 1 : \max_{1 \leq k \leq n} \sum_{i=k}^{n} \log\left(\frac{f_1(X_i)}{f_0(X_i)}\right) \geq \log\gamma\}.$$

Exercise Show that the two-sided CUSUM rule that is equivalent to the filter rule (2.24) corresponds to the two-sided SPRT for testing $H_0 : \mu = -\theta$ versus $H_1 : \mu = \theta$, with stopping rule

$$\tau^* = \inf\{n : \prod_{i=1}^{n} \frac{f_\theta(X_i)}{f_{-\theta}(X_i)} \notin (\frac{2\theta}{1+c}, 2\theta(1+c))\},$$

where f_μ denotes the $N(\mu, 1)$ density function. By considering the test $H_0 : \mu = 0$ versus $H_1 : \mu = \theta$ instead, show that the corresponding CUSUM rule is equivalent to a modified filter rule with an additional slope parameter, which Lam and Yam (1997) call the filter speed.

Exercise Read Lam and Yam (1997) and discuss how their modified filter rule that includes an additional filter speed parameter performs in comparison with the usual filter rule.

12. *Moving average and sequential GLR detection rules* Lai (1995) also applies his Lemma 1 to the Neyman-Pearson test (with fixed sample size m) to show that the moving average rule with window size m can have detection delay that is asymptotically as efficient as the CUSUM rule if m is chosen appropriately; the proof is given in his Lemma 2. His Section 3.3 describes the use of a family of window sizes, which was first proposed by Willsky and Jones (1976) for implementing generalized likelihood ratio (GLR) detection rules when the post-change parameter is not assumed known.

13. *Value and momentum everywhere* This is the title of the paper by Asness et al. (2013) who report an empirical analysis of value and momentum portfolios of individual stocks globally across four equity markets (US, UK, continental Europe, Japan) and four other asset classes (global equity indices, global government bonds, currencies, commodity futures). Their findings lead them to a 3-factor model consisting of a global market index,

a zero-cost value strategy applied across all asset classes, and a zero-cost momentum strategy across all assets. Using these "common global factors related to value and momentum", the model can "capture the comovement and the cross section of average returns both globally across asset classes and locally with an asset class." It also "does a good job capturing the returns to the Fama and French US stock portfolios as well as a set of hedge fund indices." As financial markets become "increasingly global", it is advantageous "to have a single model that describes returns across asset classes rather than specialized models for each market."

14. *Behavioral finance and portfolio selection under prospect theory*   Section 2.8.3 has referred to prospect theory, for which Daniel Kahneman was awarded the 2002 Nobel Prize in Economic Sciences. This behavioral economic theory describes how people make decisions when the outcomes of the actions to be chosen are uncertain, in contrast to expected utility theory that has been generally accepted as a normative model of rational choice in economics:

> Choices among risky prospects exhibit several pervasive effects that are inconsistent with the basic tenets of utility theory. In particular, people underweight outcomes that are merely probable in comparison with outcomes that are obtained with certainty. This tendency, called the certainty effect, contributes to risk aversion in choices involving sure gains and to risk-seeking in choices involving sure losses. In addition people generally discard components that are shared by all prospects under consideration. This tendency, called the isolation effect, leads to inconsistent preferences when the same choice is presented in different forms. An alternative theory of choice is developed, in which value is assigned to gains and losses rather than to final assets and in which probabilities are replaced by decision weights. The value function is normally concave for gains, commonly convex for losses, and is generally steeper for losses than for gains. Decision weights are generally lower than the corresponding probabilities, except in the range of low probabilities. Overweighting of low probabilities may contribute to the attractiveness of both insurance and gambling.

Benartzi and Thaler (1995) examine single-period portfolio choice for an investor with prospect-type value function. Barberis et al. (2001) bring prospect theory into asset pricing as a model for the level of average returns. Jin and Zhou (2008) formulate and study a general continuous-time behavioral portfolio selection model under prospect theory that features S-shaped value functions and probability distortions, which are non-linear transformations of the probability scale, enlarging a small probability and diminishing a large probability. These probability distortions lead to the Choquet integral[7] as a substitute for conventional expectation.

---

[7] Let $\nu : \mathcal{F} \to [0, \infty)$ be monotone in the sense that $\nu(A) \leq \nu(B)$ if $A \subset B$, where $\mathcal{F}$ is a collection of subsets of a set S. The Choquet integral of $f : S \to \mathbb{R}$ with respect to ν is defined by

$$\int_{-\infty}^{0} (\nu\{s : f(s) \geq x\}) - \nu(S))dx + \int_{0}^{\infty} \nu\{s : f(s) \geq x\}dx,$$

assuming that f is measurable with respect to ν (i.e., $\{s : f(s) \geq x\} \in \mathcal{F}$ for all $x \in \mathbb{R}$).

3

Active Portfolio Management and Investment Strategies

The cornerstones of quantitative portfolio management are prediction of asset returns from a large pool of investment possibilities, risk estimation, and portfolio optimization. The underlying methodology analyzes how wealth can be optimally invested in assets that differ in regard to their expected return and risk, and thereby also how risks can be reduced via adequate diversification. In this chapter, portfolio management refers the business of constructing and administering portfolios of assets. The typical setup of a quantitative fund in the portfolio management business consists of the following areas of expertise:

- Quantitative researchers – responsible for market forecast, risk estimation, and portfolio optimization research,

- Execution traders – responsible for daily trading and rebalancing activities, and for interfacing with prime brokers on trading related issues,

- Sales representatives – responsible for reaching out to clients to market the fund, with the goal of increasing assets under management (AUM) and diversifying the fund's client base,

- Operating managers – responsible for overseeing general administration of the fund, which includes redemption or reallocation of client wealth,

- Portfolio managers – responsible for overseeing the overall activity of the different portfolios under management.

There are two main styles of portfolio management – passive and active. Passive portfolio management constructs and administers portfolios that mimic the return of some given index. Such a portfolio is said to be tracking the index; any deviation, whether it is undershoot or overshoot, is considered a tracking error. Advocates of passive portfolio management argue that index tracking incurs low cost as it does not require much information gathering on individual stocks, and base their argument on market efficiency under which reducing investment costs is the key to improving net returns. Moreover, relatively infrequent trading results in fewer capital gains and therefore lower taxes. The goal of active portfolio management, on the other hand, is to construct portfolios that aim to outperform some index or benchmark, e.g., the S&P 500 index or the yield on the 3-month Treasury Bill; see Grinold and

Kahn (2000). An active portfolio manager is rewarded for generating additional return relative to the benchmark, and penalized for subpar return or for adding excess volatility relative to the benchmark. The additional return that a portfolio generates relative to the benchmark is commonly known as the *alpha* of the portfolio. Often, this alpha is assessed in the context of the additional risk for each unit of additional alpha. Metrics, such as the information ratio that expresses mean excess return in units of its standard deviation, are often used to measure a portfolio's performance.

If an active portfolio manager merely replicates the benchmark portfolio and uses leverage to increase return, there will be a corresponding increase in the volatility of the return and hence the risk of the portfolio. The net effect of the increase in return is then offset by the increase in risk, leading to an unchanged information ratio. Therefore, increasing leverage alone does not contribute to any incremental value added by the portfolio manager. A key to sustained profitability of active risk management is to identify, through superior information gathering, data analytics and model-based optimization, sources of alpha that boost return while minimizing additional contributions to the overall portfolio risk. In Sections 3.1 and 3.2, we discuss techniques to achieve this and also introduce the related concepts of active alpha and exotic beta, building on the foundations laid in Chapter 2. Section 3.3 considers dynamic investment strategies, which active portfolio management naturally involves, and various optimization and filtering problems that arise in this connection. In particular, incorporating transaction costs and other constraints into multiperiod portfolio optimization results in a hybrid of the passive approach and continual portfolio rebalancing that the active approach would entail in the absence of transaction costs. A review of long-standing problems and related developments is also provided, with further details in Sections 3.4 and 3.5.

3.1 Active alpha and beta in portfolio management

Following the notation in Chapter 2, let $\boldsymbol{r} \in \mathbb{R}^m$ be the vector of returns for m assets, $\boldsymbol{w}_P$ be the vector of portfolio weights, and $\boldsymbol{w}_B$ the benchmark weights. The returns of the portfolio and benchmark are then given by $r_P = \boldsymbol{w}_P^\top \boldsymbol{r}$ and $r_B = \boldsymbol{w}_B^\top \boldsymbol{r}$. Under the one-factor model $r_P - r_f = \alpha + \beta(r_B - r_f) + \epsilon$ as a "working model" to guide reductionist analysis, where r_f is the risk-free rate, subtracting $r_B - r_f$ from both sides of the equation defines the "active return" of the portfolio by the linear regression model

$$r_P - r_B = \alpha + \beta_P (r_B - r_f) + \epsilon, \tag{3.1}$$

in which the "active" α denotes the additional return of the portfolio over that of the benchmark, $\beta_P = \beta - 1$ is known as the active beta of the portfolio, and ϵ is a zero-mean random variable, commonly referred to as the residual, which is uncorrelated with $r_B - r_f$.

3.1.1 Sources of alpha

Alpha generation is a key component of active portfolio management to add value. The two main sources of alpha are (a) superior information, and (b) efficient information processing. Nowadays, the information dissemination of publicly tradable companies is under multiple layers of regulatory scrutiny, which has helped to eradicate insider trading based on private information. In order to gain a competitive advantage, alpha-generating information often needs to look beyond financial statements and news announcements, and to uncover related information that can help put the readily available material into perspective for making predictions for future price movements. For example, a visit to a factory can give a quantitative analyst a good sense of the trustworthiness and the interpretability of the company's financial statement and suggest strategies to make future projections.

Efficient use of information includes researching and developing models to describe factors that drive the underlying asset prices such that the model-based prediction can leverage the available information on the market fundamentals and anticipate market behavior. For example, a comprehensive model that captures the supply chain of a company and describes the interdependency of the chain in sufficient granularity can help to anticipate impacts on the company when components of the supply chain are disrupted. In this era of big data and vast investment universe, there are many challenges and opportunities to process and harness this rich source of information for alpha generation. Supplements 8, 9, 10 and 13 of Section 2.9 have given a brief review of modern developments in statistical modeling and data science that provide analytical tools to meet these challenges.

3.1.2 Exotic beta beyond active alpha

Litterman (2004, 2008) points out that superior risk-adjusted returns can come not only through the active alpha of a portfolio but also a low active beta. He says that in 1969, Black, Jensen and Scholes documented that "low-beta stocks seemed to be providing too high a return, relative to high-beta stocks, to be consistent with the CAPM." Since $\text{Var}(r_P - r_B) = \beta_P^2 \text{Var}(r_B) + \text{Var}(\varepsilon)$ by (3.1), a decrease in the value of β_P^2 means a smaller variance for $r_P - r_B$, which leads to a higher information ratio for a portfolio over others with similar mean returns but higher betas. In particular, since the benchmark portfolio has beta equal to 1 (and therefore zero active beta), a portfolio with positive alpha and small $|\beta| < 1$ can have a high information ratio. Such a portfolio is said to have an "exotic beta". Litterman (2004) says that the traditional CAPM framework only refers to a single beta related to a market portfolio, but today's investors "allocate weights to different asset classes and often refer to multiple betas, one for each asset class", thereby providing opportunities to "both create alpha and outperform the market portfolios." He suggests that "combining active portfolio management to generate alpha with passive exposures to exotic betas" can result in better performance of the portfolio. Litterman (2008) further points that whereas "market beta" is used to refer to the classic CAPM exposure to a market portfolio, exotic beta "can arise from a largely passive exposures to less mainstream asset classes, such as high-yield bonds, global small-cap equities, emerging market equities, emerging market debt, commodities, and global real estate", and "allows investors to think about sources of return in terms of a spectrum, beginning with pure market beta that has low cost and unlimited supply but currently has low Sharpe ratio." Exotic beta, therefore, has the advantage of relatively large capacity, enhanced return potential and low correlation with the market portfolio. Whereas he agrees that ultimately equilibrium will take place and consequently exotic beta as a source of high information ratio cannot persist, this may take many years to happen and "in the meantime, investors can benefit as the prices of the assets they hold get bid up."

3.1.3   A new approach to active portfolio optimization

An active portfolio seeks to get better risk-adjusted returns than the benchmark portfolio B to justify the fees of the portfolio manager. For a given portfolio P, the difference $\widetilde{w} = w_P - w_B$ satisfies $\widetilde{w}^\top \mathbf{1} = 0$. We can reformulate Markowitz's optimization problem (2.9) in the case of unknown $(\boldsymbol{\mu}, \boldsymbol{\Sigma})$ for the active portfolio weight vector $\widetilde{w}$ as follows:

$$\max \left\{ E(\widetilde{w}^\top r_{n+1}) - \lambda \mathrm{Var}(\widetilde{w}^\top r_{n+1}) \right\},$$
$$\text{subject to } \widetilde{w}^\top \mathbf{1} = 0 \text{ and } \widetilde{w} \in \mathcal{C}, \tag{3.2}$$

where $\mathcal{C}$ represents long-short and other constraints. Varying λ can yield the efficient frontier. Instead of solving for efficient frontier, we choose the λ that has the highest information ratio estimated from the sample data as in Section 2.7.2. Therefore we can carry out the procedure in Section 2.7, with obvious modifications, to solve this active portfolio optimization problem. A reason why this is new compared to that in Section 2.7 is that it basically tries to modify the benchmark portfolio weights, which are often updated daily and therefore already contains "free" market information; see Supplement 1 in Section 3.4. This enables one to come up with approximate solutions to (3.2) that can incorporate conveniently domain knowledge and prior beliefs about future movements of the stocks and of their associated firms and financial sectors, as we have discussed in Supplement 4 of Section 2.9. Alpha generation can be accomplished by using additional information and superior analytics, as noted in Section 3.1.1.

Another reason why we call this a new approach is that portfolio managers traditionally pick some target excess return $\widetilde{\mu}_*$, which can be used to justify their management fees, for the active portfolio, leading to the alternative optimization problem:

$$\min E(\widetilde{\boldsymbol{w}}^{\top} \boldsymbol{\Sigma} \widetilde{\boldsymbol{w}}), \quad \text{subject to } \widetilde{\boldsymbol{w}}^{\top} \boldsymbol{\mu} = \widetilde{\mu}_*, \ \widetilde{\boldsymbol{w}}^{\top} \mathbf{1} = 0 \text{ and } \widetilde{\boldsymbol{w}} \in \mathcal{C}; \qquad (3.3)$$

see Ledoit and Wolf (2004, abbreviated by LW in this section). To illustrate this point, Lai et al. (2011b, abbreviated by LXC) follow LW and use a value-weighted portfolio of 50 stocks as the benchmark portfolio and the constraint set $\mathcal{C} = \{\widetilde{\boldsymbol{w}} : -\boldsymbol{w}_B \leq \widetilde{\boldsymbol{w}} \leq c\mathbf{1} - \boldsymbol{w}_B\}$, with $c = 0.1$, i.e., the portfolio is long only and the total position in any stock cannot exceed an upper bound c. As in LW, the $m = 50$ stocks are chosen at the beginning of each month t so that they have a largest market value among those in the CRSP database that have no missing monthly pieces in the previous 120 months, which are used as the training samples. LXC uses quadratic programming to solve the optimization problem (3.3) in which $\boldsymbol{\mu}$ and $\boldsymbol{\Sigma}$ are replaced, for the plug-in active portfolio, by their sample estimates based on the training sample in the past 120 months. The covariance-shrinkage active portfolio uses a shrinkage estimator of $\boldsymbol{\Sigma}$ instead, shrinking towards a patterned matrix that assumes all pairwise correlations to be equal (Ledoit and Wolf, 2003). LXC extends the NPEB approach in Section 2.5 to construct the corresponding NPEB active portfolio, but does not incorporate domain knowledge to be comparable to the other procedures in Table 3.1.

TABLE 3.1: Means and standard deviations (in parentheses) of annualized realized excess returns over the value-based benchmark

$\widetilde{\mu}_*$	0.01	0.015	0.02	0.03
λ	2^2	2	2^{-1}	2^{-2}
(a) All test periods by re-defining portfolios in some periods				
Plug-in	0.001 (4.7e-3)	0.002 (7.3e-3)	0.003 (9.6e-3)	0.007 (1.4e-2)
Shrink	0.003 (4.3e-3)	0.004 (6.6e-3)	0.006 (8.8e-3)	0.011 (1.3e-2)
Boot	0.001 (2.5e-3)	0.001 (3.8e-3)	0.001 (5.1e-3)	0.003 (7.3e-3)
NPEB	0.029 (1.2e-1)	0.046 (1.3e-1)	0.053 (1.5e-1)	0.056 (1.6e-1)
(b) Test periods in which all portfolios are well defined				
Plug-in	0.002 (6.6e-3)	0.004 (1.0e-2)	0.006 (1.4e-2)	0.014 (1.9e-2)
Shrink	0.005 (5.9e-3)	0.008 (9.0e-3)	0.012 (1.2e-2)	0.021 (1.8e-2)
Boot	0.001 (3.5e-3)	0.003 (5.3e-3)	0.003 (7.1e-3)	0.006 (1.0e-2)
NPEB	0.282 (9.3e-2)	0.367 (1.1e-1)	0.438 (1.1e-1)	0.460 (1.1e-2)

LXC first notes that specified target excess returns $\widetilde{\mu}_*$ may be vacuous for the plug-in, covariance-shrinkage (abbreviated "shrink" in Table 3.1) and resampled (abbreviated "boot" for bootstrapping) active portfolios in a given test period. For $\widetilde{\mu}_* = 0.01, 0.015, 0.02, 0.03$, there are 92, 91, 91 and 80 test periods, respectively, for which (3.3) has solutions when $\boldsymbol{\Sigma}$ is replaced by either the sample covariance matrix or the Ledoit-Wolf shrinkage estimator of the training data from the previous 120 months. Higher levels of target returns result in even fewer of the 180 test periods for which (3.3) has solutions. On the other hand, values of $\widetilde{\mu}_*$ that are lower than 1% may be of little practical interest to active portfolio managers. When (3.3) does not have a solution to provide a portfolio of a specified type for a test period, LXC uses the value-weighted benchmark as the portfolio for the test period. Table 3.1(a) gives the actual (annualized) mean realized excess returns $12\bar{e}$ to show the extent to which they match the target value $\widetilde{\mu}_*$, and also the corresponding annualized standard deviations $\sqrt{12}s_e$, over the 180 test periods for the plug-in, covariance-shrinkage and resampled active portfolios constructed with the above modification. These numbers are very small, showing that the three portfolios differ little from the benchmark portfolio, so the realized information ratios that range from 0.24 to 0.83 for these active portfolios can be quite misleading if the actual mean excess returns are not taken into consideration. Table 3.1(a) also gives the means and standard deviations of the annualized realized excess returns of the NPEB active portfolio for four values of λ that are chosen so that the mean realized excess returns roughly match the values of $\widetilde{\mu}_*$ over a grid of the form $\lambda = 2^j$ $(-2 \leq j \leq 2)$ that LXC uses. Note that NPEB has considerably larger mean excess returns than the other three portfolios.

LXC has also tried another default option that uses 10 stocks with the largest mean returns (among the 50 selected stocks) over the training period and puts equal weights to these 10 stocks to form a portfolio for the ensuing test period for which (3.3) does not have a solution. The mean realized excess returns $12\bar{e}$ when this default option is used are all negative (between -17.4% and -16.3%) while $\widetilde{\mu}_*$ ranges from 1% to 3%. Table 3.1(b) restricts only to the 80–92 test periods in which the plug-in, covariance-shrinkage and resampled active portfolios are all well defined by (3.3) for $\widetilde{\mu}_* = 0.01, 0.015, 0.02$ and 0.03. The mean excess returns of the plug-in, covariance-shrinkage and resampled portfolios are still very small, while those of NPEB are much larger. The realized information ratios of NPEB range from 3.015 to 3.954, while those of the other three portfolios range from 0.335 to 1.214 when we restrict to these test periods. LXC also compares the S&P 500 Index with the value-weighted portfolio. The S&P 500 Index has mean 0.006 and standard deviation 0.046, while the mean of the value-weighted portfolio is 0.0137 and its standard deviation is 0.045, showing an advantage of the value-weighted portfolio over the S & P 500 Index as the benchmark portfolio for this study period.

3.2 Transaction costs, and long-short constraints

3.2.1 Cost of transactions and its components

Active portfolio management clearly outperforms the passive approach in the absence of transaction costs, but the advantage may be outweighed by the transaction costs incurred in purchase and sale of the assets constituting the portfolio. Transaction costs can be broken down into three key components: commissions, bid-ask spread, and market impact. Together, they contribute to the portfolio's overall *implementation shortfall*, which compares the actual portfolio return with that of an ideal scenario without any transaction costs.

The first component, the commission cost, consists of a number of subcomponents such as broker commission, exchange commission, and clearing house fee. These costs are usually fixed as a percentage of the transaction volume, and can vary depending on the underlying asset. For stocks, it is usually in the pennies whereas for futures and options contracts, it can be close to a dollar per contract. Being deterministic, this component is simple to account for in the portfolio construction phase. The second component, the bid-ask spread, varies depending on the market condition and on the underlying asset. Under normal market conditions, stocks listed on the US exchange have bid-ask spread typically in the region of 0.02% of the price, with a minimum set at $0.01, which is known as the minimum tick size. This minimum tick size is specified by the exchanges and is the same across all stocks. For futures, the minimum spread can vary depending on the underlying asset, and is often in the region of $10 to $25 per contract. The bid-ask spread is a function of the volatility and the overall liquidity of the asset, and is therefore a random variable. Recently, as a result of intense competition among the high-frequency market-makers in providing liquidity, the spread has diminished sharply; it now usually hovers for most assets around the minimum tick size.

The third component, the market impact, is the most difficult to quantify. It is unobservable and can only be inferred by suitable application of statistical and economic models. There is a wealth of literature on this subject, leveraging recent advances in market microstructure and behavioral finance research; see Hasbrouck (1996) and Bouchaud et al. (2002) for a comprehensive survey. A rudimentary form that is widely used is to incorporate transaction costs as a proportional cost added to the expected return. Incorporating proportional transaction cost in Markowitz's portfolio optimization problem (2.1) leads to

$$\boldsymbol{w}_{\mathrm{eff}} = \arg\min_{\boldsymbol{w}} \boldsymbol{w}^\top \boldsymbol{\Sigma}\boldsymbol{w}$$
$$\text{such that } \boldsymbol{w}^\top \boldsymbol{\mu} + \boldsymbol{\phi}^\top (\boldsymbol{w}_+ + \boldsymbol{w}_-) = \mu_*,$$
$$\boldsymbol{w}^\top \boldsymbol{1} = 1, \ \boldsymbol{w}_+ \geq \boldsymbol{0}, \ \boldsymbol{w}_- \geq \boldsymbol{0},$$

where $\boldsymbol{w}_+$ (or $\boldsymbol{w}_-$) is the vector of positive (or negative) weights, $\boldsymbol{w} = \boldsymbol{w}_+ -$

w_-, and the vector ϕ contains the transaction costs per unit of asset for all assets contained in the portfolio.

3.2.2 Long-short and other portfolio constraints

Benign constraints on portfolio weights do not substantially increase the complexity of the optimization problem. An example is the long-short constraint for which the optimization problem becomes

$$w_{\text{eff}} = \arg\min_{w} w^\top \Sigma w$$
$$\text{such that} \quad w^\top \mu + \phi^\top (w_+ + w_-) = \mu_*,$$
$$w^\top 1 = 1, \; 0 \le w_+^\top 1 \le \psi_L, \; 0 \le w_-^\top 1 \le \psi_S,$$

where ψ_L and ψ_S are the long and short portfolio weight caps, respectively. However, some practical constraints can increase the complexity of the underlying optimization problem significantly, especially when they destroy the convex nature of the problem and give rise to *NP-hard* problems for which the solution is not guaranteed in polynomial time. These problems can only be solved if the number of constraints is small or if one is willing to forgo optimality, approaching the problem approximately with some heuristics. The *cardinality constraint* is one such constraint that is often used to limit the trade size to integer shares or contracts, or to limit the number of assets that the portfolio holds. For example, for index tracking, there are cases where an index can have a large number of constituent assets that need to be tracked by a smaller portfolio using just a subset of index constituents for feasibility and transaction cost reasons. A canonical index tracking problem with cardinality constraint can be expressed as a Cardinality-Constrained Quadratic Program (CCQP).

$$w_{\text{eff}} = \arg\max_{w} w^\top \mu - \frac{1}{2}\lambda w^\top \Sigma w$$
$$\text{such that} \quad Aw \ge b, \; w^\top 1 = 1, \; \sum_i 1_{\{w_i \neq 0\}} = K,$$

where K is the constraint on the number of non-zero weighted assets in the optimal portfolio. If one is only interested in a reasonably good solution, then one can use successive truncations involving a sequence of relaxed quadratic programs so that at each iteration, assets with small or zero weights are truncated off. More sophisticated methods require algorithms that are capable of globally optimizing the objective function over the constrained parameter space, or finding a good approximation to the global solution in significantly less time. Supplements 2 and 3 in Section 3.4 give an introduction to solving CCQP and some heuristic techniques.

3.3 Multiperiod portfolio management

In his incisive discussion on smart rebalancing, diBartolomeo (2012) says that MPT as introduced by Markowitz "frames the time dimension of investing as a single period over which the parameters of the probability distribution of asset returns are both known with certainty and unchanging", but that "neither assumption is true in the real world." Although there has been much academic research in "full multi-period optimization", it has seldom been used by "investment professionals" (i.e., fund managers and investment consultants) who instead focus on when it is really necessary to bear the costs and use single-period mean-variance optimization to rebalance their portfolio weights. Although some form of these "all or nothing" rebalancing rules is commonly used in practice, he notes at least two "negative arguments" not in their favor. The first is that "when active managers are inactive because the potential benefits of rebalancing are too small, this lack of trading is perceived by clients as the manager being neglectful rather than as an analytically-driven decision to reduce trading costs." The second argument is that after a period of inactivity, the eventual rebalancing concentrates the required trading into a particular moment in time, resulting in market impact and higher transaction cost. In this section we give an overview of (a) the academic literature on multi-period portfolio theory, and (b) writings of active portfolio managers and other investment professionals on portfolio rebalancing under the constraints they encounter in implementation. This overview leads to a number of basic modeling and optimization techniques, details of which are given in Section 3.4 where we also describe some recent work that uses an interdisciplinary approach and incorporates both vantage points toward a solution of the long-standing problem of dynamic rebalancing in active portfolio management. In particular, Supplements 4–7 in Section 3.4 introduce dynamic optimization principles and methods, which will provide the background for further development in Chapters 6 and 7 for the specific quantitative trading problems studied therein.

3.3.1 The Samuelson-Merton theory of "lifetime portfolio selection" of risky assets via stochastic control

This theory begins with the first paragraph of Samuelson (1969) with a footnote acknowledging his student Merton whose "companion paper" in the same issue of the journal tackles "the much harder problem of optimal control in the presence of continuous-time stochastic variation" (which is *stochastic control* in today's terminology). Samuelson says that "most analyses of portfolio selection, whether they are of the Markowitz-Tobin mean-variance or more general type, maximize over one period," but that "a lifetime planning of consumption and investment decisions" involves a many-period generalization.

He assumes a finite-horizon discrete-time framework and uses a backward induction approach to convert multi-period optimization to "the well-known one-period portfolio problem," running the clock backwards and using an induction argument at every stage. He adds up the investor's expected utility of consumption at every stage, which is widely accepted as a normative model of rational decision making in economic theory. Although this approach only works for the finite-horizon setting, he says that letting the horizon approach infinity should also work for infinite-horizon (hence lifetime) portfolio selection.

Merton (1969) considers the continuous-time framework in which the market consists of two investment instruments: a bond paying a fixed risk-free rate $r > 0$ and a stock whose price is a geometric Brownian motion with mean rate of return $\alpha > 0$ and volatility $\sigma > 0$. Thus, the stock price at time $t \geq 0$ is given by

$$dS_t = S_t(\alpha \, dt + \sigma \, dB_t), \tag{3.4}$$

where $\{B_t : t \geq 0\}$ is a standard Brownian motion. The investor's position is denoted by (X_t, Y_t), where

$$
\begin{aligned}
X_t &= \text{dollar value of investment in bond,} \\
Y_t &= \text{dollar value of investment in stock,} \\
y_t &= \text{number of shares held in stock.}
\end{aligned}
\tag{3.5}
$$

Note in particular that $Y_t = y_t S_t$. Supplements 4 and 5 in Section 3.4 provides some basic background for discrete-time and continuous-time dynamic programming.

The investment and consumption decisions of an investor comprise three non-negative stochastic processes C, L, and M, such that C is integrable on each finite time interval, and L and M are nondecreasing and right-continuous with left-hand limits. Specifically, the investor consumes at rate C_t from the bond, and L_t (respectively, M_t) represents the cumulative dollar value of stock bought (respectively, sold) within the time interval $[0, t]$, $0 \leq t \leq T$. Thus, the investor's position (X_t, Y_t) satisfies

$$dX_t = (rX_t - C_t) \, dt - dL_t + dM_t, \tag{3.6}$$
$$dY_t = \alpha Y_t \, dt + \sigma Y_t \, dB_t + dL_t - dM_t. \tag{3.7}$$

By requiring the investor to remain solvent (i.e., to have non-negative net worth) at all times, the investor's position is constrained to lie in the solvency region $\mathcal{D}$ which is a closed convex set bounded by the line segments

$$
\begin{aligned}
\partial_\lambda \mathcal{D} &= \{(x, y) : x > 0, \ y < 0 \text{ and } x + y = 0\}, \\
\partial_\mu \mathcal{D} &= \{(x, y) : x \leq 0, \ y \geq 0 \text{ and } x + y = 0\}.
\end{aligned}
\tag{3.8}
$$

Denote by $\mathcal{A}(t, x, y)$ the class of admissible policies, for the position $(X_t, Y_t) = (x, y)$, satisfying $(X_s, Y_s) \in \mathcal{D}$ for $t \leq s \leq T$, or equivalently, $Z_s \geq 0$ for

$t \leq s \leq T$ where $Z_s = X_s + Y_s$. At time t, the investor's objective is to maximize over $\mathcal{A}(t, x, y)$ the expected utility

$$J(t, x, y) =$$
$$E\left[\int_t^T e^{-\beta(s-t)} U_1(C_s) \, ds + e^{-\beta(T-t)} U_2(Z_T) \middle| X_t = x, Y_t = y\right], \quad (3.9)$$

in which $\beta > 0$ is a discount factor and U_1 and U_2 are concave utility functions of consumption and terminal wealth. Suppose U_1 and U_2 are chosen from the so-called CRRA (constant relative risk aversion)[1] class:

Assume that U_1 is differentiable and that the inverse function $(U_1')^{-1}$ exists. Often

$$U(c) = c^\gamma / \gamma \quad \text{if } \gamma < 1, \gamma \neq 0; \qquad U(c) = \log c \quad \text{if } \gamma = 0, \quad (3.10)$$

which has constant relative risk aversion $-cU''(c)/U'(c) = 1 - \gamma$. The value function is defined by

$$V(t, x, y) = \sup_{(C,L,M) \in \mathcal{A}(t,x,y)} J(t, x, y). \quad (3.11)$$

Summing (3.6) and (3.7) yields for the investor's total wealth $Z_t = X_t + Y_t$:

$$dZ_t = \{rZ_t + (\alpha - r)\theta_t Z_t - C_t\} \, dt + \sigma \theta_t Z_t \, dB_t, \quad (3.12)$$

where $\theta_t = Y_t/(X_t + Y_t)$ is the proportion of the investment held in stock. Using the reparameterization $z = x + y$, define the value function

$$V(t, z) =$$
$$\sup_{(C,L,M) \in \mathcal{A}(t,z)} \mathbb{E}\left[\int_t^T e^{-\beta(s-t)} U_1(C_s) \, ds + e^{-\beta(T-t)} U_2(Z_T) \middle| Z_t = z\right], \quad (3.13)$$

where $\mathcal{A}(t, z)$ denotes the class of admissible policies (C, θ) for which $Z_s \geq 0$ for all $t \leq s \leq T$. Let

$$p = \frac{\alpha - r}{(1 - \gamma)\sigma^2}, \qquad c = \frac{1}{1 - \gamma}\left[\beta - \gamma r - \frac{\gamma(\alpha - r)^2}{2(1 - \gamma)\sigma^2}\right],$$
$$C_i(t) = c/\{1 - \phi_i e^{c(t-T)}\} \ (i = 1, 2), \qquad \phi_1 = 1, \ \phi_2 = 1 - c. \quad (3.14)$$

Assuming the CRRA form (3.10) for the utility functions U_1 and U_2 in the

---

[1] Risk aversion in economics refers to human preference for choices with more certain outcomes. An agent with utility function $U(x)$, where x represents the value that the agent might receive in money or goods, is said to be *risk-averse* if U is concave, *risk-seeking* if U is convex, and *risk-neutral* if U is linear. The coefficient, also called the Arrow-Pratt measure, of absolute risk aversion is $-U''(x)/U'(x)$, and the coefficient of relative risk aversion is $-xU''(x)/U'(x)$.

stochastic control problem (3.13), Merton (1969) finds that the optimal strategy is to devote a constant proportion (the Merton proportion) p of the investment to the stock and to consume at a rate proportional to wealth. Details of his solution are given in Supplement 5 of Section 3.4. Similar results for more general utility functions U_1 and U_2 have been given by Karatzas et al. (1987) and Cox and Huang (1989).

3.3.2   Incorporating transaction costs into Merton's problem

Since Merton does not consider transaction costs, trading occurs continuously so that dL_t and dM_t are multiples of dt and (3.6)–(3.7) can be combined into (3.12). The reason why we have gone through the more detailed dynamics captured by (3.6) and (3.7) is that it can be easily extended to the setting of proportional transaction costs that we now consider. Suppose the investor pays fractions $0 \leq \lambda < 1$ and $0 \leq \mu < 1$ of the dollar value transacted on purchase and sale of a stock, respectively. Then, the investor's position (X_t, Y_t) satisfies

$$dX_t = (rX_t - C_t)\, dt - (1 + \lambda)\, dL_t + (1 - \mu)\, dM_t, \qquad (3.15)$$
$$dY_t = \alpha Y_t\, dt + \sigma Y_t\, dB_t + dL_t - dM_t. \qquad (3.16)$$

The factor $1 + \lambda$ in (3.15) reflects the fact that a transaction fee in the amount of λdL_t needs to be paid from the bond when purchasing a dollar amount dL_t of the stock. Similarly, the factor $1 - \mu$ in (3.15) is due to the transaction fee in the amount of μdM_t when selling a dollar amount dM_t of the stock. Define the investor's dollar value of net worth as

$$Z_t = X_t + (1 - \mu)Y_t \quad \text{if } Y_t \geq 0; \qquad Z_t = X_t + (1 + \lambda)Y_t \quad \text{if } Y_t < 0.$$

Including transaction costs also changes the definition (3.8) to

$$\partial_\lambda \mathcal{D} = \{(x,y) : x > 0,\ y < 0 \text{ and } x + (1 + \lambda)y = 0\},$$
$$\partial_\mu \mathcal{D} = \{(x,y) : x \leq 0,\ y \geq 0 \text{ and } x + (1 - \mu)y = 0\}.$$

With these changes, the problem is still to find $(C, L, M) \in \mathcal{A}(t,x,y)$ that maximizes the expected utility $J(0,x,y)$ defined in (3.9). This is a *singular stochastic control* problem, characterized by a no-trade region that is bounded between a buy-boundary $\mathcal{B}_t$ and a sell boundary $\mathcal{S}_t$, buying stock immediately when the stock price S_t is at or below the buy-boundary $\mathcal{B}_t$, and selling stock immediately when $S_t \geq \mathcal{S}_t$. Details about singular stochastic control and the buy- and sell-boundaries are given in Supplement 6 of Section 3.4. Since there is no analytic solution for the buy- and sell-boundaries, numerical methods have been developed to compute them by Gennotte and Jung (1994), Zakamouline (2005), Dai and Zhong (2010), Chen and Dai (2013) and Palczewski et al. (2015). The problem is simpler in the infinite-horizon case $(T = \infty)$ because the buy- and sell- boundaries do not depend on t. Magill and Constantinides (1976) considered this case heuristically and arrived at the fundamental insight that there is a no-trading region which is the "wedge" that Davis and Norman (1990) later proved rigorously by using the principle of smooth fit and deriving the ordinary differential equation that determine the wedge. Extensions to multiple risky assets have been considered by Akian et al. (1996), Leland (2000), and Liu (2004).

3.3.3 Multiperiod capital growth and volatility pumping

The multiperiod setting and logarithmic utility have led to an alternative approach to "lifetime portfolio selection" considered in Section 3.3.1, dating back to Kelly (1956), Latane (1959), and Breiman (1961). This approach basically deals with the discrete-time version of (3.13) with $U_1 = 0$ and $U_2(c) = \log(c)$. Let Z_t denote the capital (which is the same as wealth since there is no consumption) at time $t = 0, 1, \ldots, T$. Let P_t^i denote the price of the ith security in the portfolio and let $\boldsymbol{w}_t = (w_t^1, \ldots, w_t^m)^\mathsf{T}$ be the portfolio weights for the next period based on the the information set $\mathcal{F}_t$ up to time t. Let $\mathcal{R}_t^i = P_{t+1}^i / P_t^i$ denote the next period's gross return. Then $Z_{t+1} = Z_t \boldsymbol{w}_t^\mathsf{T} \mathcal{R}_t$, where $\mathcal{R}_t = (\mathcal{R}_t^1, \ldots, \mathcal{R}_t^m)^\mathsf{T}$. Assuming $\mathcal{R}_t$ to be i.i.d., Breiman (1961) shows that the *log-optimum strategy* which chooses at every t the portfolio weight vector

$$\boldsymbol{w}_t^* = \arg\max_{\boldsymbol{w}} E\left[\log(\boldsymbol{w}^\mathsf{T} \mathcal{R}_t) | \mathcal{F}_t\right] \tag{3.17}$$

attains the asymptotically maximal growth rate of $(\log Z_T)/T$ as $T \to \infty$. Algoet and Cover (1988) show that this result holds with no restrictions on the joint distribution of $\{\mathcal{R}_t, 0 \leq t < T\}$, as a consequence of the following supermartingale property of the wealth process Z_t^* of a log-optimum strategy: For the wealth process Z_t of any multiperiod portfolio selection strategy, $\{Z_t/Z_t^*, t \geq 0\}$ is a non-negative supermartingale and

$$\limsup_{T \to \infty} T^{-1} \log(Z_T/Z_T^*) \leq 0 \qquad \text{with probability 1.} \tag{3.18}$$

In Supplement 8 of Section 3.4 we provide further background between log-optimum strategies and information theory.

Samuelson (1969, p. 245–246), however, argues that although "there is nothing wrong with the logical deduction from premise to theorem", the premise — that with high probability the portfolio which maximizes the geometric mean of the gross return at each stage (in the sense of (3.17)) will end up with larger terminal wealth than any other portfolio selection strategy — is "faulty to begin with" because it does not yield a better expected utility (which incorporates the investor's risk aversion as pointed out in footnote 1) except for the special case of log utility. Ignoring questions of optimality, Dempster et al. (2007) have shown that for time-invariant weight vectors $\boldsymbol{w} \geq \boldsymbol{0}$ (no short selling) that characterize a constant proportions strategy, the growth rate $\lim_{T \to \infty} T^{-1} \log Z_T$ exists with probability 1 when the random vectors $\mathcal{R}_t$, $t = 1, 2, \ldots$, are nondegenerate, stationary and ergodic;

moreover, the growth rate is almost surely larger than $(\rho_1, \ldots, \rho_m)\boldsymbol{w}$, where $\rho_i = \lim_{T\to\infty} T^{-1} \log(\prod_{t=1}^{\top} \mathcal{R}_t^i)$ is the growth rate of the ith security. The positive variance (representing uncertainty) in a non-degenerate random variable W coupled with Jensen's inequality $E(\log W) > \log(EW)$ is often used to explain this "volatility-induced" excess growth rate (or "volatility pumping"). Fernholz and Shay (1982) and Luenberger (2013) have shown further that the constant proportions strategy in the case of $m = 2$ risky assets with diffusion-type price processes not only has the volatility-induced growth rate but also a simultaneous decrease of the volatility of the portfolio. Dempster et al. (2007) show that the volatility-induced excess growth rate still holds when there are proportional transaction costs for rebalancing, provided that the proportional costs are sufficiently small.

3.3.4 Multiperiod mean-variance portfolio rebalancing

Even after they incorporate proportional transaction costs, multiperiod expected utility maximization theory and capital growth theory described in the preceding subsections have ignored an important practical considerations of the active (or passive) portfolio manager of an investment fund, namely that the fund's performance is evaluated periodically according to some risk-adjusted measure such as information ratio as described in Sections 1.4 and 3.1.3. Accordingly most active portfolio managers still prefer to use Markowitz's framework of mean-variance optimization. Moreover, Levy and Markowitz (1979) and Kroll et al. (1984) have shown how expected utility can be approximated by a function of mean and variance in applications to portfolio selection. However, mean-variance portfolio rebalancing in a multiperiod dynamic framework has been a long-standing problem, and Chapter 21 of Grinold and Kahn (2000) lists when rebalancing should occur as one of the open questions: "Active portfolio management is a dynamic problem,.... With a proper frame, managers should make decisions now, accounting for these dynamics and interactions now and in the future. One simple open question is, when should we trade (in the presence of transaction costs)?"

Merton (1990) has introduced a continuous-time analog of mean-variance optimization and Zhou and Li (2000) provide a linear-quadratic (LQ) stochastic control framework for the problem. Pliska (1997) and Li and Ng (2000) consider the discrete-time version of this multiperiod problem. However, because of the "curse of dimensionality" in dynamic programming, this approach has to be limited to relatively few assets and is seldom used in practice; see Kritzman et al. (2007). Instead, heuristic procedures extending single-period mean-variance analysis to multiple periods in a changing world, which are scalable to higher dimensions, are often used. In particular, Markowitz and van Dijk (2003) propose to use (a) a "mean-variance surrogate", which is a linear combination of the mean and variance of the portfolio return, to substitute for the discounted sum of the conditional expected future utilities given the information up to the present in an infinite-horizon setting, and (b) approximate dynamic programming (ADP). Although Markowitz and van Dijk do not use the term ADP and specifically illustrate their idea with a concrete example, their procedure is actually an example of ADP that will be described in greater detail in Supplement 7 of Section 3.4, where we also provide more details about the "mean-variance surrogate" and show how it is used in conjunction with ADP. Kritzman et al. (2007) report some simulation studies, using real world scenarios, comparing the Markowitz-van-Dijk (MvD) approach with the dynamic programming (DP) solution that is feasible in these cases and several standard industry heuristics. They find that the performance of MvD is "remarkably close" to that of DP in cases with DP can be computed with reliable numerical accuracy, and "far superior to solutions based on standard industry heuristics." They also highlight the issue of "a changing world" in MvD's title, saying: "Almost immediately upon implementation (of mean-variance analysis to determine optimal portfolio weights), however, the portfolio's weights become sub-optimal as changes in asset prices cause the portfolio to drift away from the optimal targets."

3.3.5 Dynamic mean-variance portfolio optimization in the presence of transaction costs

As pointed out by Grinold (2005), implementation costs are a barrier to adoption of these strategies which has been developed without assuming these costs. Kroner and Sultan (1993), Perold and Sharpe (1995), and Engle et al. (1998) propose rules to avoid unnecessary turnover. Markowitz and van Dijk (2003) and Kritzman et al. (2007) actually incorporate linear transaction costs

$$C_t = \sum_{j=1}^{m} \kappa_j |w_t^j - w_{t-1}^j| \tag{3.19}$$

in carrying out the MvD heuristic procedure, in which $\kappa_1, \ldots, \kappa_m$ are prescribed constants, and their portfolios do not include short selling. Gârleanu and Pedersen (2013) consider more tractable transaction costs that are quadratic functions of $\Delta u_t = u_t - u_{t-1}$, where u_t^j is the number of shares of stock j in the portfolio at time t. They assume a multi-factor model for the vector $\tilde{R}_{t+1}$ of next period's excess returns over the risk-free rate R_f, multiplied by current stock prices, so that the components of $\tilde{R}_{t+1}$ are $\tilde{R}_{t+1}^j = P_t^j (R_{t+1}^j - R_f)$, $j = 1, \ldots, m$:

$$\tilde{R}_{t+1} = Bf_t + \varepsilon_{t+1}, \tag{3.20}$$

with observed factor f_t at time t and known covariance matrix Σ for the i.i.d. random disturbances ε_{t+1}. Assuming a vector autoregressive model $f_{t+1} = (I - \Phi)f_t + w_{t+1}$ for the factors, they consider the optimization problem of sequential choice of u_t to maximize

$$E\sum_{t=1}^{\infty}(1-\rho)^t \left[(1-\rho)(u_t^\mathsf{T}\tilde{R}_{t+1} - \gamma u_t^\mathsf{T}\Sigma u_t) - \frac{1}{2}(\Delta u_t)^\mathsf{T}\Lambda\Delta u_t \right], \tag{3.21}$$

in which the transaction cost matrix $\mathbf{\Lambda}$ is proportional to $\mathbf{\Sigma}$, with $\lambda > 0$ being the constant of proportionality; their Appendix B gives a microeconomic foundation for this assumption. The parameter $0 < \rho < 1$ is a discount factor in the infinite-horizon problem (3.21), in which γ either a Lagrange multiplier when a risk constraint is imposed or a measure of the investor's risk aversion in the absence of risk constraints. This is a Markov decision problem with state $\boldsymbol{f}_t$ and control $\Delta \boldsymbol{u}_t$ at time t. Because of the linear dynamics and quadratic costs consisting of both $\boldsymbol{u}_t^\mathsf{T} \mathbf{\Sigma} \boldsymbol{u}_t$ and $(\Delta \boldsymbol{u}_t)^\mathsf{T} \mathbf{\Lambda} \Delta \boldsymbol{u}_t$, (3.21) can be solved explicitly by dynamic programming. Letting $\bar{\rho} = 1 - \rho$, their Proposition 1 yields an explicit formula for the value function and Propositions 2–4 characterize the optimal policy, which will be described in Supplement 9 of Section 3.4 where its connection to the classical LQG problem in stochastic control will also be discussed. They give an empirical application of this approach using 15 different commodity futures in the period January 1, 1996 – January 23, 2009, showing superior net returns relative to several benchmarks. The factors they choose are (a) mean divided by standard deviation of the past 5 days' price changes for each commodity future, stacked into a vector, (b) vector of an analogous quantities for past year's price changes, and (c) corresponding vector for the past 5 years' price changes. For the transition matrix $\mathbf{\Lambda} = \lambda \mathbf{\Sigma}$, they use an estimate of λ proposed by Engle et al. (2012). They also use a shrinkage estimator of $\mathbf{\Sigma}$ for their empirical study.

3.3.6 Dynamic portfolio selection in the presence of parameter uncertainty

In Chapter 2, we have described the Markowitz optimization enigma when the means and covariances of the underlying asset returns that are assumed known in the single-period optimization are replaced by their sample estimates. We have also reviewed different approaches in the literature to address these difficulties and in particular the approach proposed by Lai et al. (2011b, LXC), that treats the unknown parameters as random variables in a Bayesian or empirical Bayesian reformulation of the problem. For fully dynamic (continuous-time) portfolio optimization, the role of sequential learning has been recognized by Williams (1977), Detemple (1986), Dothan and Feldman (1986), Gennotte (1986), and Brennan (1998) who treat the unknown drift parameters as state variables and estimate them sequentially by a filtering approach; see Supplement 9 of Section 3.4. These sequential estimates are used to substitute for the parameters in the procedures of Merton (1971) and others that assume known parameters in the asset price models.

In particular, Brennan (1998) considers an investor who can invest in a risk-free asset with instantaneous rate of return r and a risky asset whose price process follows geometric Brownian motion $dP_t/P_t = \mu dt + \sigma dB_t$ in which σ is known to the investor but μ is unknown. Treating μ as a state variable with prior distribution $N(m_0, v_0)$, he notes that the investor can learn sequentially about μ through the posterior distribution $N(m_t, v_t)$, in which m_t is given by the Kalman-Bucy filter

$$dm_t = \frac{v_t}{\sigma^2}\left(\frac{dP_t}{P_t} - m_t dt\right), \quad \text{with } v_t = \frac{v_0 \sigma^2}{v_0 t + \sigma^2}; \tag{3.22}$$

see Supplement 9 of Section 3.4 which also gives the dynamic programming equation for choosing w_t, $0 \le t \le T$, to minimize the expected utility of the investor's terminal wealth Z_T. He applies this approach to an empirical study with the S&P 500 index as the risky asset for the 69-year period 1926–1994, assuming $r = 5\%$ and $m_0 = 13\%$, two values of $\sqrt{v_0} = 0.0243$ or 0.0452, and two values of $\sigma = 20.2\%$ (standard deviation of the 69 annual returns) and $\sigma = 14\%$ (that corresponds "more closely to recent experience"). He uses a finite difference approximation to the dynamic programming equation for the case of power utility $U(z) = z^\gamma/\gamma$, with $\gamma < 1$ and $\gamma \ne 0$. The empirical study shows that imperfect knowledge about the parameters gives rise to two quite distinct phenomena which he labels as "estimation risk" and "learning". He points out that although increased risk due to uncertainty about the mean return tends to reduce the fraction of the portfolio allocated to the risky asset, "the prospect of learning more about the true value of the mean parameter, μ, as more returns are observed, induces an additional hedging demand for the risky asset."

Xia (2001) examines the effects of estimation risk and learning in a more general framework and shows that "the hedge demand associated with the uncertain parameters plays a predominant role in the optimal strategy", which is "horizon dependent: the optimal stock allocation can increase, decrease or vary non-monotonically with the horizon, because parameter uncertainty induces a state-dependent hedge demand that may increase or decrease with the horizon." Instead of the geometric Brownian motion model for the asset prices considered by Brennan (1998), she considers a price process of the form $dP_t/P_t = \mu_t dt + \sigma dB_t$, in which $\mu_t = \alpha + \boldsymbol{\beta}^\top \boldsymbol{f}_t$ and $\boldsymbol{f}_t$ is a $k \times 1$ vector of observable factors that undergo linear stochastic dynamics $d\boldsymbol{f}_t = (\boldsymbol{A}_{0,t} + \boldsymbol{A}_{1,t}\boldsymbol{\beta})dt + \boldsymbol{\Sigma}_t^{1/2}d\tilde{\boldsymbol{B}}_t$, where $\boldsymbol{A}_{1,t}$ and $\boldsymbol{\Sigma}_t$ are given $(k \times k)$ matrices and $\boldsymbol{A}_{0,t}$ is a $k \times 1$ vector that may depend on $(\boldsymbol{P}_t, \boldsymbol{f}_t)$, and $\tilde{\boldsymbol{B}}_t$ is a k-dimensional Brownian motion that is independent of B_t. Assuming a normal prior distribution for $(\alpha, \boldsymbol{\beta}^\top)^\top$, the posterior distribution given $(\boldsymbol{P}_s, \boldsymbol{f}_s)$, $0 \le s \le t$, is normal and is given by the Kalman-Bucy filter; see Supplement 9 of Section 3.4. Her analysis of this setting shows that "the relation between the optimal portfolio allocation and the predictor variable depends crucially on future learning and investment horizon."

Cvitanić et al. (2006) follow up on Brennan's work but extend to m risky assets plus a market or benchmark portfolio with price process $P_{0,t}$:

$$\frac{dP_{0,t}}{P_{0,t}} = \mu_0 dt + \sigma_0 dB_t^0, \qquad \frac{dP_{i,t}}{P_{i,t}} = \mu_i dt + \beta_i \sigma_0 dB_t^0 + \sigma_i dB_t^i, \tag{3.23}$$

$i = 1, \ldots, m$, where $\mu_i - r = \alpha_i + \beta_i(\mu_0 - r)$, in which $\mu_i - r$ represents the mean excess return over the risk-free asset that is also available for inclusion in the portfolio. Assuming that $\sigma_0, \ldots, \sigma_m$ are known and letting

$$\theta_0 = (\mu_0 - r)/\sigma_0, \quad \theta_i = \alpha_i/\sigma_i \quad (i = 1, \ldots, m), \tag{3.24}$$

and putting a prior distribution $N(\bar{\boldsymbol{\theta}}(0), \boldsymbol{Q}^\top \boldsymbol{D} \boldsymbol{Q})$ on $\boldsymbol{\theta} = (\theta_0, \ldots, \theta_m)^\top$, they use the Kalman-Bucy filter to show that the posterior distribution of $\boldsymbol{\theta}$ given the observed prices up to time t is $N(\bar{\boldsymbol{\theta}}(t), \boldsymbol{Q}^\top \boldsymbol{D}(t) \boldsymbol{Q})$, where $\bar{\boldsymbol{\theta}}(t) = \boldsymbol{Q}^\top \bar{\boldsymbol{D}}(t) \{\boldsymbol{Q} \boldsymbol{B}^*(t) + (\bar{\boldsymbol{D}}(0))^{-1} \boldsymbol{Q} \bar{\boldsymbol{\theta}}(0)\}$, $\bar{\boldsymbol{D}}(t) = diag(\delta_0(t), \ldots, \delta_m(t))$, in which the $\delta_i(t)$ and $\boldsymbol{B}^*(t)$ are defined as follows. First note that $\boldsymbol{Q}^\top \boldsymbol{D} \boldsymbol{Q}$ is the singular value decomposition of the covariance matrix of the prior distribution of $\boldsymbol{\theta}$, with $\boldsymbol{D} = diag(d_0, \ldots, d_m)$ consisting of the eigenvalues d_i, and $\boldsymbol{Q}$ being the orthogonal matrix consisting of the corresponding eigenvectors of unit length. Define $\delta_i(t) = d_i/(1 + d_i t)$, $i = 0, \ldots, m$, and $\boldsymbol{B}^*(t) = \bar{\boldsymbol{B}}(t) + \int_0^t \bar{\boldsymbol{\theta}}(s) \, ds$, where $\bar{\boldsymbol{B}}(\cdot)$ is the innovation process in filtering theory (Cvitanić et al., 2006, p. 1147, in particular (A1) and (A2) there). For the power utility $U(z) = z^\gamma/\gamma$ with $\gamma < 0$ and the log utility $U(z) = \log(z)$ that corresponds to the case $\gamma = 0$ in (3.10), Cvitanić et al. (2006) derive an explicit formula for the optimizing portfolio weight vector for the $m + 1$ risky assets:

$$\boldsymbol{\pi}_t = \boldsymbol{V}^{-1/2} \boldsymbol{Q}^\top \boldsymbol{A}^{-1}(t) \boldsymbol{Q} \bar{\boldsymbol{\theta}}(t), \tag{3.25}$$

where $\boldsymbol{V}$ is the covariance matrix of $(B_1^0, B_1^1, \ldots, B_1^m)^\top$ and $\boldsymbol{A}(t)$ is the diagonal matrix with diagonal elements $(1 - \gamma) - \gamma(T - t)\delta_i(t)$, $i = 0, \ldots, m$. The optimal portfolio vector $\boldsymbol{w}_t$ consists of the subvector $\boldsymbol{\pi}_t$ and an additional component $1 - \boldsymbol{\pi}_t^\top \boldsymbol{1}$, which is the weight assigned to the risk-free asset.

Thus, filtering (specifically the Kalman-Bucy filter) has played a prominent role in dynamic portfolio optimization when there is uncertainty about the drift parameters of the price processes. Uncertainty about the volatility parameters, however, has not received much attention in the literature and previous works all assumed them to be known. On the other hand, Chapter 2 has shown that the important role played by volatility estimation even in the single-period portfolio optimization theory. The main reason for the lack of similar treatment of uncertainty about the volatility parameters is that it would involve nonlinear filtering methods. These methods have been recently developed and are summarized in Supplement 10 of Section 3.4, which also describes some recent work to solve the long-standing problem of multiperiod portfolio optimization in the presence of transaction costs and parameter uncertainty.

3.4   Supplementary notes and comments

1. **SPX, SPY and other indices and ETFs as benchmarks**   SPX refers

to the S&P 500 index, which is calculated from the stocks of 500 US large-cap corporations. The index is owned and maintained by Standard & Poor's Financial Services LLC (S&P), a division of McGraw-Hill. Although 500 companies are represented, at any time it may contain more than 500 stocks as some companies may have multiple share class lines in the index. Since the index aims at reflecting various sectors in the US economy, a company that satisfies all the qualifying characteristics to be included may not be selected if the sector to which it belongs is already well represented in the index. The index is updated daily using a methodology described in *S&P Dow Jones Indices: Index Mathematics Methodology* which can be accessed at `www.spdji.com` and which says: "The simplest capitalization weighted index can be thought of as a portfolio of available shares of stocks in the index, which is a large unwieldy number in trillions of US dollars and needs to be scaled by a suitably chosen divisor ... To assure the index's value, or level, does not change when stocks are added or deleted, the divisor is adjusted to offset the change in market value of the index." The 500 large-cap US companies in the S&P 500 Index are represented by 11 sectors listed in Table 3.2 which presents also the sectorial weights as of October 17, 2016.

Many index funds and exchange traded funds track the performance of the S&P 500 index by holding the same stocks as the index, in the same or modified proportions, to match or improve its performance (before fees and expenses). Partly because of this, the index is often used as a benchmark against whose performance portfolio managers are assessed. Passive portfolio management attempts to track the index, while active portfolio managers try to outperform it. Because the index weights and returns are updated daily, S&P already carries out extensive data collection and analytics and provides a good benchmark.

SPY is the stock symbol for the Exchange Traded Fund (ETF) which is designed to track the S&P 500 Index. The inception date of the ETF is January 22, 1993. For a long time, this fund was the largest ETF in the world. Its expense ratio is among the lowest and is currently at 0.09%. SPY offers investors diversified exposure to US large-cap stocks. Whereas S&P 500 is the most often cited proxy for the US equity market, SPY is the most heavily traded security in the world. This level of liquidity makes it very inexpensive to trade in large amounts, which is an attractive feature for traders and institutions with relatively short anticipated holding periods. Depending on one's investment preferences, other indices and associated ETFs are also available. For example, the Dow Jones industrial Average (DJIA) is another index owned by S&P Dow Jones Indices. It involves stocks of 30 large US companies that are traded at NYSE or NASDAQ, and DIA is the stock symbol for the associated ETF, which was introduced in 1998 and is part of the SPDR family of ETFs from State Street Global Advisors.

TABLE 3.2: The sectors of S&P 500 Index and corresponding sectorial weights as of October 17, 2016.

Sector	Weight	Main Companies
Information Technology	21.42%	Apple Microsoft Facebook
Health Care	14.36%	Johnson & Johnson Pfizer Merck
Financials	13.09%	Berkshire Hathaway Class B JPMorgan Chase Wells Fargo
Consumer Discretionary	12.57%	Amazon Comcast Home Depot
Consumer Staples	9.94%	Procter & Gamble Coca-Cola PepsiCo Inc.
Industrials	9.71%	General Electric 3M Company Honeywell International
Energy	7.26%	Exxon Mobil Corporation Chevron Corporation Schlumberger
Utilities	3.23%	American Electric Power Activision Blizzard PG&E Corp.
Real Estate	2.96%	Simon Property Group Inc. American Tower Crown Castle
Materials	2.83%	E. I. du Pont Dow Chemical Monsanto Company
Telecommunication Services	2.59%	AT&T Verizon Level 3 Communications
Cash	0.03%	

The October 21 2016 issue of *The Wall Street Journal* has an article on "the active-passive powerhouse" Dimensional Fund Advisors LP (DFA), saying that this "fastest-growing major mutual fund company in the US isn't strictly an active or passive investor. It is both." Its founders are "pioneers of index funds" but "concluded long ago that investors who respected efficient markets (or the EMH) would nevertheless achieve better returns than plain index funds deliver" by modifying them using an active approach. The article says: "Here is how DFA invests: It designs its own indexes, often of small-capitalization stocks, then waits — for weeks, if necessary — until an eager seller is willing to unload shares at below the prevailing asking price in the market. Such tactics can minimize and in some cases even erase transaction costs, providing a small but meaningful boost to returns. The firm often emphasizes cheap, small-cap stocks that are cumbersome for active managers to buy," making use of the research by their advisers Fama and French that we have referred to in Supplement 6 of Section 2.9. The article reports: "Since its launch in late 1981, DFA's flagship fund has returned an average of 11.8% annually; the Russell 2000 index of small-cap stocks has gained an average of 10.3% annually. Two-thirds of the firm's approximately 50 stock funds have beaten their benchmark since inception, according to data from DFA. Fees (are) on average about a third of a percentage point."

2. **Solving CCQP with a global smoothing algorithm** Murray and Ng (2008) propose an algorithm for nonlinear optimization problems with discrete variables. Shek (2010) extends that framework by casting the problem in the portfolio optimization setting. There is an equivalence of the CCQP problem and that of finding the global minimizer of a function of continuous variables (see, e.g., Ge and Huang, 1989). However, it is insufficient to introduce only penalty functions to enforce the integer constraint as it risks introducing a large number of local minima which can significantly increase the complexity of the original problem. The idea of a global smoothing algorithm is to add a strictly convex function to the original objective, together with a suitable penalty function. The algorithm then iteratively increases the penalty on non-integer values and decreases the amount of smoothing introduced.

Consider the following CCQP problem:

$$\min_{\boldsymbol{x}} f(\boldsymbol{x}) = \boldsymbol{c}^\top \boldsymbol{x} + \frac{1}{2} \boldsymbol{x}^\top \boldsymbol{H} \boldsymbol{x} \tag{3.26}$$

$$\text{such that } \boldsymbol{A}\boldsymbol{x} = \boldsymbol{b}, \tag{3.27}$$

$$y_i \geq x_i \geq 0, \ i = 1, \ldots, n, \tag{3.28}$$

$$\sum_i y_i = K, \ y_i \in \{0, 1\}, \tag{3.29}$$

where $\boldsymbol{c}, \boldsymbol{x} \in \mathbb{R}^{n \times 1}$, $\boldsymbol{H} \in \mathbb{R}^{n \times n}$, $\boldsymbol{A} \in \mathbb{R}^{m \times n}$, $\boldsymbol{b} \in \mathbb{R}^{m \times 1}$, and $m \leq n$. The transformed problem becomes:

$$\min_{\boldsymbol{x}, \boldsymbol{y} \in \mathbb{R}^n} F(\boldsymbol{x}, \boldsymbol{y}) = f(\boldsymbol{x}) + \underbrace{\frac{\gamma}{2} \sum_i y_i (1 - y_i)}_{\text{penalty term}} - \underbrace{\mu \sum_i (\log s_i + \log x_i)}_{\text{smoothing term}}$$

$$\text{such that} \quad \begin{bmatrix} A & 0 & 0 \\ 0 & e^\top & 0 \\ -I & I & -I \end{bmatrix} \begin{bmatrix} x \\ y \\ s \end{bmatrix} = \begin{bmatrix} b \\ K \\ 0 \end{bmatrix}, \quad (3.30)$$

where $e \in \mathbb{R}^{n \times 1}$ is a column vector of ones, s is the slack variable, γ is the parameter that governs the degree of penalty on non-integer values, and μ is the parameter that governs the amount of global smoothing introduced into the problem. Let

$$\check{A} = \begin{bmatrix} A & 0 & 0 \\ 0 & e^\top & 0 \\ -I & I & -I \end{bmatrix}, \quad \check{x} = \begin{bmatrix} x \\ y \\ s \end{bmatrix}, \quad \text{and} \quad \check{b} = \begin{bmatrix} b \\ K \\ 0 \end{bmatrix}.$$

Then the gradient and Hessian can be expressed as

$$\check{g} = \nabla F(\boldsymbol{x}, \boldsymbol{y}) = \begin{bmatrix} g_1 - \frac{\mu}{x_1} \\ \vdots \\ g_n - \frac{\mu}{x_n} \\ \frac{\gamma}{2}(1 - 2y_1) \\ \vdots \\ \frac{\gamma}{2}(1 - 2y_n) \\ -\frac{\mu}{s_1} \\ \vdots \\ -\frac{\mu}{s_n} \end{bmatrix}, \quad \check{H} = \nabla^2 F(\boldsymbol{x}, \boldsymbol{y}) = \begin{bmatrix} \check{H}_{11} & 0 & 0 \\ 0 & \check{H}_{22} & 0 \\ 0 & 0 & \check{H}_{33} \end{bmatrix},$$

where

$$g = c + Hx, \quad \check{H}_{11} = H + \text{diag}\left(\frac{\mu}{x_i^2}\right),$$

$$\check{H}_{22} = \text{diag}(-\gamma_i), \quad \check{H}_{33} = \text{diag}\left(\frac{\mu}{s_i^2}\right).$$

Shek (2010) uses an adaptation of the conjugate gradient method to compute the solution of (3.30), thereby introducing a global smoothing CCQP algorithm.

3. **Solving CCQP with local relaxation approach**   Murray and Shek (2012) propose an algorithm that solves the following CCQP problem

$$\min_{\boldsymbol{x}} f(x) = -c^\top x + \lambda x^\top H x$$

$$\text{such that} \quad \sum_i x_i = 0, \quad \sum_i I_{\{x_i \neq 0\}} = K,$$

by exploring the inherent similarity of asset returns, and relies on solving a series of relaxed problems in a lower-dimensional space by first projecting and clustering returns in this space. This approach mimics the line-search and trust-region methods for solving continuous problems, and it uses local approximations to determine an improved solution at each iteration. Murray and Shanbhag (2006, 2007) have used a variant of the local relaxation approach to solve for a grid-based electrical substation site. Instead of a predefined two-dimensional grid search, Murray and Shek (2012) project asset returns onto a multi-dimensional space after reducing dimensionality via factor models. Moreover, instead of using physical distance, they use a function of factor loadings, risk aversion and expected returns to define a metric. They also use a probabilistic procedure to assign members to each cluster at each iteration. The proposed framework can be divided into three key steps:

- Identifying an initial set of starting centroids $\mathcal{S}_0$
  - A successive truncation algorithm works by discarding, at each iteration, a portion of assets which have small weights, until the appropriate number of assets in the portfolio is equal to the desired cardinality value.
- Iterative local relaxation such that at each iteration, the algorithm solves the following quadratic programs (QPs) for subsets $\mathcal{S}_r$ of the original set S of assets, with $\mathcal{S}_0 \subset \mathcal{S}_r \subseteq \mathcal{S}$:
  - *Globally relaxed QP*

  $$\min_{\boldsymbol{x}} \{-\boldsymbol{c}^\top \boldsymbol{x} + \frac{1}{2} \boldsymbol{x}^\top \boldsymbol{H} \boldsymbol{x}\}$$
  $$\text{such that } \boldsymbol{e}^\top \boldsymbol{x} = 0, \ \boldsymbol{x} \in \mathcal{S},$$

  where $\boldsymbol{c}$ and $\boldsymbol{H}$ are the expected return and covariance matrix for the set $\mathcal{S}$ of assets and $\boldsymbol{e}$ is a column vector of 1's.
  - *Centroid-asset QP*

  $$\min_{\boldsymbol{x}_0} \{-\boldsymbol{c}_0^\top \boldsymbol{x}_0 + \frac{1}{2} \boldsymbol{x}_0^\top \boldsymbol{H}_0 \boldsymbol{x}_0\}$$
  $$\text{such that } \boldsymbol{e}^\top \boldsymbol{x}_0 = 0, \ \boldsymbol{x}_0 \in \mathcal{S}_0,$$

  where $\boldsymbol{c}_0$ and $\boldsymbol{H}_0$ are expected return and covariance matrix for the set of centroid assets $\mathcal{S}_0$ only. This gives an optimal weight vector $\boldsymbol{x}_0^*$, where $x_{0,i}^*$ is the ith element of this vector corresponding to ith centroid asset, $s_{0,i}$.
  - *Relaxed-neighborhood QP*

  $$\min_{\boldsymbol{x}_r} \{-\boldsymbol{c}_r^\top \boldsymbol{x}_r + \frac{1}{2} \boldsymbol{x}_r^\top \boldsymbol{H}_r \boldsymbol{x}_r\}$$
  $$\text{such that } \boldsymbol{e}_i^\top \boldsymbol{x}_r = x_{0,i}^*, \ i = 1, \ldots, K, \ \boldsymbol{x}_r \in \mathcal{S}_r,$$

where e_i is a column vector with 1's at positions that correspond to the ith centroid cluster assets and are zero elsewhere, c_r is the vector of expected returns, and H_r is the corresponding covariance matrix of the sub-universe of assets.

- Identifying new centroids to update the clusters.

4. **Dynamic programming: Discrete-time case** Consider the problem of choosing the actions (controls) $u_t \in U$ sequentially so as to minimize $E\{\sum_{t=1}^{N-1} c_t(x_t, u_t) + c_N(x_N)\}$, with state variables x_t. Here c_t are given functions, and N is called the "horizon" of the problem. The control u_t affects the dynamics of the future states $x_{t+1}, \ldots, x_N$, and depends on the information set $\mathcal{F}_t$ consisting of $x_t, u_{t-1}, x_{t-1}, \ldots, u_1, x_1$. The summand $c_t(x_t, u_t)$ represents the immediate cost and $c_N(x_N)$ the terminal cost; no control is exercised at time N because it only affects the states after N. Noting that the choice of u_t can only depend on $\mathcal{F}_t$, we use the "tower property" of conditional expectations to write

$$E\left\{c_t(x_t, u_t) + V_{t+1}\right\} = E\left\{E\left[c_t(x_t, u_t) + V_{t+1}|\mathcal{F}_t\right]\right\}, \qquad (3.31)$$

where V_{t+1} is the *value function* at time $t + 1$ that will be defined below. The tower property suggests the following *backward induction* algorithm to choose u_t for $t = N - 1, N - 2, \ldots, 1$, and to define V_t, initializing with $V_N = c_N(x_N)$:

$$u_t = \arg\min_{u \in U} \left\{c_t(x_t, u) + E(V_{t+1}|\mathcal{F}_t)\right\},$$
$$V_t = c_t(x_t, u_t) + E(V_{t+1}|\mathcal{F}_t). \qquad (3.32)$$

When x_t is a controlled Markov chain so that the conditional probability of x_{t+1} given x_s, u_s ($s \le t$) depends only on x_t, u_t as in (3.33), the optimal control u_t can be chosen to depend only on x_t.

Infinite-horizon discounted problem

Letting $N \to \infty$ in the finite-horizon case leads to the infinite-horizon problem of minimizing $E\{\sum_{t=1}^{\infty} c_t(x_t, u_t)\}$ when the infinite series is summable. The case $c_t(x, u) = \beta^t c(x, u)$ is called the discounted problem, with discount factor $0 < \beta < 1$. When x_t is a controlled Markov chain with state space S and stationary transition probabilities

$$P\{x_{t+1} \in A | x_t = x, \ u_t = u\} = P_{x,u}(A) \qquad (3.33)$$

for all $t \ge 0$, $u \in U$, $x \in S$, and $A \subset S$, the value function

$$V(x) := \sup_{u_1, u_2, \ldots} E\left\{\sum_{t=1}^{\infty} \beta^t c(x_t, u_t) \middle| x_0 = x\right\}$$

is the solution of the dynamic programming equation

$$V(x) = \inf_{u \in U} \left\{ c(x, u) + \beta \int_S V(y) \, dP_{x,u}(y) \right\}. \tag{3.34}$$

The sequence $\boldsymbol{u} = (u_1, u_2, \dots)$ is called a control policy. From (3.34), it follows that the optimal control policy is *stationary* in the sense that $u_t = g(x_t)$ for some time-invariant function g.

Value and policy iterations

There are two commonly used methods to solve (3.34) for V and for the stationary policy. One is *value iteration*. Starting with an initial guess v_0 of V, it uses successive approximations

$$v_{k+1}(x) = \min_{u \in U} \left\{ c(x, u) + \beta \int_S v_k(y) \, dP_{x,u}(y) \right\}.$$

The minimizer $u = u_k(x)$ of the right-hand side yields an approximation to the stationary policy at the kth iteration. The other is *policy iteration*, which can be applied when the state space S is finite as follows. Denote the states of S by $1, \dots, m$, and let

$$\boldsymbol{c}_g = [c(1, g(1)), \dots, c(m, g(m)]^\top,$$
$$\boldsymbol{P}_g = \left[P_{x, g(x)}(y) \right]_{1 \le x \le m, \, 1 \le y \le m}.$$

Note that $\boldsymbol{P}_g$ is the transition matrix of the controlled Markov chain with stationary control policy g that is specified by the vector $[g(1), \dots, g(m)]^\top$. Minimization over g, therefore, is the same as minimization of $\mathbb{R}^m$. In view of (3.34), we define the affine transformation $T_g(\boldsymbol{x}) = \boldsymbol{c}_g + \beta \boldsymbol{P}_g(\boldsymbol{x})$ for $\boldsymbol{x} \in \mathbb{R}^m$. Letting $T(\boldsymbol{x}) = \min_g T_g(\boldsymbol{x})$ and $\boldsymbol{v} = [V(1), \dots, V(m)]^\top$, note that (3.34) can be written as $\boldsymbol{v} = T(\boldsymbol{v})$. This suggests the following policy iteration scheme that initializes with a preliminary guess g_0 of the optimal g^*. At the kth iteration, after determining g_k, solve the linear system $\boldsymbol{v} = T_{g_k}(\boldsymbol{v})$ and denote the solution by $\boldsymbol{v}_k$. If $T(\boldsymbol{v}_k) = \boldsymbol{v}_k$, stop and set $g^* = g_k$. Otherwise solve the linear system $T_g(\boldsymbol{v}_k) = T(\boldsymbol{v}_k)$ for $(g(1), \dots, g(m))^\top$ and set $g_{k+1} = g$.

Optimal stopping

When the control u_t consists of whether to stop at time t, the stochastic control problem is called optimal stopping. Note that unlike more general stochastic control problems, the stopping rule does not change the dynamics of x_t. The control policy reduces to a stopping rule τ in this case. Suppose the costs are $c(x_t)$ prior to stopping and $h(x_\tau)$ upon stopping, and $c_t(x_t) = 0$ for $t > \tau$. The transition matrix (3.33) now has the form

$$P\{x_{t+1} \in A | x_t = x\} = P_x(A) \qquad \text{for } 0 \le t < \tau, \; x \in S, \text{ and } A \subset S.$$

The value function $V(x) := \sup_\tau E\{\sum_{t<\tau} c(x_t) + h(x_\tau) | x_0 = x\}$ (with supremum over all stopping rules τ) satisfies the dynamic programming equation

$$V(x) = \min\left\{ h(x), c(x) + \int_S V(y) \, dP_x(y) \right\}. \tag{3.35}$$

The right-hand side of (3.35) refers to either stopping when state x is observed and incurring cost $h(x)$, or continuing after paying cost $c(x)$ and then proceeding optimally thereafter. The minimum in (3.35) refers to choosing continuation or stopping according to which has a smaller expected cost. The optimal stopping rule is a stationary policy of the form $\tau = \inf\{n \geq 0 : h(x_t) = V(x_t)\}$.

5. **Dynamic programming: Continuous-time case**   Consider a continuous-time Markov process X_t with state space $\mathbb{R}$ and *infinitesimal generator* (the continuous-time analog of transition probability) $\mathcal{A}$ defined by

$$\mathcal{A}f(x) = \lim_{t\downarrow 0} \frac{1}{t} E[f(X_t) - f(x)], \qquad \text{with } X_0 = x,$$

for any bounded function f with bounded first and second derivative. In particular, for a diffusion process defined by the stochastic differential equation

$$dX_t = \mu(X_t)dt + \sigma(X_t)dB_t, \tag{3.36}$$

with $|\mu(x)-\mu(y)|+|\sigma(x)-\sigma(y)| \leq K|x-y|$ and $\mu^2(x)+\sigma^2(x) \leq K^2(1+x^2)$, its generator $\mathcal{L}$ can be expressed as

$$\mathcal{L}f(x) = \mu(x)\frac{\partial}{\partial x}f(x) + \frac{1}{2}\sigma^2(x)\frac{\partial^2}{\partial x^2}f(x). \tag{3.37}$$

Such expression is motivated by applying the Taylor's expansion

$$f(X_t) \approx f(x) + f'(X_t)(X_t - x) + \frac{1}{2}f''(X_t)(X_t - x)^2, \tag{3.38}$$

in which the first-order term gives

$$\lim_{t\downarrow 0} \frac{1}{t} E_x\left[f'(X_t)(X_t - x)\right] = f'(x)\mu(x)$$

and the second-order term gives

$$\lim_{t\downarrow 0} \frac{1}{t} E_x\left[f''(X_t)(X_t - x)^2\right] = f''(x)\sigma^2(x),$$

yielding the generator (3.37), where E_x denotes expectation under the initial state $X_0 = x$. To model how the control process U_t could affect the drift and volatility and to incorporate the time-inhomogeneity of the process, the preceding idea can be generalized to the controlled diffusion process

$$dX_t = \mu(t, X_t, U_t)dt + \sigma(t, X_t, U_t)dB_t, \qquad (3.39)$$

in which U_t takes values in $\mathbb{U}$ and is *non-anticipative* with respect to B_t in the sense that $B_{t'} - B_t$ is independent of $\{U_s, 0 \leq s \leq t\}$ for $t' > t$. The corresponding generator is a differential operator of the form

$$
\begin{aligned}
\mathcal{A}^u &= \frac{\partial}{\partial t} + \mu(t, x, u)\frac{\partial}{\partial x} + \frac{1}{2}\sigma^2(t, x, u)\frac{\partial^2}{\partial x^2} \\
&= \frac{\partial}{\partial t} + \mathcal{L}^u,
\end{aligned}
\qquad (3.40)
$$

defined on a suitably chosen class of functions f so that $\mathcal{A}^u f(t, x)$ is a linear combination of $\partial f/\partial t$, $\partial f/\partial x$, and $\partial^2 f/\partial x^2$ given by (3.40). In particular, if $U_t = u(t, X_t)$ for some measurable function u, then U_t is called a *Markov control rule* and u is called a *Markov control function*.

Infinite-horizon discounted problem

Let g be the running cost function. Then the infinite-horizon discounted cost of a Markov control rule is given by

$$J(x, u) = E_x \left[\int_0^\infty e^{-\gamma s} g(X_s, u(X_s)) \, ds \right].$$

The value function of the stochastic control problem is given by

$$V(x) = \inf_u J(x, u), \qquad (3.41)$$

with the infimum taken over all Markov control functions. Using the following backward induction argument, $V(x)$ can be expressed in the recursive form

$$V(x) = \inf_u E_x \left[\int_0^t e^{-\gamma s} g(X_s^u, u(t, X_s^u)) \, ds + e^{-\gamma t} V(X_t^u) \right], \qquad (3.42)$$

in which X_s^u refers to the controlled diffusion process associated with the Markov control function u. For the optimal control function u^*,

$$
\begin{aligned}
V(x) &= E_x \left[\int_0^\infty e^{-\gamma s} g(X_s, u^*(X_s^{u^*})) \, ds \right] \\
&= E_x \left[\int_0^t e^{-\gamma s} g(X_s, u^*(X_s^{u^*})) \, ds \right] \\
&\quad + E_x \left[\int_t^\infty e^{-\gamma s} g(X_s, u^*(X_s)) \, ds \right].
\end{aligned}
$$

The second term can be further simplified to

$$e^{-\gamma t} E_x \left[\int_t^\infty e^{-\gamma(s-t)} g(X_s, u^*(X_s)) \, ds \right]$$
$$= e^{-\gamma t} E_x \left\{ E \left[\int_0^\infty e^{-\gamma v} g(X_{t+v}, u^*(X_{t+v})) \, dv \Big| X_t \right] \right\}$$
$$= e^{-\gamma t} E_x \{ V(X_t^{u^*}) \}.$$

A similar argument shows that for $u \neq u^*$,

$$V(x) \le E \left[\int_0^t e^{-\gamma s} g(X_s^u, u(X_s^u)) \, ds + e^{-\gamma t} V(X_t^u) \right]. \tag{3.43}$$

Assuming smoothness of V and applying Itô's formula to $f(t, x) = e^{-\gamma t} V(x)$, we obtain

$$d(e^{-\gamma t} V(X_t^{u^*})) = \left\{ e^{-\gamma t} [-\gamma V(X_t^{u^*}) + \mathcal{A}^{u^*} V(X_t^{u^*})] \right\} dt + d\xi_t, \tag{3.44}$$

where $d\xi_t = e^{\gamma t} \sigma(X_t^{u^*}) V'(X_t^{u^*}) dB_t$. Hence $\int_0^t d\xi_s$ is a martingale and $e^{-\gamma t} V(X_t^{u^*})$ is equal to

$$V(x) + \int_0^t \left\{ e^{-\gamma s} \left[-\gamma V(X_s^{u^*}) + \mathcal{A}^{u^*} V(X_s^{u^*}) \right] \right\} ds + \int_0^t d\xi_s.$$

Adding $\int_0^t e^{-\gamma s} g(X_s^{u^*}, u^*(X_s^{u^*})) \, ds$ to both sides of the equality and taking expectation yields

$$E_x \left[\int_0^t \left\{ e^{-\gamma s} \left[-\gamma V(X_s^{u^*}) + \mathcal{A}^{u^*} V(X_s^{u*}) + g(X_s^{u^*}, u^*(X_s^{u^*})) \right] \right\} ds \right] = 0. \tag{3.45}$$

Letting $t \downarrow 0$ then gives

$$-\gamma V(x) + \mathcal{A}^{u^*} V(x) + g(x, u^*(x)) = 0. \tag{3.46}$$

For any $u \neq u^*$, a similar argument leads to

$$-\gamma V(x) + \mathcal{A}^u V(x) + g(x, u(x)) \ge 0. \tag{3.47}$$

Combining (3.46) and (3.47) gives

$$\inf_u [-\gamma V(x) + \mathcal{A}^u V(x) + g(x, u(x)] = 0, \tag{3.48}$$

with the infimum taken over all Markov control functions. The nonlinear ordinary differential equation (3.48) is known as the *Hamilton-Jacobi-Bellman* (HJB) equation of the stochastic control problem.

HJB equation for finite-horizon case

The finite-horizon cost function of a non-anticipative control rule U is

$$J(t, x, U) = E\left[\int_t^T g(s, X_s, U_s) \, ds + h(X_T) \Big| X_t = x\right], \qquad (3.49)$$

where g is the running cost function and h is the terminal cost function, and X_t is the controlled diffusion process (3.39). The HJB equation is

$$0 = \inf_u [\mathcal{A}^u V(t, x) + g(t, x, u)], \qquad 0 \le t < T, \qquad (3.50)$$

in which $\mathcal{A}^u$ is defined by (3.40), with boundary condition

$$V(T, x) = h(x); \qquad (3.51)$$

see Øksendal (2003) and Karatzas and Shreve (1991).

To illustrate, consider the Merton problem of maximizing (3.13) in Section 3.3.1, in which the total wealth process satisfies the SDE (3.12). The HJB equation[2] for the value function is:

$$\max_{C, \theta} \{(\partial/\partial t + \mathcal{L}) V(t, z) + U_1(C) - \beta V(t, z)\} = 0, \qquad (3.52)$$

subject to the terminal condition $V(T, z) = U_2(z)$, where $\mathcal{L}$ is the infinitesimal generator of (3.12):

$$\mathcal{L} = \frac{\sigma^2 \theta^2 z^2}{2} \frac{\partial^2}{\partial z^2} + [rz + (\alpha - r)\theta z - C] \frac{\partial}{\partial z}.$$

Formal maximization with respect to C and θ yields $C = (U_1')^{-1}(V_z)$ and $\theta = -(V_z/V_{zz})(\alpha - r)/\sigma^2 z$ (in which subscript denotes partial derivative, e.g., $V_z = \partial V/\partial z$). Substituting for C and θ in (3.52) leads to the PDE

$$\frac{\partial V}{\partial t} - \frac{(\alpha - r)^2}{2\sigma^2} \frac{(\partial V/\partial z)^2}{\partial^2 V/\partial z^2} + (rz - C^*)\frac{\partial V}{\partial z} + U_1(C^*) - \beta V = 0, \quad (3.53)$$

where $C^* = C^*(t, z) = (U_1')^{-1}[V_z(t, z)]$.

6. **Singular stochastic control and transaction costs** The HJB equation can be extended to a class of singular stochastic problems such as the

---

[2] Section 7.3.2 will give further discussion of the HJB equation and its interpretation as the "infinitesimal version" of the dynamic programming principle. The HJB equation implicitly assumes that the value function is sufficiently smooth in t and x. Relaxing this assumption leads to "viscosity solutions" of the HJB equation. When the state space is discrete as in the case of a continuous-time controlled Markov chain in Appendix B or its discrete-time counterpart (3.34), the value function also satisfies a similar equation which is often called the "Bellman equation".

Merton problem in the presence of transaction costs as discussed in Section 3.3.2, which will be considered in the rest of this supplement to illustrate the main ideas. The solvency region can be partitioned into three regions corresponding to "buy stock" ($\mathcal{B}$), "sell stock" ($\mathcal{S}$), and "no transactions" ($\mathcal{N}$). Instantaneous transition from $\mathcal{B}$ to the buy boundary $\partial\mathcal{B}$ or from $\mathcal{S}$ to the sell boundary $\partial\mathcal{S}$ takes place and moving the portfolio parallel to $\partial_\lambda\mathcal{D}$ or $\partial_\mu\mathcal{D}$ (i.e., in the direction of $(-1, (1+\lambda)^{-1})^\top$ or $(1, -(1-\mu)^{-1})^\top$). This suggests that the value function satisfies $V(t, x, y) = V(t, x + (1 - \mu)\delta y, y - \delta y)$ for $(t, x, y) \in \mathcal{S}$ and $V(t, x, y) = V(t, x - (1 + \lambda)\delta y, y + \delta y)$ for $(t, x, y) \in \mathcal{B}$ so that as $\delta y \to 0$,

$$V_y(t, x, y) = (1 - \mu)V_x(t, x, y), \qquad (t, x, y) \in \mathcal{S}, \qquad (3.54)$$
$$V_y(t, x, y) = (1 + \lambda)V_x(t, x, y), \qquad (t, x, y) \in \mathcal{B}. \qquad (3.55)$$

In $\mathcal{N}$ the value function satisfies (3.53). Section 7.3.2 considers another class of singular stochastic control problems that are associated with market making strategies, for which the HJB equation such as (3.53) has to be replaced by variational, or quasi-variational, inequalities.

Infinite-horizon case with stationary policy

To simplify the calculations, consider the infinite-horizon case $T = \infty$ so that the value function and the buy/sell boundaries do not depend on t. For the power utility $U(z) = z^\gamma/\gamma$ with $\gamma < 1$, $\gamma \neq 0$, we can reduce the HJB equation to an ordinary differential equation (ODE) with one state variable by using the homotheticity property $V(x, y) = y^\gamma \psi(x/y)$, where $\psi(x) = V(x, 1)$. The system (3.53)–(3.55) now reduces to the ODE

$$b_3 x^2 \frac{d^2\psi}{dx^2} + b_2 x \frac{d\psi}{dx} + \frac{1 - \gamma}{\gamma}\left(\frac{d\psi}{dx}\right)^{\gamma/(\gamma-1)} + b_1\psi = 0, \qquad x_* < x < x^*, \tag{3.56}$$

$$\psi(x) = \begin{cases} \gamma^{-1}A_*(x + 1 - \mu)^\gamma, & x \leq x_* \\ \gamma^{-1}A^*(x + 1 + \lambda)^\gamma, & x \geq x^*, \end{cases} \tag{3.57}$$

where

$$b_1 = \sigma^2\gamma(\gamma - 1)/2 + \alpha\gamma - \beta, \qquad b_2 = \sigma^2(1 - \gamma) + r - \alpha, \; b_3 = \sigma^2/2. \tag{3.58}$$

Principle of smooth fit

This principle pertains to the existence and uniqueness of convex C^2-solutions to optimal stopping and singular stochastic control problems associated with diffusion processes; see Benes et al. (1980), Karatzas (1983), and Ma (1992). It determines the "free boundary" in these problems by matching the first derivatives across the boundaries. In particular, to find the constants x_*, x^*, A_*, A^*, and the function ψ in (3.56), Davis and Norman (1990) apply the principle of smooth fit to ψ'' at x_* and x^* to solve for A_* and A^* (which depend on x_* and x^* respectively) and x_* and x^*.

7. **Approximate dynamic programming (ADP)** Although dynamic programming (DP) provides a standard tool for stochastic optimization to minimize the running and terminal expected costs sequentially over time, its practical implementation is limited by the exponential rise in computational complexity as the dimension of the state space increases, which is often called the "curse of dimensionality" (Bellman, 1957, p. ix). For example, Åström (1983, p. 478) used the backward induction algorithm in DP to compute the optimal policy that minimizes

$$\int E_\beta \left\{ \sum_{t=2}^{T} (y_t - y_t^*)^2 \right\} d\pi(\beta) \qquad (3.59)$$

in the regression model $y_t = y_{t-1} + \beta u_t + \varepsilon_t$, where $\varepsilon_t \sim N(0,1)$, $y_t^* \equiv 0$, and π is a normal prior distribution on β. With $T = 30$, the numerical computations took 180 CPU hours and showed that the optimal policy takes relatively large and irregular control actions u_t to probe the system when the Bayes estimate β_{t-1} of β has poor precision, but is well approximated by the *certainty equivalence* rule $\beta_{t-1} u_t = -y_{t-1}$. Because of the computational complexity of DP, no simulation studies have been performed on the Bayes risk (3.59) of the computationally formidable optimal policy to compare it with that of the much simpler myopic or certainty equivalence rule. In addition, instead of Bayes risks, one often considers risk functions evaluated at different values of α and β.

As shown by Han et al. (2006), important advances in ADP "in the machine learning/artificial intelligence literature, motivated by large combinatorial optimization problems in traffic and communication networks and computer algorithms to play games such as backgammon and solitaire" can be modified for applications to adaptive control of regression models such as (3.59). In particular, Han et al. describe how the rollout algorithm, which arises from the game of backgammon, can be used to derive nearly optimal policies for adaptive control in regression models.

Approximate policy iteration via rollout

To begin with, consider a controlled Markov chain $\{X_t\}$ on a finite state space $\mathcal{X}$, whose control u_t at time t belongs to a finite set U. Let $[p_{xy}(u)]_{x,y \in \mathcal{X}}$ denote the transition probability matrix corresponding to control $u \in U$. An infinite-horizon Markov decision problem is an optimization problem that chooses a sequence $\boldsymbol{u} = (u_1, u_2, \dots)$ of controls to minimize the total expected discounted cost

$$c_{\boldsymbol{u}}(x) = E \left\{ \sum_{t=1}^{\infty} \delta^{t-1} g(X_t, u_t) \,\middle|\, X_0 = x \right\}, \qquad (3.60)$$

where $0 < \delta < 1$ is a discount factor and g is a cost function. The minimum $V(x)$ of $c_u(x)$ over all control policies u satisfies the Bellman equation

$$V(x) = \inf_{u \in U} \left\{ g(x, u) + \delta \sum_{y \in \mathcal{X}} p_{xy}(u) V(y) \right\}. \tag{3.61}$$

The minimum is attained by a stationary policy, i.e., $u_t = \mu(x_t)$, where $\mu(x)$ is the minimizer u of the right-hand side of (3.61). The *policy iteration* method in DP uses (3.61) at the kth iteration a "policy evaluation" step that computes the cost $c_{\mu^{(k)}}(x)$ of a stationary policy (associated with $\mu^{(k)} : \mathcal{X} \to U$) and a "policy improvement" step that defines $\mu^{(k+1)}(x)$ as the minimizer u of

$$g(x, u) + \delta \sum_{y \in \mathcal{X}} p_{xy}(u) c_{\mu^{(k)}}(y). \tag{3.62}$$

The cost-to-go function c_μ can be computed by solving the linear system

$$c_\mu(x) = g(x, \mu(x)) + \delta \sum_{y \in \mathcal{X}} p_{xy}(\mu(x)) c_\mu(y), \qquad x, y \in \mathcal{X}. \tag{3.63}$$

Note that (3.62) in the policy improvement step can be expressed as

$$g(x, u) + E \left\{ \sum_{t=2}^{\infty} \delta^{t-1} g(X_t, u_t^{(k)}) \,\middle|\, X_1 = x, \ u_1 = u \right\}. \tag{3.64}$$

When the state space $\mathcal{X}$ is very large, solving the linear system (3.63) becomes computationally expensive. Also storing $\mu(x)$, $x \in \mathcal{X}$, for a stationary policy requires exorbitant memory. Instead of solving (3.63), one can use Monte Carlo simulations to compute the cost-to-go function of a stationary policy u for those states that are visited during the execution of the policy, as is done in a "rollout", which corresponds to a single iteration of policy improvement via minimization of (3.64) with $k = 0$ and which uses a good heuristic policy for $u_t^{(0)}$. An "iterated rollout" involves $k \geq 1$, where $k = 1$ corresponds to the rollout of a heuristic policy and $k = 2$ corresponds to the rollout of that rollout policy, etc. Since the calculation of $u_t^{(k)}$ in (3.64) requires optimization of the cost-to-go that is computed by Monte Carlo simulations, an iterated rollout requires layers of Monte Carlo simulations, and it is therefore important to find a good heuristic policy to initialize because at most a few iterations are computationally tractable. To compute the cost-to-go by Monte Carlo, the infinite series in (3.64) has to be replaced by a finite sum over $2 \leq t \leq n_\delta$. Unless δ is very near 1, n_δ can be of moderate size (< 100) for δ^{n_δ} to be reasonably small. Details of the Monte Carlo procedure are given in Section 4 of Han et al. (2006).

Rollout for finite-horizon problems

Section 3 of Han et al. (2006) modifies the preceding procedure for the finite-horizon case, for which DP chooses controls $u_1^*, \ldots, u_n^*$ using backward induction that determines u_k^* by minimizing

$$h_{k-1}(u) + E\left[\sum_{i=k+1}^{n} h_{i-1}(u_i^*)\bigg|\mathcal{F}_{k-1}, \ u_k = u\right] \tag{3.65}$$

after determining the future controls $u_{k+1}^*, \ldots, u_n^*$. For $i \geq k+1$, since u_i^* is a complicated nonlinear function of the past observations and of $y_k, u_{k+1}^*, y_{k+1}, \ldots, x_{i-1}^*, y_{i-1}$ that are not yet observed, evaluation of the aforementioned conditional expectation is a formidable task. To overcome this difficulty, rollout approximates the optimal policy u_k^* by minimizing (3.65) with $u_{k+1}^*, \ldots, u_n^*$ replaced by some known *base policy* $\hat{u}_{k+1}, \ldots, \hat{u}_n$, which ideally can be easily computed and is not far from the optimum. Specifically, given a base policy $\hat{\boldsymbol{u}} = (\hat{u}_1, \ldots, \hat{u}_n)$, let $\hat{u}_1^{(1)}$ be the u that minimizes

$$h_{k-1}(u) + E\left[\sum_{i=k+1}^{n} h_{i-1}(\hat{u}_i)\bigg|\mathcal{F}_{k-1}, \ \hat{u}_k = u\right], \tag{3.66}$$

where the expectation in the second term in (3.66) is typically evaluated by Monte Carlo simulation. The policy $\hat{\boldsymbol{u}}^{(1)} = (\hat{u}_1^{(1)}, \ldots, \hat{u}_n^{(1)})$ is called the *rollout* of $\hat{\boldsymbol{u}}$. The rollout $\hat{\boldsymbol{u}}^{(1)}$ may itself be used as a base policy, yielding $\hat{\boldsymbol{u}}^{(2)}$, and in theory, this process may be repeated an arbitrary number of times, yielding $\hat{\boldsymbol{u}}^{(1)}, \hat{\boldsymbol{u}}^{(2)}, \hat{\boldsymbol{u}}^{(3)}, \ldots$ Letting $R(\boldsymbol{u}) = E[\sum_{i=1}^{n} h_{i-1}(u_i)]$, Bayard (1991) has shown that, regardless of the base policy, rolling out n times yields the optimal design and that rolling out always improves the base design, i.e.,

$$R(\hat{\boldsymbol{u}}) \geq R(\hat{\boldsymbol{u}}^{(1)}) \geq R(\hat{\boldsymbol{u}}^{(2)}) \geq \cdots \geq R(\hat{\boldsymbol{u}}^{(n)}) = R(\boldsymbol{u}^*) \tag{3.67}$$

for any policy $\hat{\boldsymbol{u}}$, where $\boldsymbol{u}^*$ denotes the optimal policy. Although (3.67) says that rolling out a base design can improve it and rolling out n times yields the dynamic programming solution, in practice it is difficult to use a rollout as the base policy for another rollout (which is defined by a backward induction algorithm that involves Monte Carlo simulations followed by numerical optimization at every stage). Han et al. (2006) have shown how this difficulty can be resolved in adaptive control of regression models by restricting the sum $\sum_{i=k+1}^{n}$ in (3.66) to $\sum_{i=k+1}^{\min(n,k+m)}$ that involves at most m summands in a tractable limited look-ahead scheme.

Approximate value iteration via LS-MC (regression and Monte Carlo)

The conditional expectation in (3.65), as a function of u, is called the *cost-to-go function* in DP. An ADP method, which grew out of the machine learning literature, is based on two statistical concepts concerning the conditional expectation. First, for given u and the past information $\mathcal{F}_{k-1}$, the conditional expectation is an expectation and therefore can be evaluated by Monte Carlo simulations, if one knows how $h_k(u_{k+1}^*), \ldots, h_{n-1}(u_n^*)$ are generated. The second concept is that, $h_i(u_{i+1})$ is a conditional expectation given $\mathcal{F}_i$ and is therefore a regression function with regressors generated from $\mathcal{F}_i$. Based on a large sample (generated by Monte Carlo), the regression function can be estimated by least squares using basis function approximations, as is typically done in nonparametric regression. Combining least squares (LS) regression with Monte Carlo (MC) simulations yields the following LS-MC method for Markov decision problems. Let $\{X_t, t \geq 0\}$ be a Markov chain whose transition probabilities from state X_t to X_{t+1} depend on the action u_t at time t, and let $f_t(x, u)$ denote the cost function at time t, incurred when the state is x and the action u is taken. Consider the statistical decision problem of choosing x at each stage k to minimize the cost-to-go function

$$Q_k(x, u) = E\left\{ f_k(x, u) + \sum_{t=k+1}^n f_t(X_t, u_t) \Big| X_k = x, \ u_k = u \right\}, \quad (3.68)$$

assuming that $u_{k+1}, \ldots, u_n$ have been determined. Note that Q_k is a function of both the state x and the action u, which is the basic ingredient in "Q-learning" techniques in the machine learning literature. Let

$$V_k(x) = \min_x Q_k(x, u), \qquad u_k^* = \arg\min_u Q_k(x, u). \quad (3.69)$$

These functions can be evaluated by the backward induction algorithm of dynamic programming: $V_n(x) = \min_u f_n(x, u)$, and for $n > k \geq 1$,

$$V_k(x) = \min_u \left\{ f_k(x, u) + E\left[V_{k+1}(X_{k+1}) | X_k = x, \ u_k = u \right] \right\}, \quad (3.70)$$

in which the minimizer yields u_k^*. Assuming the state space to be finite-dimensional, the LS-MC method first uses basis functions ϕ_j, $1 \leq j \leq J$, to approximate V_{k+1} by $\hat{V}_{k+1} = \sum_{j=1}^J a_{k+1,j}\phi_j$, and then uses this approximation together with B Monte Carlo simulations to approximate

$$E\left[V_{k+1}(X_{k+1}) | X_k = x, \ u_k = u \right]$$

for every u in a grid of representative values. This yields an approximation $\tilde{V}_k$ to V_k and also $\hat{u}_k$ to u_k^*. Moreover, using the sample

$$\left\{ \left(X_{k,b}, \tilde{V}_k(u_{k,b}) \right) : 1 \leq b \leq B \right\} \quad (3.71)$$

generated by the control action $\hat{u}_k$, we can perform least squares regression of $\tilde{V}_k(u_{k,b})$ on $[\phi_1(X_{k,b}), \ldots, \phi_J(X_{k,b})]$ to approximate $\tilde{V}_k$ by $\hat{V}_k = \sum_{j=1}^J a_{k,j}\phi_j$. The LS-MC method has also useful applications in the valuation of complex path-dependent financial derivatives; see Longstaff and Schwartz (2001), Tsitsiklis and Van Roy (1999), Lai and Wong (2004), and the references therein.

Reinforcement learning

As pointed out by Powell and Ryzhov (2013), "ADP (including research under names such as reinforcement learning, adaptive dynamic programming and neuro-dynamic programming) has become an umbrella for a wide range of algorithmic strategies", most of which involve the learning functions (or policies) of some form "quickly and effectively" and Monte Carlo sampling. They say: "Learning arises in both offline settings (training an algorithm within the computer) and online settings (where we have to learn as we go). Learning also arises in different ways within algorithms, including learning the parameters of a policy, learning a value function and learning how to expand the branches of a tree." They also point out that *passive learning* policies collect information and update beliefs about functions, without making any explicit attempt at collecting information in a way that would accelerate the learning policy. In contrast, *"active learning* refers to policies where we are willing to make suboptimal actions explicitly because the information gained will add value later in the process."

Reinforcement learning is an area of active learning that has been widely studied in computer science and artificial intelligence and has applications in other disciplines such as neuroscience, game theory, multi-agent systems, operations research and control systems. As pointed out in the survey article of Kaelbling et al. (1996), "its promise is beguiling — a way of programming (systems or) agents by rewards and punishment without needing to specify how the task is to be achieved"; note the role of "algorithms" here. Learning is "through trial-and-error interactions with a dynamic environment." Specifically, "an agent is connected to its environment via perception and action" and receives on each step of interaction an input, i, some indication after the current state, s, of the environment, and chooses an action, a, to generate as output. "The action changes the state of the environment, and the value of this state transition is communicated to the agent through a scalar *reinforcement signal*." Active learning takes the form of learning from the reinforcement signals to improve the outcomes of the actions sequentially over time, guided by a wide variety of algorithms including LS-MC, certainty equivalence, adaptive heuristic critic and temporal differences, Q-learning; see Kaelbling et al. (1996), Barto et al. (1983), Sutton (1988), Watkins and Dayan (1992), and Kumar and Varaiya (1986).

Monte Carlo tree search, policy and value networks, deep learning

Algorithms and optimization in the book's subtitle feature prominently in ADP, and in particular, reinforcement learning for which there have been major advances in recent years. These advances are reviewed by Silver et al. (2016) in connection with their applications to programming computers to play the game of Go, which "has long been viewed as the most challenging of classic games for artificial intelligence owing to its enormous search space and the difficulty of evaluating board positions and moves." Among the advances reviewed are (a) machine learning from human expert games, (b) reinforcement learning of value networks to evaluate board positions, (c) reinforcement learning of policy networks to select moves, (d) Monte Carlo tree search (MCTS) that uses Monte Carlo rollouts to estimate the value of each state and action pair in a search tree so that the relevant values become more accurate as the search tree grows larger. The policy and value networks are used to narrow the search to a beam of high-probability actions and to sample actions during rollouts, and "deep convolutional neural networks" are used to "reduce the effective depth and breath of the search tree: evaluating positions using a value network and sampling actions using a policy network." The computer program AlphaGo thereby efficiently combines the policy and value networks with MCTS. It defeated the human European Go champion by 5 games to 0 in 2015 and a 9-dan Korean champion by 4 games to 1 in 2016.

8. **Information theory and universal portfolios** Suppose $\mathcal{R}_t$ are i.i.d. Then because $\mathcal{R}_t$ is independent of $\mathcal{F}_t$, the optimal portfolio weight (3.17) becomes $\boldsymbol{w}^* = \arg\max_{\boldsymbol{w}\in\mathcal{W}} E[\log(\boldsymbol{w}^\mathsf{T}\mathcal{R}_t)]$, which is independent of t. Hence the asymptotically optimal strategy is the *best constant rebalancing portfolio* (BCRP), i.e., the optimal choice for $\boldsymbol{w}_t$ is constant for all t. For finite T, $\boldsymbol{w}_T^*$ is given by

$$\boldsymbol{w}_T^* = \arg\max_{\boldsymbol{w}\in\mathcal{W}} \prod_{t=1}^{T} \boldsymbol{w}^\mathsf{T}\mathcal{R}_t \tag{3.72}$$

which gives the optimal wealth Z_T^*. However, this also means that $\boldsymbol{w}_T^*$ is a function of all $\mathcal{R}_t$ for $t = 1,\ldots,T$, making BCRP anticipative and therefore not implementable without hindsight. To resolve this difficulty, Cover and Ordentlich (1996) propose to use a Dirichlet-weighted portfolio

$$\hat{\boldsymbol{w}}_{t+1} = \frac{\int_{\mathcal{W}} \boldsymbol{w} \left[\prod_{i=1}^{t} \boldsymbol{w}^\mathsf{T}\mathcal{R}_i\right] \mu_{\boldsymbol{\alpha}}(d\boldsymbol{w})}{\int_{\mathcal{W}} \left[\prod_{i=1}^{t} \boldsymbol{w}^\mathsf{T}\mathcal{R}_i\right] \mu_{\boldsymbol{\alpha}}(d\boldsymbol{w})}, \qquad \text{for } t = 1, 2, \ldots, \tag{3.73}$$

where the weight is the Dirichlet distribution

$$\mu_{\boldsymbol{\alpha}}(d\boldsymbol{w}) = \frac{\Gamma(\sum_{j=1}^{m} \alpha_j)}{\prod_{j=1}^{m} \Gamma(\alpha_j)} \left(\prod_{j=1}^{m} w_j^{\alpha_j - 1}\right) d\boldsymbol{w}. \tag{3.74}$$

over the m-dimensional simplex $\{\boldsymbol{w} = (w_1,\ldots,w_m)^\mathsf{T} : w_j \geq 0$ and $\sum_{j=1}^{m} w_j = 1\}$ with parameter vector $\boldsymbol{\alpha} = (\alpha_1,\ldots,\alpha_m)^\top$ such that $\alpha_j > 0$ for all $j = 1,\ldots,m$.

For $t = 0$, $\boldsymbol{w}_T^*$ is given by the mean vector of $\mu_\alpha(d\boldsymbol{w})$. To prove the "universality" of such a scheme, Ordentlich and Cover (1998)[3] show that for any T,

$$\min_{\{\tilde{\boldsymbol{w}}_t\}_{t=1}^T} \max_{\{\boldsymbol{\mathcal{R}}_t\}_{t=1}^T} \left(Z_T^* / \tilde{Z}_T \right) = U_T, \tag{3.75}$$

where the minimum is over all non-anticipative portfolios $\{\tilde{\boldsymbol{w}}_t\}_{t=1}^T$, $\tilde{Z}_T = Z_0 \prod_{t=1}^T \tilde{\boldsymbol{w}}_t^\mathsf{T} \boldsymbol{\mathcal{R}}_t$ is the terminal wealth and

$$U_T = \sum_{n_1 + \cdots + n_m = T} \binom{T}{n_1, \ldots, n_m} 2^{-T\mathcal{I}(n_1/T, \ldots, n_m/T)}, \tag{3.76}$$

in which $\mathcal{I}(p_1, \ldots, p_m) = -\sum_{j=1}^m p_j \log_2 p_j$ is Shannon's entropy in information theory and the term in parentheses is the multinomial coefficient. The minimax value (3.75) is the value of a game between an investor, who has to use a non-anticipative strategy $\{\tilde{\boldsymbol{w}}_t\}_{t=1}^T$, and an adversarial (worst-case) market that selects a sequence $\{\boldsymbol{\mathcal{R}}_t\}_{t=1}^T$ to maximize the ratio $Z_T^*/\tilde{Z}_T$. That is, the investor aims at minimizing the maximum regret for not being able to use the optimal anticipative portfolio with weights (3.72) that are available only at time T. It follows from (3.76) that $U_T = O(T^{(m-1)/2})$. By choosing $\boldsymbol{\alpha} = (1/2, \ldots, 1/2)^\mathsf{T}$, Cover and Ordentlich (1996) show that $Z_T^*/\tilde{Z}_T \le (T+1)^{m+1}$ for all T and all choices of $\{\boldsymbol{\mathcal{R}}_t\}_{t=1}^T$. Combining this result with (3.75) yields

$$\lim_{T \to \infty} \sup_{\{\boldsymbol{\mathcal{R}}_t\}_{t=1}^T} \frac{1}{T} \log \frac{Z_T^*}{\tilde{Z}_T} = 0. \tag{3.77}$$

Cover (1996) calls the portfolios attaining the minimax value of (3.75) "universal" because they are the counterpart of the universal data compression algorithms in information theory. He shows that (3.75) is the same as "the value of the minimax redundancy" in data compression algorithms that "achieve the entropy limit for the unknown source within a minimax redundancy" over the set of unknown distributions (and is therefore "universal"). Note that (3.73) basically replaces the maximization procedure in (3.72) by a sequence of weighted averages. This is similar to the pseudo-maximization method in Chapters 11 and 17 of de la Peña et al. (2009) who use pseudo-maximization to connect likelihood (and in particular, generalized likelihood) inference with Bayesian methods.

---

[3] Instead of using the minimax approach, Ordentlich and Cover (1998) adopt the maximin approach by considering the reciprocal of $\left(Z_T^* / \tilde{Z}_T \right)$. Under such approach, the equivalent result for (3.75) is a maximin statement on $\left(\tilde{Z}_T / Z_T^* \right)$ with the term on the right-and-side being U_T^{-1}. Ordentlich and Cover (1998) call the maximin approach as the process of finding the non-anticipative strategy that achieves the best portfolio with hindsight (i.e., the wealth of BCRP Z_T^*), while Cover and Ordentlich (1996) and Cover (1996) phrase the equivalent strategy in (3.75) as the procedure that minimizes the maximum regret.

9. **Kalman filter, LQG control and separation principle**   In Section 3.3.6 that reviews the literature on dynamic portfolio selection in the presence of parameter uncertainty, we have introduced the Bayesian approach which puts a normal prior distribution on the unknown parameters (or some transformation thereof) and have used Kalman filters (or Kalman-Bucy filters) to estimate the posterior means and covariance matrices recursively. The literature reviewed emphasizes the sequential learning aspect of multiperiod portfolio selection, i.e., the aspect that involves "estimation risk" for portfolio selection under power utility and "learning while doing" to reduce parameter uncertainty. It does not consider transaction costs which Section 3.3.5 has treated, by reviewing the work of Gârleanu and Pedersen (2013) who assume quadratic transaction costs and a multi-factor regression model (3.20) on excess returns. Assuming autoregressive dynamics for the factors, their multiperiod portfolio optimization problem (3.21) turns out to be related to Kalman filters for linear state-space models that will be discussed below.

Linear state-space models with inputs and the Kalman filter

Section 5.3 of Lai and Xing (2008) gives an introduction to linear state-space models $x_{t+1} = F_t x_t + w_{t+1}$, $y_t = G_t x_t + \varepsilon_t$, in which x_t is an unobserved state vector undergoing linear dynamics and y_t is the observation that is some linear transformation of the state x_t plus additive noise ε_t. Assuming independent zero-mean Gaussian vectors w_{t+1} and ε_t that are independent and are also independent of the state vectors, with $\mathrm{Cov}(w_t) = \Sigma_t$ and $\mathrm{Cov}(\varepsilon_t) = V_t$, the conditional distribution of x_t given the observation y_s, $s \le t$, is normal and the Kalman filter provides recursive formulas to calculate the mean $\hat{x}_{t|t}$ and the covariance matrix of the estimation error $x_t - \hat{x}_{t|t}$. Chapter 7 of Kumar and Varaiya (1986) shows that these formulas can be easily extended to the linear control systems

$$\begin{aligned}
x_{t+1} &= F_t x_t + \Gamma_t u_t + w_{t+1}, \\
y_t &= G_t x_t + \Lambda_t u_t + \varepsilon_t,
\end{aligned} \tag{3.78}$$

in which u_t is the non-anticipative input (control) at time t. In fact, Section 3 of that chapter shows how to use linearity of the state and measurement equations to obtain the following Kalman recursions from those in the case $u_t \equiv 0$:

$$\begin{aligned}
\hat{x}_{t+1|t} &= F_t \hat{x}_{t|t-1} + \Gamma_t u_t + K_t(y_t - G_t \hat{x}_{t|t-1} - \Lambda_t u_t), \\
P_{t+1} &= (F_t - K_t G_t) P_t (F_t - K_t G_t)^\top + \Sigma_{t+1} + K_t V_t K_t^\top,
\end{aligned} \tag{3.79}$$

in which $K_t = (F_t P_t G_t^\top)(G_t P_t G_t^\top + V_t)^{-1}$ is the Kalman gain matrix. The second recursion, for $P_t = \mathrm{Cov}(x_t - \hat{x}_{t|t-1})$, is called a Riccati equation, and the first recursion is about the Kalman predictor

$E[\boldsymbol{x}_{t+1}|(\boldsymbol{y}_s, \boldsymbol{u}_s), s \leq t]$. The Kalman filter $\hat{\boldsymbol{x}}_{t|t}$ is given by

$$\hat{\boldsymbol{x}}_{t|t} = \hat{\boldsymbol{x}}_{t|t-1} + \boldsymbol{P}_t\boldsymbol{G}_t^\top(\boldsymbol{G}_t\boldsymbol{P}_t\boldsymbol{G}_t^\top + \boldsymbol{V}_t)^{-1}(\boldsymbol{y}_t - \boldsymbol{G}_t\hat{\boldsymbol{x}}_{t|t-1} - \boldsymbol{\Lambda}_t\boldsymbol{u}_t).$$

LQG control scheme

The L and G stand for the linear and Gaussian model (3.78) and Q stands for quadratic costs for control errors and control costs. Specifically, the stochastic control problem for the MIMO system (multiple-input and multiple-output, hence multivariate $\boldsymbol{u}_t$ and $\boldsymbol{y}_t$) is to choose non-anticipative controls $\boldsymbol{u}_t$, $1 \leq t \leq T$, to minimize

$$E\left\{\sum_{t=1}^{T}\boldsymbol{x}_t^\top\boldsymbol{Q}_t\boldsymbol{x}_t + \sum_{t=1}^{T-1}\boldsymbol{u}_t^\top\boldsymbol{R}_t\boldsymbol{u}_t\right\}. \tag{3.80}$$

Backward induction can be used to solve the associated dynamic programming problem. Kumar and Varaiya (1986, Sect. 7.5) derive explicit backward recursion formulas for the optimal $\boldsymbol{u}_t^*$:

$$\boldsymbol{u}_t^* = -\boldsymbol{L}_t\hat{\boldsymbol{x}}_{t|t-1}, \qquad \text{where } \boldsymbol{L}_t = (\boldsymbol{\Gamma}_t^\top\boldsymbol{S}_{t+1}\boldsymbol{\Gamma}_t + \boldsymbol{R}_t)^{-1}\boldsymbol{\Gamma}_t^\top\boldsymbol{S}_{t+1}\boldsymbol{F}_t, \tag{3.81}$$

in which $\boldsymbol{S}_t$ is determined by the backward Riccati equation $\boldsymbol{S}_T = \boldsymbol{Q}_T$, and for $1 \leq t < T$,

$$\boldsymbol{S}_t = \boldsymbol{F}_t^\top\left(\boldsymbol{S}_{t+1} - \boldsymbol{S}_{t+1}\boldsymbol{\Gamma}_t[\boldsymbol{\Gamma}_t^\top\boldsymbol{S}_{t+1}\boldsymbol{\Gamma}_t + \boldsymbol{R}_t]^{-1}\boldsymbol{\Gamma}_t^\top\boldsymbol{S}_{t+1}\right)\boldsymbol{F}_t + \boldsymbol{Q}_t.$$

Separation principle and certainty equivalence

Note that the covariance matrices $\boldsymbol{\Sigma}_t$ and $\boldsymbol{V}_t$ of the random noise $\boldsymbol{w}_t$ and $\boldsymbol{\varepsilon}_t$ do not appear in the optimal control $\boldsymbol{u}_t^*$. Hence the control law is the same when there is no noise, or as if the estimate were without error. It is a certainty equivalence rule, and is also a case in stochastic control where the separation principle is valid. The separation principle separates estimation and control in the design of an optimal feedback control law for a stochastic system. Specifically, the control problem can be solved when there are no unknown parameters or unobserved states. When there are unobserved states, the separation principle first estimates them and then substitute the unobserved states in the optimal control law with known parameters and observed states by their estimates.

Continuous-time extensions and the Kalman-Bucy filter

For simplicity, assume $\boldsymbol{\Lambda}_t = \boldsymbol{0}$, which is the case in practice. The continuous-time extension of (3.78) is

$$\begin{aligned}d\boldsymbol{x}_t &= (\boldsymbol{F}_t\boldsymbol{x}_t + \boldsymbol{\Gamma}_t\boldsymbol{u}_t)dt + \boldsymbol{\Sigma}_t^{1/2}d\boldsymbol{B}_t, \\ d\boldsymbol{y}_t &= \boldsymbol{G}_t\boldsymbol{x}_tdt + \boldsymbol{V}_t^{1/2}d\boldsymbol{B}_t^*,\end{aligned} \tag{3.82}$$

in which $\boldsymbol{B}_t$ and $\boldsymbol{B}_t^*$ are independent Brownian motions. The Kalman-Bucy filter, which is the continuous-time analog of (3.79), is defined by the ordinary differential equations (ODEs)

$$d\hat{\boldsymbol{x}}_t = (\boldsymbol{F}_t\hat{\boldsymbol{x}}_t + \boldsymbol{\Gamma}_t\boldsymbol{u}_t)dt + \boldsymbol{K}_t(d\boldsymbol{y}_t - \boldsymbol{G}_t\hat{\boldsymbol{x}}_t dt),$$
$$\frac{d}{dt}\boldsymbol{P}_t = \boldsymbol{F}_t\boldsymbol{P}_t + \boldsymbol{P}_t\boldsymbol{F}_t^\top + \boldsymbol{\Sigma}_t - \boldsymbol{P}_t\boldsymbol{G}_t^\top\boldsymbol{V}_t^{-1}\boldsymbol{G}_t\boldsymbol{P}_t, \tag{3.83}$$

with $\boldsymbol{K}_t = \boldsymbol{P}_t\boldsymbol{G}_t^\top\boldsymbol{V}_t^{-1}$. The LQG problem in continuous time is to find non-anticipative controls to minimize

$$E\left\{\boldsymbol{x}_T^\top\boldsymbol{Q}_T\boldsymbol{x}_T + \int_0^T (\boldsymbol{x}_t^\top\boldsymbol{Q}_t\boldsymbol{x}_t + \boldsymbol{u}_t^\top\boldsymbol{R}_t\boldsymbol{u}_t)\,dt\right\}. \tag{3.84}$$

The optimal control law is $\boldsymbol{u}_t^* = -\boldsymbol{L}_t\hat{\boldsymbol{x}}_t$, where $\boldsymbol{L}_t = \boldsymbol{R}_t^{-1}\boldsymbol{\Gamma}_t^\top\boldsymbol{S}_t$ and $\boldsymbol{S}_t$ is defined by an ODE with terminal condition $\boldsymbol{S}_T = \boldsymbol{Q}_T$.

10. **A new approach to multiperiod portfolio optimization**   In Section 3.3, we have reviewed the extensive literature on multiperiod portfolio optimization. Unlike the single-period case treated in Chapter 2, for which the issue of modeling and predicting the future covariance matrix of the assets in a portfolio has received much attention and undergone major recent development, this issue seems to have been ignored in the works on the multi-period problem. In particular, it has been noted in Sections 1.4 and 2.7 that investment funds are periodically evaluated on some risk-adjusted performance measure and therefore the choice of λ in Section 2.7 can be targeted towards optimizing that measure along the efficient frontier, and it is natural to do something similar for the multiperiod setting. However, most of the papers reviewed in Section 3.3 do not consider predicting future covariances. For example, in Section 3.3.6 on dynamic portfolio selection in the presence of parameter uncertainty, the parameter considered is the mean return and the Kalman filter is used to handle this parameter, while the covariance parameter is assumed to be known or reliably estimated from the available data. The reason why the filtering approach has not been used to model uncertainties in the volatility parameters is that it will entail nonlinear filtering, as mentioned in the last paragraph of Section 3.3. The preceding supplement has highlighted the power of the Kalman filter in both state estimation and control of linear state-space models, which are characterized by linear state and measurement equations. If either of these equations is nonlinear, then one encounters a much harder filtering problem. However, there are important recent advances in this problem by using sequential Monte Carlo methods called adaptive particle filters that can be used to handle volatility estimation; see Chan and Lai (2013, 2016), Lai and Bukkapatanam (2013), and Chapter 6 of Lai and Xing (2016).

We have used nonlinear filtering only as an illustration of the new tools that are now available to tackle some of the long-standing problems in multiperiod portfolio optimization. Besides giving additional details and discussion of the literature reviewed in Section 3.3, we have also introduced new techniques that pave the way for innovations in multiperiod portfolio optimization. In particular, singular stochastic control, approximate dynamic programming, adaptive filtering, and Monte Carlo tree search are used in Lai and Gao (2016a,b) and Lai et al. (2016a) to develop a new approach to multiperiod portfolio optimization that addresses some of the long-standing issues raised in the first paragraph of Section 3.3.

3.5 Exercises

1. Explain why imposing proportional transaction costs leads to a bang-bang-type control problem, i.e., a no-trade region when prices stay in the region, and pushing back into the region with full trade intensity when prices are too cheap (signaling "buy") or too high (signaling "sell"). Read Magill and Constantinides (1976) and Davis and Norman (1990) and describe the no-trade zone they have found.

2. Read Markowitz and van Dijk (2003) and explain their heuristic approach to modifying the single-period Markowitz's mean-variance portfolio rebalancing to multiperiod rebalancing with many securities and transaction costs. Pinpoint some ADP ideas underlying their approach.

3. Read Cvitanić et al. (2006) and explain how they derive the optimal portfolio weight vector (3.25). Use Supplement 9 in Section 3.4 to derive an alternative optimal portfolio that involves the Kalman-Bucy filter but uses the quadratic cost criterion (3.83) in that supplement.

4

Econometrics of Transactions in Electronic Platforms

As reviewed in Sections 1.1 and 1.7, the evolution of bid and ask for assets to a continuous low-latency double auction system, hitherto inaccessible, has occurred in the 1980s, capitalizing on major advances in exchange technology which not only substantially increased the accessibility of high-resolution data but also greatly enhanced data quality. High-frequency econometrics, which is concerned with econometric analysis of these high-resolution time series data and modeling of transactions in electronic platforms, has emerged as an active area of research in financial econometrics. This chapter reviews some major methods and models in high-frequency econometrics. We begin in Section 4.1 with an introduction to transactions in electronic exchanges, describing in particular the limit order book, how buy (bid) and sell (ask) orders are placed, and when transactions occur. The bulk of high-frequency econometrics is about the last aspect (i.e., the price, quantity and time when a transaction occurs) and the time series of these multivariate observations. High-frequency trading strategies that involve predictions of future transactions involve high-resolution data from the limit order book, and the statistical analysis and modeling of limit order book data is the theme of the next chapter. Section 4.2 introduces some basic econometric models of transaction prices for a single asset. An important problem in financial econometrics is estimation of the quadratic variation, also called realized variance, of the asset's underlying efficient price process, and this problem is studied in Section 4.3. Section 4.4 considers multivariate extensions to high-frequency joint models for multiple assets. Besides quadratic variation, instantaneous volatility and co-volatility are additional parameters that play pivotal roles in quantitative finance. In Section 4.5, the theory of Fourier estimator is introduced as an alternative approach to estimating the quadratic variation as well as these instantaneous measures. Section 4.6 describes econometric models of inter-transaction durations and other summary statistics of transactions data. It also discusses the relationship between high-frequency fluctuations and the longer-horizon (e.g., daily) volatility of asset prices. Supplements and problems are given in Sections 4.7 and 4.8.

4.1   Transactions and transactions data

A trade is an exchange of assets between two parties. In principle, any data and other information generated from the trading activities are transactions data and have been used by traders since ancient times. According to Lo and Hasanhodzic (2010), as early as the period from 747 B.C. to 651 B.C., transactions data together with astronomical observations, recorded on clay Babylonian tablets, had been used by commodity traders to derive strategies based upon seasonal patterns. As mentioned in Chapter 1, since the invention of stock tickers, transaction-by-transaction (or the "tick-by-tick") data have been disseminated from the trading pits. However, a full record of tick-by-tick data is costly (Freeman, 2006) and may be flawed because of human errors in the process of transaction and data transferal under an open outcry system (Gorham and Singh, 2009). In sharp contrast are the transactions and transactions data from today's electronic exchange.

To illustrate, buying one share of AAPL (stock for Apple Inc) for \$114.50 is a transaction that involves exchanging the most liquid asset of \$114.50, cash, for a less liquid asset of one share of AAPL. On most order- and quote-driven[1] electronic exchanges, transactions occur when a matching algorithm used by the exchange matches the buying and selling sides of a specific asset for a specific amount, with one side having had posted a limit order in the limit order book (LOB). The LOB consists of untransacted limit bid and ask orders for a specific asset, e.g., AAPL stock or E-Mini futures, at discrete prices levels. Table 4.1 illustrates a limit order to buy 150 shares of AAPL on NASDAQ with a price of \$114.50, showing the LOB before and after the order is matched. For liquid assets, there can be multiple transactions within a second, hence a potentially rich source of data for modeling. Chapter 5 and Chapter 8 provide detailed descriptions of the data, matching, and statistical models of the LOB. In this chapter, we only consider the transactions data broadcast by an exchange immediately after each matched transaction.

4.2   Models for high-frequency data

We begin with Roll's microstructure model, which was introduced to reconcile the EMH (efficient market hypothesis) with the negative lag-1 autocorrelations of the time series of transaction price changes that had been used as evidence

---

[1] *Order-driven* means that any trader can submit orders and untransacted orders stay in the limit order book. *Quote-driven* means the limit order book consists of only orders submitted by the dealers. These dealers are also known as specialists or market makers and are obligated to post quotes even under drastic market conditions. Most electronic exchanges can be considered as a hybrid of order-driven and quote-driven platforms.

TABLE 4.1: Limit order to buy 150 shares with limit price of \$114.50, GTC,[2] resulting in 100 shares of the limit order traded at \$114.50, and with the remaining 50 shares of the original order now becoming the best bid in the LOB.

		114.52	145			114.53	350
		114.51	51			114.52	145
		114.50	100	$\implies$		114.51	51
80	114.49				50	114.50	
150	114.48				80	114.49	

against EMH. We then generalize Roll's model by adding microstructure noise to the underlying efficient price process after taking logarithms, and relate the models proposed by Aït-Sahalia et al. (2005) and Zhang et al. (2005) to this framework.

4.2.1 Roll's model of bid-ask bounce

Roll (1984) uses the bid-ask bounce feature of market microstructure to derive under EMH the stylized MA(1) autocorrelation of price changes in high-frequency transactions data noted in Section 2.1.2. His model is

$$P_i = P_i^* + cI_i, \tag{4.1}$$

where $2c$ is the bid-ask spread in the market, $I_i = 1$ (or -1) for bid (or ask), and P_i^* and P_i are the efficient price under EMH and the transacted price at the time of the ith transaction, respectively. Roll's model assumes that in equilibrium the I_i are i.i.d. and

$$P\{I_i = 1\} = P\{I_i = -1\} = 1/2,$$

and the P_i^* have independent increments and are independent of I_i. Let

$$D_i = P_i - P_{i-1} = P_i^* - P_{i-1}^* + c(I_i - I_{i-1}) \tag{4.2}$$

denote the price change at the ith transaction. Then

$$\text{Cov}(D_i, D_{i-1}) = -c^2, \quad \text{Cov}(D_i, D_{i-j}) = 0 \text{ with } j \geq 2, \tag{4.3}$$

which agrees with the stylized MA(1) feature of transaction price. In Supplement 6 of Section 4.7, we describe alternative models that use other features of market microstructure to explain the MA(1) autocorrelation under EMH.

---

[2]GTC stands for "good till canceled".

4.2.2   Market microstructure model with additive noise

We begin by the following modification and generalization of (4.1). Let X_t be the logarithm of the efficient price at time t, and let $t_1 < \cdots < t_n$ be the set of transaction times belonging to $[0, T]$, and let Y_{t_i} be the logarithm of the transaction price at time t_i. Other than replacing P_i and P_i^* in (4.1) by their logarithms (which has the advantage of interpretability of the multiplication effect of market frictions on prices, e.g., bid-ask spread being some percentage of the transaction price), we can greatly broaden Roll's model into

$$Y_{t_i} = X_{t_i} + \varepsilon_i, \tag{4.4}$$

where ε_i represents the market microstructure noise, with $E(\varepsilon_i) = 0$, that comes from the following sources:

- Bid-ask bounce, but without Roll's overly restrictive assumption of constant bid-ask spread over time.

- Aggregation across different electronic communication networks that have synchronicity issues.

- Delay of recording because the timestamp for a transaction can lag behind the true transaction time due to latency issues, and other recording errors.

- Gradual instead of instantaneous response to block trades.[3]

- Discrete transaction prices Y_{t_i} whereas X_t is a continuous random variable.

- Strategic components of order flow and inventory control effects.

- Post-processing adjustments during periods of subdued activity, at the exchange or at the data vendor, such as extrapolation of last period's price.

Assuming X_t to have independent increments and ε_i to be i.i.d. with mean zero and variance v_ε and independent of X_t, we still have $\text{Cov}(D_i, D_{i-1}) = -v_\varepsilon < 0$ and $\text{Cov}(D_i, D_{i-j}) = 0$ for $j \geq 2$ in the generalization (4.4) of Roll's model, where $D_i = Y_{t_i} - Y_{t_{i-1}}$.

The continuous-time efficient price process P_t^*, particularly its volatility, is a cornerstone of option pricing theory and risk management; see Chapters 8, 10, and 12 of Lai and Xing (2008). Although high-frequency trading seems to have brought market data into the continuous-time framework in the sense that the set of transaction times t_i tends to be a dense partition of $[0, T]$ in

---

[3]A block trade is a privately negotiated transaction at or exceeding an exchange-determined minimum quantity of shares, which is executed apart from the electronic markets. For example, to sell a large number of shares of a stock (at least 10,000), a hedge fund can arrange a block trade with another fund through a broker-dealer so that the selling company can get a better purchase price while the purchasing company can negotiate a discount off market rates.

the sense that its mesh[4] is close to 0, the presence of microstructure noise as first formalized by Roll's model (4.1) to account for bid-ask bounce shows that the efficient price dynamics cannot be retrieved by direct statistical methods applied to the transaction prices P_{t_i}. Aït-Sahalia et al. (2005) and Zhang et al. (2005) studied the estimation problem in the context of geometric Brownian motion (GBM) $dP_t^*/P_t^* = \mu dt + \sigma dB_t$, and more generally of a diffusion process for $X_t = \log P_t^*$:

$$dX_t = \mu_t dt + \sigma_t dB_t, \tag{4.5}$$

where B_t is standard Brownian motion. Assuming the presence of additive microstructure noise ε_i as in (4.4), they show that estimating the GBM parameters μ and σ directly from Y_{t_i} would lead to inconsistent estimates. While it is possible to estimate them consistently by throwing away a large fraction of the available data and using instead time-aggregated (e.g., daily or weekly) returns, based on closing prices, it does not make efficient use of all available data. This is particularly relevant if one considers the more general model (4.5) in which μ_t and σ_t are time-varying parameters. In the following section, we describe three methods that have been developed to overcome this difficulty.

4.3 Estimation of integrated variance of X_t

To begin with, note that (4.4) does not specify a model for the efficient continuous-time price process X_t. Estimating the unobserved states X_t, $0 \le t \le T$, from the observations Y_{t_i} is a filtering problem when the dynamics of the unobserved process X_t are specified; see Section 3.4. Instead of estimating the trajectory of this theoretical price X_t, a quantity that is of much greater interest in finance theory is the *integrated variance* (or *quadratic variation*)

$$[X]_T = \underset{\text{mesh}(\Pi) \to 0}{\text{p-lim}} \sum_{i=1}^{n} (X_{t(i)} - X_{t(i-1)})^2, \tag{4.6}$$

in which Π denotes a partition $0 = t_0 < \cdots < t_n = T$ of $[0,T]$ and $\text{mesh}(\Pi) = \max_{1 \le i \le n}(t_i - t_{i-1})$. The existence of the p-limit (i.e., limit in probability) in (4.6) comes from theory of continuous-time semimartingales, which are widely regarded as the most general model for efficient price processes. Appendix A summarizes the basic semimartingale theory that is used in this and subsequent chapters.

In the absence of microstructure noise (i.e., $\varepsilon_i = 0$ in (4.4) so that $X_{t_i} =$

---

[4]See the first paragraph of Section 4.3 for a formal definition.

Y_{t_i}), a consistent estimator of the integrated variance (4.6) is the realized variance

$$RV^{(n)} = \sum_{i=2}^{n} \left(Y_{t_i} - Y_{t_{i-1}} \right)^2, \tag{4.7}$$

noting that $\max_{2 \le i \le n}(t_i - t_{i-1})$ is typically very small in high-frequency trading. However, $RV^{(n)}$ is no longer consistent when the ε_i are independent non-degenerate random variables. In fact, assuming ε_i to be i.i.d. with mean 0 and variance v_ε and $E(\varepsilon_i^4) < \infty$, and that X_t is a diffusion process independent of the ε_i, Zhang et al. (2005) and Hansen and Lunde (2006) have shown that

$$RV^{(n)}/n \xrightarrow{P} 2v_\varepsilon. \tag{4.8}$$

Therefore, the microstructure noise ε_i makes $RV^{(n)}$ a biased estimator of the integrated variance, with bias increasing linearly with the sample size n. To prove (4.8) for general semimartingales which include the diffusion processes (4.5) considered by these authors, note that

$$
\begin{aligned}
\sum_{i=2}^{n}(Y_{t_i} - Y_{t_{i-1}})^2 &= \sum_{i=2}^{n} \left(X_{t_i} + \varepsilon_i - X_{t_{i-1}} - \varepsilon_{i-1} \right)^2 \\
&= \sum_{i=2}^{n}(X_{t_i} - X_{t_{i-1}})^2 + 2\sum_{i=2}^{n}(X_{t_i} - X_{t_{i-1}})(\varepsilon_i - \varepsilon_{i-1}) \\
&\quad + \sum_{i=2}^{n}(\varepsilon_i - \varepsilon_{i-1})^2.
\end{aligned}
$$

The first two summands are $O_p(1)$ in view of (4.6) and the martingale convergence theorem (see Appendix A). Hence, (4.8) follows from $n^{-1}\sum_{i=2}^{n}(\varepsilon_i - \varepsilon_{i-1})^2 \xrightarrow{P} 2v_\varepsilon$, which is the consequence of the law of large numbers.

The rest of this subsection reviews various estimators of $[X]_T$ under the microstructure noise model (4.4) together with the diffusion model (4.5) for the dynamics of X_t, which implies $[X]_T = \int_0^T \sigma_t^2 dt$. Sections 4.3.1–4.3.3 introduce methods that make use of coarser sampling frequencies. Sections 4.3.4 and 4.3.5 introduce the realized kernel estimator and the method of pre-averaging, respectively. The quasi-maximum likelihood approach is described in Section 4.3.6.

4.3.1   Sparse sampling methods

Sparse sampling refers to aggregating the returns over time so that the sum of squares of lower-frequency returns is used to estimate $[X]_T$. Specifically, instead of $RV^{(n)}$, we use

$$[Y]_T^{(k)} = \sum_{0 \le k(i-1) \le n-k} (Y_{t_{ki}} - Y_{t_{k(i-1)}})^2 \tag{4.9}$$

based on the subsample of size $n_k \sim n/k$ consisting of the returns $Y_{t_{ki}} - Y_{t_{k(i-1)}}$. Aït-Sahalia et al. (2005) and Zhang et al. (2005) have shown that

$$[Y]_T^{(k)} - [X]_T - 2n_k v_\varepsilon \overset{\mathcal{L}}{=} (1 + o_p(1))\gamma(n_k)N(0,1), \qquad (4.10)$$

where $\overset{\mathcal{L}}{=}$ denotes equivalence in distribution and $\gamma^2(n_k)$ is a sum of two terms depending on the sample size n_k, with one representing the source of error due to the market microstructure noise ε_i and the other representing the discretization error due to replacing the continuous time t in $[X]_T$ by the discrete time t_{ki}. They make use of (4.10) to come up with an asymptotically optimal choice of n_k when $v_\varepsilon \to 0$. Under certain assumptions, they show that the optimal subsample size n^* is of the order $v_\varepsilon^{-2/3}$.

4.3.2   Averaging method over subsamples

Zhang et al. (2005) comment that a drawback of sparse sampling methods is that they do not use the data to the full extent. To improve on sparse sampling, they propose to divide $\{t_1, \ldots, t_n\}$ into k disjoint subsets of the form $S_j = \{t_{j-1+ki} : i = 1, \ldots, n_j\}$, $j = 1, \ldots, k$. Note that the case $j = 1$ reduces to the subsample in the preceding section and that $n_j \sim n_k \sim n/k$. Define $[Y]_T^{(j,k)}$ as in (4.9) but with S_j replacing S_1. Essentially all data points are used in their average estimator $[Y]_T^{\text{avg}}$ of $[X]_T$:

$$[Y]_T^{\text{avg}} = \frac{1}{k} \sum_{j=1}^{k} [Y]_T^{(j,k)}. \qquad (4.11)$$

Letting $\bar{n} = k^{-1} \sum_{j=1}^{k} n_j$, they show that under (4.5) and some other conditions,

$$[Y]_T^{\text{avg}} - [X]_T - 2\bar{n}v_\varepsilon \overset{\mathcal{L}}{=} (1 + o_p(1)) \left[4\frac{n_k}{k}E(\varepsilon^4) + \frac{4}{3n_k}\int_0^T \sigma_t^4 \, dt \right]^{1/2} N(0,1).$$
$$(4.12)$$

Since $\bar{n} \sim n_k \sim n/k$, it follows that the optimizing k that gives the smallest standard deviation of the distribution on the right-hand side of (4.12) is of the order $n^{2/3}$, for which n_k has order $n^{1/3}$.

4.3.3   Method of two time-scales

A major issue with the preceding methods is the bias $2n_k v_\varepsilon$ in (4.10) and $2\bar{n}v_\varepsilon$ in (4.12), which is unbounded since the asymptotic theory requires $n \to \infty$ and $n_k \to \infty$. To address this issue, Zhang et al. (2005) use the method of two time-scales, the first of which is the "sparse" time-scale, yielding the subsample sizes $n_j \sim n/k$ in the preceding subsection that averages $[Y]_T^{(j,k)}$ over these k subsamples to arrive at the biased estimate (4.11). The second

time-scale is the original time-scale of the data and is used in conjunction with
(4.8) to estimate v_ε consistently by

$$\hat{v}_\varepsilon = \frac{RV^{(n)}}{2n}, \tag{4.13}$$

and also the bias by $2\bar{n}\hat{v}_\varepsilon = (\bar{n}/n)RV^{(n)}$. The method of two time-scales leads
to the bias-corrected estimator:

$$\widehat{[X]}_T = [Y]_T^{\text{avg}} - (\bar{n}/n)\, RV^{(n)}, \tag{4.14}$$

for which Zhang et al. (2005) show under certain conditions that

$$(k/n_k)^{1/2} \, (\widehat{[X]}_T - [X]_T) \xrightarrow{\mathcal{L}} N(0, 8v_\varepsilon^2), \tag{4.15}$$

with the choice of $k \sim n^{2/3}$ yielding $n^{1/6}(\widehat{[X]}_T - [X]_T) \xrightarrow{\mathcal{L}} N(0, 8v_\varepsilon^2)$.

4.3.4 Method of kernel smoothing: Realized kernels

The additive noise ε_i in (4.4) suggests that smoothing techniques in nonparametric regression (see, e.g., Chapter 7 of Lai and Xing, 2008) can be
used to remove its effect in estimation of $[X]_T$. The problem is, however,
considerably more difficult than estimating the conditional mean $E(Y|X)$ in
nonparametric regression because (a) here X is not observed and (b) the
problem is to estimate $[X]_T$ rather than $f(X) = E(Y|X)$. But the principle
of averaging out the noise locally and discretizing the continuous distribution X in nonparametric regression (e.g., kernel smoothing) still works here.
This idea was first tried by Zhou (1996) and Hansen and Lunde (2006), but
their proposed estimators are inconsistent. A definitive solution was given by
Barndorff-Nielsen et al. (2008). They call their kernel estimator of $[X]_T$ *realized kernel*, following the widely used terminology of realized variance for
(4.7). The realized kernel estimator uses time-aggregated log returns data
$Y^{(\delta)} = \{Y_{i\delta} - Y_{(i-1)\delta}, i = 1, \ldots, n\}$, partitioning the time interval $[0, T]$ into
$n = T/\delta$ grids and considering the change in log-price from beginning to close
of each grid. For $h = -H, \ldots, -1, 0, 1, \ldots, H$, define

$$\gamma_h(Y^{(\delta)}) = \sum_j (Y_{j\delta} - Y_{(j-1)\delta})(Y_{(j-h)\delta} - Y_{(j-h-1)\delta}), \tag{4.16}$$

in which the sum Σ_j is over $(h+1) \le j \le n$ if $h \ge 0$, and over $1 \le j \le n - |h|$
if $h < 0$. Note that for $h = 0$, (4.16) reduces to (4.7). The realized kernel,
which is used to estimate $[X]_T$, is defined by

$$K(Y^{(\delta)}) = \underbrace{\gamma_0(Y^{(\delta)})}_{\text{realized variance}} + \underbrace{\sum_{h=1}^{H} \kappa\left(\frac{h-1}{H}\right)\left(\gamma_h(Y^{(\delta)}) + \gamma_{-h}(Y^{(\delta)})\right)}_{\text{realized kernel correction}}, \tag{4.17}$$

where κ is a twice continuously differentiable function on $[0, 1]$ with $\kappa(0) = 1$ and $\kappa(1) = 0$ (called the "flat-top" kernels).

Assuming the diffusion model (4.5) for X_t and independence of ε_i and the process $\{X_t\}$, Barndorff-Nielsen et al. (2008) show that for the choice $H = cn^{2/3}$,

$$n^{1/6}\{K(Y^{(\delta)}) - [X]_T\} \overset{\mathcal{L}}{\to} N\left(0, 4c^*T \int_0^T \sigma_t^4 \, dt\right) \tag{4.18}$$

as $n \to \infty$ (and therefore $\delta = T/n \to 0$), where c^* depends on c, $(\kappa'(0))^2 + (\kappa'(1))^2$, $\int_0^1 \kappa^2(x) \, dx$ and $v_\varepsilon^2/(T \int_0^T \sigma_t^4 \, dt)$. Moreover, in the case of $\kappa'(0) = \kappa'(1) = 0$, the choice $H = cn^{1/2}$ gives a faster rate of convergence:

$$n^{1/4}\{K(Y^{(\delta)}) - [X]_T\} \overset{\mathcal{L}}{\to} N\left(0, 4\tilde{c}T \int_0^T \sigma_t^4 \, dt\right), \tag{4.19}$$

where $\tilde{c}$ depends also on $\int_0^1 [\kappa'(x)]^2 \, dx$ and $\int_0^1 [\kappa''(x)]^2 \, dx$. They also consider the case where the ε_i are generated by an autoregressive process in Section 5.4 of their paper. They show that the same realized kernel under the working model of i.i.d. ε_i can still be used, and that "the rate of convergence of the realized kernel is not changed by this form of serial dependence, but the asymptotic distribution is altered."

Fan and Wang (2008) use an alternative kernel smoothing method to derive consistent estimators of $[X]_T$ from transactions data under the market microstructure noise model (4.4) in which X_t is the diffusion model (4.5) that has to satisfy much stronger regularity conditions.

4.3.5 Method of pre-averaging

Using the average $\bar{Y}_{t_i}$ of $Y_{t_i}, Y_{t_{i+1}}, \ldots, Y_{t_{i+k-1}}$ can substantially reduce the variance of the additive microstructure noise and therefore replacing Y_{t_i} by $\bar{Y}_{t_i}$ in (4.7) may yield a better approximation to the underlying semimartingale X_t. Assuming the diffusion model (4.5) for X_t and additional regularity conditions, Jacod et al. (2009) implement this idea by using the "pre-averaging" method described as follows. They assume evenly spaced times $t_i = i\delta_n$ (with $n = \lceil T/\delta_n \rceil$) and choose an integer $k_n = \theta/\sqrt{\delta_n} + o(\delta_n^{-1/4})$ for some $\theta > 0$. With this choice of t_i and $k = k_n$, they denote Y_{t_i} by Y_i^n and $\bar{Y}_{t_i}$ by $\bar{Y}_i^n$. Let g be a continuously differentiable weight function on $[0, 1]$ with a piecewise Lipschitz derivative and $g(0) = g(1) = 0$. Let $\psi_1 = \int_0^1 [g'(u)]^2 \, du$ and

$\psi_2 = \int_0^1 g^2(u)\,du > 0$. They propose to estimate $[X]_T$ by

$$\hat{C}_T = \frac{\sqrt{\delta_n}}{\theta\psi_2} \sum_{i=1}^{n-k_n+1} \left\{ \sum_{j=1}^{k_n-1} g\left(\frac{j}{k_n}\right) (\bar{Y}_{i+j}^n - \bar{Y}_{i+j-1}^n) \right\}^2$$
$$- \frac{\psi_1\delta_n}{2\theta^2\psi_2} \sum_{i=1}^{n}(Y_i^n - Y_{i-1}^n)^2, \quad (4.20)$$

and show that $n^{-1/4}(\hat{C}_T - [X]_T)$ has a limiting normal distribution as $\delta_n \sim T/n \to 0$. The second sum in (4.20) is used for bias correction, and pre-averaging via the moving averages $\bar{Y}_{i+j}^n$ in the first sum results in the $n^{-1/4}$ rate of convergence to $[X]_T$ after bias correction.

4.3.6   From MLE of volatility parameter to QMLE of $[X]_T$

Although the asymptotic theory and simulation studies of the nonparametric methods reviewed in Sections 4.3.3–4.3.5 have shown them to perform well, Gatheral and Oomen (2010) have not found the theoretical advantages to translate into superior performance in a comparative study of these methods and other ad hoc modifications or the realized variance (4.7) using simulated data from what they call a "zero-intelligence" limit order book market that mimics some key properties of actual markets. In particular, they have shown that the maximum likelihood estimator of Hansen et al. (2008) is the best performer even though the model (4.5) for X_t is misspecified as $dX_t = \sigma dB_t$. Xiu (2010) applies the statistical theory of quasi-maximum likelihood (QML), which will be reviewed in Supplement 4 of Section 4.7, to study the properties of the MLE of the parameter σ and v_ε in the model (4.4) under the parametric assumptions that ε_i are i.i.d. $N(0, v_\varepsilon)$ and $dX_t = \sigma dB_t$. He also assumes for simplicity that the times t_i are regularly spaced in $[0, T]$ so that $t_i = i\delta$, with $T = n\delta$.

Let $\Delta_i = Y_{i\delta} - Y_{(i-1)\delta}$ and $\mathbf{\Delta} = (\Delta_2, \ldots, \Delta_n)^\top$. Under this parametric model, $\Delta_i = \sigma(B_{i\delta} - B_{(i-1)\delta}) + \varepsilon_i - \varepsilon_{i-1}$, $\mathrm{Var}(\Delta_i) = \sigma^2\delta + 2v_\varepsilon$ and the log-likelihood function of the model is given by

$$l(v_\varepsilon, \sigma^2) = -\frac{1}{2}\log\det(\mathbf{\Omega}) - \frac{n}{2}\log(2\pi) - \frac{1}{2}\mathbf{\Delta}^\top\mathbf{\Omega}^{-1}\mathbf{\Delta}, \quad (4.21)$$

where $\mathbf{\Omega}$ is the MA(1) tri-diagonal band matrix:

$$\mathbf{\Omega} = \begin{pmatrix} \sigma^2\delta + 2v_\varepsilon & -v_\varepsilon & 0 & \cdots & 0 \\ -v_\varepsilon & \sigma^2\delta + 2v_\varepsilon & -v_\varepsilon & \cdots & \vdots \\ 0 & -v_\varepsilon & \sigma^2\delta + 2v_\varepsilon & \ddots & 0 \\ \vdots & \ddots & \ddots & \ddots & -v_\varepsilon \\ 0 & \cdots & 0 & -v_\varepsilon & \sigma^2\delta + 2v_\varepsilon \end{pmatrix}. \quad (4.22)$$

Xiu first applies the central limit theorem to show that the MLE $(\hat{v}_\varepsilon, \hat{\sigma}^2) = \arg\max_{(v_\varepsilon, \sigma^2)} l(v_\varepsilon, \sigma^2)$ is still consistent when the actual distribution of ε_i is not normal. Specifically, as $n \to \infty$ (and therefore $\delta \sim T/n \to 0$),

$$\begin{pmatrix} n^{1/4}(\hat{\sigma}^2 - \sigma^2) \\ n^{1/2}(\hat{v}_\varepsilon - v_\varepsilon) \end{pmatrix} \xrightarrow{\mathcal{L}} N(\mathbf{0}, \mathbf{\Sigma}) \qquad (4.23)$$

when $dX_t = \sigma dB_t$ and ε_i are i.i.d. with mean 0 and variance v_ε, where

$$\mathbf{\Sigma} = \begin{pmatrix} 8\sqrt{v_\varepsilon}\sigma^3 T^{-1/2} & 0 \\ 0 & 2v_\varepsilon^2 + \mathrm{cum}_4[\varepsilon] \end{pmatrix}, \qquad (4.24)$$

in which $\mathrm{cum}_4[\varepsilon]$ is the fourth cumulant of ε_i. The only change from the assumed normal distribution for ε_i is that $\mathrm{cum}_4[\varepsilon] = E(\varepsilon^4) - 3v_\varepsilon^2$ may not be equal to 0. Note that $\hat{\sigma}^2$ converges to σ^2 at the rate of $n^{-1/4}$ (instead of the usual rate of $n^{-1/2}$) because $X_{i\delta} - X_{(i-1)\delta}$ has variance $\sigma^2\delta \sim \sigma^2 T/n$. Xiu then studies how well $\hat{\sigma}^2$ estimates $(1/T)\int_0^T \sigma_t^2 \, dt$ when X_t is the diffusion process (4.5). Under certain regularity conditions, he shows that as $n \to \infty$,

$$\begin{pmatrix} n^{1/4}\left(\hat{\sigma}^2 - \frac{1}{T}\int_0^T \sigma_t^2 \, dt\right) \\ n^{1/2}(\hat{v}_\varepsilon - v_\varepsilon) \end{pmatrix} \xrightarrow{\mathcal{L}} N(\mathbf{0}, \mathbf{V}),$$

where

$$\mathbf{V} = \begin{pmatrix} \dfrac{5\sqrt{v_\varepsilon}\int_0^T \sigma_t^4 \, dt}{T\left(\int_0^T \sigma_t^2 \, dt\right)^{1/2}} + \dfrac{3\sqrt{v_\varepsilon}\left(\int_0^T \sigma_t^2 \, dt\right)^{3/2}}{T^2} & 0 \\ 0 & 2v_\varepsilon^2 + \mathrm{cum}_4[\varepsilon] \end{pmatrix};$$

see Sections 4.2 and 4.3.1 of Xiu (2010), whose Sections 4.3.2 and 4.3.3 show that the results can also be applied to random sampling times t_i and non-Gaussian and serial-dependent ε_i.

4.4 Estimation of covariation of multiple assets

The continuous-time martingale theory in Appendix A shows how the quadratic variation $[X]_T$ can be generalized to the quadratic covariation $[X, \tilde{X}]_T$ of two semimartingales X_t and $\tilde{X}_t$, which correspond to the time series of efficient log prices of two risky assets in a portfolio. In this section, we study how to estimate $[X, \tilde{X}]_T$ in the presence of microstructure noise and asynchronicity in transactions of multiple assets.

4.4.1 Asynchronicity and the Epps effect

How asynchronicity in trading (which takes place as a function of the supply and demand of the underlying asset, the news impact and the transmission mechanism of information) should factor into the analysis of multivariate financial transactions data has been an active area of research in financial econometrics since 1960s; see Fisher (1966) and the review by Campbell et al. (1997). Using the fact that the last transaction of two assets on a trading day do not happen at the same time and the theory of subordinated stochastic processes (see Section 2.6.2), Scholes and Williams (1977) demonstrated the downward bias of the least squares estimator of beta in the CAPM model computed from the daily closing prices. By making use of the method of instrumental variables, they developed a consistent estimator of beta to address the asynchronicity in daily closing prices. Subsequently, Epps (1979) documented empirically that shortening the width of the sampling interval resulted in the shrinkage of the contemporaneous correlations of the log returns of different stocks towards 0. This stylized fact, called the *Epps effect*, has also been found in exchange rate returns; see Guillaume et al. (1997) and Muthuswamy et al. (2001). The time-aggregation technique that has been used to estimate $[X]_T$ in the microstructure noise model (4.4) can be extended to synchronize the transactions data of different assets. We next describe these synchronization procedures and then introduce estimators of $[X, \tilde{X}]_T$ to circumvent the Epps effect.

4.4.2   Synchronization procedures

For $i = 1, \ldots, M$, let $Y_{i,t}$ be the logarithm of the transaction price of the ith asset at time $t \in [0, T]$. To synchronize the high-frequency data of these M assets, Aït-Sahalia et al. (2010) define *generalized sampling times* τ_j, $1 \leq j \leq n$, such that

(a) $0 = \tau_0 < \tau_1 < \cdots < \tau_n = T$,

(b) $(\tau_{j-1}, \tau_j] \cap \{t_{i,k} : k = 1, \ldots, n_i\} \neq \varnothing$ for all $i = 1, \ldots, M$,

(c) $\max_{1 \leq j \leq n} \delta_j \to 0$ in probability where $\delta_j = \tau_j - \tau_{j-1}$.

In view of (b), for $i = 1, \ldots, M$, there exists $\tilde{t}_{i,j} \in (\tau_{j-1}, \tau_j]$ for which $Y_{i,\tilde{t}_{i,j}}$ is observed, and this yields a synchronized dataset $\{Y_{i,\tau_j}^\tau : i = 1, \ldots, M; j = 1, \ldots, n\}$ such that $Y_{i,\tau_j}^\tau = Y_{i,\tilde{t}_{i,j}}$. Several synchronization methods have been introduced to define τ_j. The Previous Tick approach proposed by Zhang (2011) corresponds to the case that $\delta_j \equiv \delta$ for all j. The Refresh Time scheme of Barndorff-Nielsen et al. (2011) defines τ_j recursively by

$$\tau_{j+1} = \max_{1 \leq i \leq M} t_{i, N_i(\tau_j)+1},$$

where $\tau_1 = \max\{t_{i,1} : i = 1, \ldots, M\}$ and $N_i(t)$ is the number of observations for the ith asset up to time t. Both the Previous Tick and Refresh Time methods choose the previous tick for the ith asset at or before τ_j as Y_{i,τ_j}^τ.

As pointed out in Section 4.4.1, the usual daily closing price data are obtained by setting τ_j to be the market closing time.[5] All synchronization methods inevitably leave out a portion of the data. The loss proportion is particularly pronounced if one of the M assets is illiquid. Among all methods that use generalized sampling times, the Refresh Time scheme includes the highest amount of data. Also, attention should be paid to the somewhat arbitrary selection mechanisms of the generalized synchronized methods; see Supplement 9 in Section 4.7 for further discussion.

4.4.3 QMLE for covariance and correlation estimation

Consider the case $M = 2$. Suppose the synchronized data $\{(Y_{1,\tau_j}^\tau, Y_{2,\tau_j}^\tau)\}_{j=1}^n$ are obtained by the generalized sampling times described Section 4.4.2. Then

$$Y_{i,\tau_j}^\tau = X_{i,\tau_j}^\tau + \varepsilon_{i,j} \tag{4.25}$$

for $i = 1, 2$, $j = 1, \ldots, n_i$, in which the $\varepsilon_{i,j}$ corresponds to the i.i.d. microstructure noise with mean 0 and variance $v_{\varepsilon,i}$ such that $\{\varepsilon_{1,j}\}$ and $\{\varepsilon_{2,j}\}$ are independent of each other. Noting that

$$[X_1, X_2]_T = \frac{1}{4}\{[X_1 + X_2]_T + [X_1 - X_2]_T\}, \tag{4.26}$$

Aït-Sahalia et al. (2010) apply the QMLE method in Section 4.3.6 to $Z_{1,\tau_j} = Y_{1,\tau_j}^\tau + Y_{2,\tau_j}^\tau$ and $Z_{2,\tau_j} = Y_{1,\tau_j}^\tau - Y_{2,\tau_j}^\tau$, thereby obtaining

$$\widehat{[X_1, X_2]}_T = \frac{1}{4}\left\{\widehat{[X_1 + X_2]}_T + \widehat{[X_1 - X_2]}_T\right\}.$$

Similar to (4.5), suppose the efficient price processes are diffusions $dX_{i,t} = \mu_{it}dt + \sigma_{it}dB_{it}$ for $i = 1, 2$, where B_{1t} and B_{2t} are correlated Brownian motions such that $E(dB_{1t}dB_{2t}) = \rho_t dt$. Then, under the assumption of τ_j being evenly spaced (i.e., $\tau_j - \tau_{j-1} \equiv \delta$ for all j), the asymptotic theory in Section 4.3.6 gives

$$n^{1/4}\left(\frac{1}{T}\widehat{[X_1, X_2]}_T - \frac{1}{T}\int_0^T \rho_t\sigma_{1t}\sigma_{2t}\,dt\right) \xrightarrow{\mathcal{L}} N(0, v_1),$$

which together with the delta method (Lai and Xing, 2008, p. 59) yields

$$n^{1/4}\left(\frac{\widehat{[X_1, X_2]}_T}{\sqrt{\widehat{[X_1]}_T \widehat{[X_2]}_T}} - \frac{\int_0^T \rho_t\sigma_{1t}\sigma_{2t}\,dt}{\sqrt{\int_0^T \sigma_{1t}^2\,dt \int_0^T \sigma_{2t}^2\,dt}}\right) \xrightarrow{\mathcal{L}} N(0, v_2)$$

---

[5]Instead of choosing the previous tick, the Hong Kong Stock Exchange sets the closing price as the median of the last 5 snapshots (15 seconds each) of the nominal price in the last minute of the trading day. The nominal price, rather than the transaction price, is used so that the closing price of illiquid stocks can still be determined when no transaction is observed in the last minute.

as $n \to \infty$; see Aït-Sahalia et al. (2010) whose Appendix also gives explicit formulas for v_1 and v_2. For the case of random τ_j such that $\delta_j = \tau_j - \tau_{j-1}$ are i.i.d. with mean $\bar{\delta}$, Aït-Sahalia et al. (2010) show that $\widehat{[X_1, X_2]}_T$ and $\widehat{[X_1, X_2]}_T / \sqrt{\widehat{[X_1]}_T \widehat{[X_2]}_T}$ are consistent estimates of the covariation process $[X_1, X_2]_T$ and the correlation process $[X_1, X_2]_T / \sqrt{[X_1]_T [X_2]_T}$, respectively, with convergence rate $O_p(\bar{\delta}^{1/4})$. Supplements 1–4 of Section 4.7 give further discussion of how QMLE works in practice and its connection to Hansen et al. (2008) using the MA(1) correlation structure of transactions data.

4.4.4 Multivariate realized kernels and two-scale estimators

Zhang (2011) synchronizes the data by using the Previous Tick approach as described to obtain $\mathbf{Y}_\tau = \{\mathbf{Y}_{\tau_j} : j = 1, \dots, n\}$, where $\mathbf{Y}_{\tau_j} = (Y_{1,\tau_j}^\tau, Y_{2,\tau_j}^\tau)$. In analogy with Sections 4.3.1–4.3.3, define the sparse-sampling estimator

$$[Y_1^\tau, Y_2^\tau]_T^{(k)} = \sum_{0 \leq k(i-1) \leq n-k} \left(Y_{1,\tau_{ki}}^\tau - Y_{1,\tau_{k(i-1)}}^\tau \right) \left(Y_{2,\tau_{ki}}^\tau - Y_{2,\tau_{k(i-1)}}^\tau \right),$$

the average estimator

$$[Y_1^\tau, Y_2^\tau]_T^{\text{avg}} = \frac{1}{k} \sum_{j=1}^k [Y_1^\tau, Y_2^\tau]_T^{(j,k)},$$

and the two-scales realized covariance (TSRC) estimator

$$\widehat{[X_1, X_2]}_T^{\text{TSRC}} = [Y_1^\tau, Y_2^\tau]_T^{\text{avg}} - (\bar{n}/n) \operatorname{Cov}^{(n)} \tag{4.27}$$

where $\bar{n} = k^{-1} \sum_{j=1}^k n_j$ and

$$\operatorname{Cov}^{(n)} = \sum_{j=1}^n \left(Y_{1,\tau_j}^\tau - Y_{1,\tau_{j-1}}^\tau \right) \left(Y_{2,\tau_j}^\tau - Y_{2,\tau_{j-1}}^\tau \right).$$

Similar to the univariate case, she shows that TSRC is consistent and that, by choosing $k \sim n^{2/3}$, the rate of convergence is of order $n^{1/6}$.

Barndorff-Nielsen et al. (2011) use the Refresh Time scheme to synchronize the data into $\mathbf{Y}_\tau = \{\mathbf{Y}_{\tau_j} : j = 1, \dots, n\}$. They define the matrix generalization of (4.16) by

$$\mathbf{\Gamma}_h (\mathbf{Y}_\tau) = \sum_{j=h+1}^n \left(\mathbf{Y}_{\tau_j} - \mathbf{Y}_{\tau_{j-1}} \right) \left(\mathbf{Y}_{\tau_{j-h}} - \mathbf{Y}_{\tau_{j-h-1}} \right)^\top \quad \text{for } h \geq 0 \tag{4.28}$$

and $\mathbf{\Gamma}_h(\mathbf{Y}_\tau) = [\mathbf{\Gamma}_{-h}(\mathbf{Y}_\tau)]^\top$ for $h < 0$. The multivariate realized kernel (MRK) estimator for $[X_1, X_2]_T$ is defined by

$$\mathbf{K} (\mathbf{Y}_\tau) = \sum_{h=-n}^n \mathcal{K} \left(\frac{h}{H} \right) \mathbf{\Gamma}_h (\mathbf{Y}_\tau), \tag{4.29}$$

in which the kernel function $\mathcal{K}(\cdot)$ is similar to $\kappa(\cdot)$ in Section 4.3.4 on the univariate case. In particular, to ensure $\boldsymbol{K}(\boldsymbol{Y}_\tau)$ to be positive definite, Barndorff-Nielsen et al. only use $\mathcal{K}(\cdot)$ that satisfies the following conditions:

(a) $\mathcal{K}(0) = 1, \mathcal{K}'(0) = 0$;

(b) $\mathcal{K}$ is twice differentiable with continuous derivatives;

(c) $\int_0^\infty \mathcal{K}^2(x)\,dx < \infty$, $\int_0^\infty [\mathcal{K}'(x)]^2\,dx < \infty$, and $\int_0^\infty [\mathcal{K}''(x)]^2\,dx < \infty$;

(d) $\int_{-\infty}^\infty \mathcal{K}(x)\exp\{\sqrt{-1}x\lambda\}\,dx \geq 0$ for all $\lambda \in \mathbb{R}$.

Condition (d) guarantees $\boldsymbol{K}(\boldsymbol{Y}_\tau)$ to be positive definite according to Bochner's theorem; see Supplement 11 in Section 4.7. In condition (a), $\mathcal{K}(0) = 1$ implies that $\boldsymbol{\Gamma}_0(\boldsymbol{Y}_\tau)$ gets unit weight, while $\mathcal{K}(0) = 1$ and $\mathcal{K}'(0) = 0$ imply that $\boldsymbol{\Gamma}_h(\boldsymbol{Y}_\tau)$ receives some weight close to 1 if $|h|$ is small. Under assumptions similar to those in Section 4.3.4, Barndorff-Nielsen et al. (2011) prove asymptotic normality and consistency of the MRK estimator and show that the rate of convergence is of order $n^{1/5}$ for the choice of $H = cn^{3/5}$. They also show that MRK is robust against possible serial correlations in $(\varepsilon_{1,j}, \varepsilon_{2j})$.[6]

Whereas the preceding estimators rely on explicit synchronization schemes described in Section 4.4.2, Hayashi and Yoshida (2005) propose the following estimator of $[X_1, X_2]_T$ that seemingly can make use of all available data:

$$\sum_{i=1}^{n_1} \sum_{j=1}^{n_2} \left(Y_{1,t_{1,i}} - Y_{1,t_{1,i-1}}\right) \left(Y_{2,t_{2,j}} - Y_{2,t_{2,j-1}}\right) \mathbf{1}_{\{O_{ij} \neq \varnothing\}}, \qquad (4.30)$$

where $O_{ij} = (t_{1,i-1}, t_{1,i}] \cap (t_{2,j-1}, t_{2,j}]$. Aït-Sahalia et al. (2010) have pointed out that (4.30) can be expressed as

$$\sum_{i=1}^{n_1} \left(Y_{1,t_{1,i}} - Y_{1,t_{1,i-1}}\right) \left(Y_{2,t_{2,i+}} - Y_{2,t_{2,(i-1)-}}\right),$$

where $t_{2,i+} = \min\{t_{2,k} : t_{2k} \geq t_{1,i}\}$ and $t_{2,i-} = \max\{t_{2,k} : t_{2k} \leq t_{1,i}\}$. That is, there is an implicit synchronization procedure that removes data in the interval $(t_{2,(j-1)+}, t_{2,j-})$. Moreover, Hayashi and Yoshida (2005, 2008) require the absence or negligibility of microstructure noise in establishing their asymptotic theory that shows the estimator (4.30) to be $\sqrt{n}$-consistent under the assumption of independent Poisson arrivals of nonsynchronous transactions. However, as shown by Voev and Lunde (2007), not only is (4.30) biased under the microstructure noise model (4.25), but its variance also diverges as the number of observations becomes infinite.

---

[6] Barndorff-Nielsen et al. (2011) suggest to use averaging for controlling the edge effect in the initial and final time points. This requires an adjustment of the upper limit of the sum in (4.28).

4.5   Fourier methods

As discussed in Section 4.4.2, synchronization procedures may result in significant loss of information if any one of the M assets is illiquid. In addition, since a synchronization procedure uses time blocks of the form $(\tau_j, \tau_{j+1}]$ with $\tau_{j+1} - \tau_j = \delta$, the logarithmic returns computed from the synchronized transaction prices are mostly zeroes when δ is small. To estimate the covariation directly from the asychronous transactions data, Malliavin and Mancino (2002) propose to use Fourier methods. We first introduce their approach and the underlying theory for the case of $M = 1$ and then present the corresponding multivariate version.

4.5.1   Fourier estimator of $[X]_T$ and spot volatility

Malliavin and Mancino (2002, 2009) assume (4.5) as the price dynamics, in which μ_t and σ_t are adapted processes satisfying

$$E\left[\int_0^T \mu_t^2 \, dt\right] < \infty \quad \text{and} \quad E\left[\int_0^T \sigma_t^4 \, dt\right] < \infty. \tag{4.31}$$

They assume no microstructure noise so that the observations are $Y_t = X_t$ at possibly irregularly spaced times $t_1 < \cdots < t_n$ in the interval $[0, T]$. Whereas $[X]_T = \int_0^T \sigma_t^2 \, dt$ is called the integrated variance (or the quadratic variation) of X_t, σ_t is called the *spot volatility* or the *instantaneous volatility*. Letting $\Sigma_t = \sigma_t^2$, their main idea consists of (a) using Itô's formula for stochastic integrals to express the Fourier coefficients of Σ_t in terms of those of dX_t, (b) using a piecewise constant function to approximate the integrands in the stochastic integrals defining these Fourier coefficients, and (c) using a finite-sum approximation to the Fourier series of Σ_t.

To begin with, let the terminal time T be 2π (by rescaling if necessary). Then the Fourier coefficients are given by

$$a_0(dX) = \frac{1}{2\pi} \int_0^{2\pi} dX_t,$$

$$a_k(dX) = \frac{1}{\pi} \int_0^{2\pi} \cos(kt) \, dX_t \qquad \text{for } k > 0, \tag{4.32}$$

$$b_k(dX) = \frac{1}{\pi} \int_0^{2\pi} \sin(kt) \, dX_t \qquad \text{for } k \geq 0.$$

Since the X_t are observed only at a discrete set of times t_i, a Lebesgue-Stieltjes approximation (with $\tilde{X}_s = X_{t_i}$ for all $s \in (t_i, t_{i+1}]$) to the integrals in (4.32) yields the estimates

$$\hat{a}_0(dX) = \frac{X_{2\pi} - X_0}{2\pi}, \tag{4.33}$$

$$\hat{a}_k(dX) = \frac{X_{2\pi} - X_0}{\pi} + \frac{1}{\pi} \sum_{i=1}^{n-1} [\cos(kt_i) - \cos(kt_{i+1})] X_{t_i}, \quad \text{for } k > 0, \tag{4.34}$$

$$\hat{b}_k(dX) = \frac{1}{\pi} \sum_{i=1}^{n-1} [\sin(kt_i) - \sin(kt_{i+1})] X_{t_i}, \qquad \text{for } k \geq 0. \tag{4.35}$$

Application of Itô's formula to the diffusion process (4.5) gives the following representation of the Fourier coefficients of $\Sigma_t = \sigma_t^2$:

$$a_0(\Sigma) = \text{p-lim}_{N_x \to \infty} \frac{\pi}{N_X + 1} \sum_{j=0}^{N_X} \left[a_j^2(dX) + b_j^2(dX) \right], \tag{4.36}$$

$$a_k(\Sigma) = \text{p-lim}_{N_x \to \infty} \frac{2\pi}{N_X + 1} \sum_{j=0}^{N_X} [a_j(dX)a_{j+k}(dX)], \qquad \text{for } k > 0 \tag{4.37}$$

$$b_k(\Sigma) = \text{p-lim}_{N_x \to \infty} \frac{2\pi}{N_X + 1} \sum_{j=0}^{N_X} [a_j(dX)b_{j+k}(dX)], \qquad \text{for } k \geq 0, \tag{4.38}$$

in which p-lim corresponds to convergence in probability; the key steps in deriving (4.36)–(4.38) are given in Problem 4 in Section 4.8. Replacing $a_i(dX)$ and $b_j(dX)$ by $\hat{a}_i(dX)$ and $\hat{b}_j(dX)$ and using the finite-sum approximation $\sum_{i=0}^{N_X}$ in (4.36)–(4.38) gives the estimates of $\hat{a}_0(\Sigma)$, $\hat{a}_k(\Sigma)$, and $\hat{b}_k(\Sigma)$ of the Fourier coefficient of Σ_t and, in particular, the Fourier estimator of the integrated variance $\widehat{[X]}_T^F = 2\pi\hat{a}_0(\Sigma)$. Noting that the Fourier-Fejér inversion formula

$$\Sigma_t = \text{p-lim}_{N_S \to \infty} \sum_{k=0}^{N_S} \left(1 - \frac{k}{N_S}\right) [a_k(\Sigma)\cos(kt) + b_k(\Sigma)\sin(kt)] \tag{4.39}$$

can be used to retrieve the spot volatilities for $t \in (0, 2\pi)$, Malliavin and Mancino (2002) estimate Σ_t by

$$\hat{\Sigma}_t = \sum_{k=0}^{N_S} \left(1 - \frac{k}{N_S}\right) \left[\hat{a}_k(\Sigma)\cos(kt) + \hat{b}_k(\Sigma)\sin(kt) \right] \tag{4.40}$$

in the absence of market microstructure noise. It is shown that N_X and N_S can be chosen appropriately so that the estimator $\hat{\Sigma}_t$ is robust against microstructure noise added to X_t, as will be explained below.

4.5.2   Statistical properties of Fourier estimators

Malliavin and Mancino (2009) have shown that $\widehat{[X]}_T^F$ is associated with $RV^{(n)}$ in the following way. Note that $\widehat{[X]}_T^F$ can be expressed as

$$\widehat{[X]}_T^F = \sum_{i=1}^n \sum_{j=1}^n D_{N_X}(t_i - t_j)(X_{t_i} - X_{t_{i-1}})(X_{t_j} - X_{t_{j-1}}), \qquad (4.41)$$

where

$$D_{N_X}(t) = \frac{1}{2N_X + 1} \frac{\sin\left[(N_X + 1/2)\,t\right]}{\sin\,(t/2)} \qquad (4.42)$$

is the *rescaled Dirichlet kernel*. By separating the terms with $i = j$ from those with $i \neq j$, (4.41) can be rewritten as

$$\widehat{[X]}_T^F = RV^{(n)} + \sum_{i=1}^n \sum_{\substack{j=1; \\ i \neq j}}^n D_{N_X}(t_i - t_j)(X_{t_i} - X_{t_{i-1}})(X_{t_j} - X_{t_{j-1}}),$$

in which the double sum can be expressed, when $t_i - t_{i-1} = \delta$ for all i, as

$$\sum_{i=1}^n \sum_{\substack{j=1; \\ i \neq j}}^n D_{N_X}((i-j)\delta)(X_{i\delta} - X_{(i-1)\delta})(X_{j\delta} - X_{(j-1)\delta})$$

$$= \sum_{h=1}^H D_{N_X}(h\delta)\left[\gamma_h(X) + \gamma_{-h}(X)\right],$$

where γ_h is the autocovariance at lag h defined in (4.16). Thus, the double sum is similar to the correction term of the realized kernel estimator in (4.17). The main difference is that the Dirichlet kernel in the Fourier estimator depends on the number N_X of frequencies included, besides the time gap between the observations. Hence $\widehat{[X]}_T^F$ is analogous to the realized kernel estimator but does not require evenly spaced data.

The Dirichlet kernel can also be used to study the bias of $\widehat{[X]}_T^F$ when there is microstructure noise as in (4.4), under which Mancino and Sanfelici (2008) have shown that

$$E\left(\widehat{[X]}_T^F - [X]_T\right) = 2nv_\varepsilon\left[1 - D_{N_X}\left(\frac{2\pi}{n}\right)\right]. \qquad (4.43)$$

That is, $\widehat{[X]}_T^F$ is biased under (4.4), but if $N_X^2/n \to 0$ and $n \to \infty$, then the bias tends to zero. They also derive a similar identity for MSE $= E(\widehat{[X]}_T^F - [X]_T)^2$, evaluated first under the microstructure noise model (4.4) and then under the case of no microstructure noise. They also show that as $N_X^2/n \to 0$ and

$n \to \infty$, the first MSE exceeds the second one by $C + o(1)$, where C is a positive constant that depends on v_ε, $E(\varepsilon^4)$, and $[X]_T$. To enhance the finite-sample performance of the estimator, they propose to select N_X by an empirical criterion that is similar to the choice of bandwidth in realized kernel estimates by Bandi and Russell (2006) and Barndorff-Nielsen et al. (2008).

Since the Fourier spot volatility estimator is inspired by the Fourier-Féjer inversion formula (4.39), various asymptotic results can be obtained by making use of this linkage. For example, Malliavin and Mancino (2009) have shown that under some regularity conditions,

$$\sup_{t \in [0,2\pi]} |\hat{\Sigma}_t - \Sigma_t| \to 0$$

if Σ_t is continuous over time; see their Corollary 3.5 for details. Mancino and Recchioni (2015) have also proved the asymptotic normality $\hat{\Sigma}_t$ (for fixed t) under additional assumptions. There are other theoretical results and simulation studies showing that Fourier estimator is robust even against endogenous and correlated microstructure noise and that it can give comparable or better forecasts relative to other volatility estimators; see Mancino and Sanfelici (2008, 2011a,b) and Mancino and Recchioni (2015).

4.5.3 Fourier estimators of spot co-volatilities

Suppose the logarithmic prices X_{it} of a portfolio of assets, $i = 1, \ldots, M$, $0 \le t \le T$, satisfy

$$dX_{it} = \mu_{i,t} dt + \sum_{k=1}^{d} \sigma_{k,t}^{(i)} dB_{k,t}, \tag{4.44}$$

where $\boldsymbol{B}_t = (B_{1,t}, \ldots, B_{d,t})^\top$ is d-dimensional Brownian motion. Similar to (4.31), assume that for $i = 1, \ldots, M$ and $k = 1, \ldots, d$

$$E\left[\int_0^T (\mu_{i,t})^2 \, dt\right] < \infty \quad \text{and} \quad E\left[\int_0^T (\sigma_{k,t}^{(i)})^2 \, dt\right] < \infty.$$

Let $\Sigma_t^{1/2}$ be the spot co-volatility matrix of the vector logarithmic price process $\boldsymbol{X}_t = (X_{1,t}, \ldots, X_{M,t})^\top$ and $\Sigma_t^{(i,j)}$ be the (i,j)th element of $\boldsymbol{\Sigma}_t$ for $i, j = 1, \ldots, d$; the square root of a non-negative definite matrix can be defined by its singular value decomposition (Lai and Xing, 2008, p. 43). Under model (4.44), $\Sigma_t^{(i,j)} = \sum_{k=1}^{d} \sigma_{k,t}^{(i)} \sigma_{k,t}^{(j)}$ and the corresponding quadratic covariance is

$$[X_i, X_j]_T = \int_0^T \Sigma_t^{(i,j)} \, dt.$$

Similar to the univariate case, we take $T = 2\pi$. Thus, the Fourier coefficients of dX_i can be computed by applying (4.33), (4.34) and (4.35) for each $i =$

$1, \ldots, M$. Likewise, the Fourier coefficients of the spot covariation process $\Sigma_t^{(i,j)}$ can be calculated from the Fourier coefficients of dX_i's as follows:

$$a_0(\Sigma^{(i,j)}) = \text{p-lim}_{N \to \infty} \frac{\pi}{N_X + 1} \sum_{q=0}^{N_X} [a_q(dX_i)a_q(dX_j) + b_q(dX_i)b_q(dX_j)],$$

$$a_r(\Sigma^{(i,j)}) = \text{p-lim}_{N \to \infty} \frac{\pi}{N_X + 1} \sum_{q=0}^{N_X} [a_q(dX_i)a_{q+r}(dX_j) + a_q(dX_j)a_{q+r}(dX_i)],$$

$$b_s(\Sigma^{(i,j)}) = \text{p-lim}_{N \to \infty} \frac{\pi}{N_X + 1} \sum_{q=0}^{N_X} [a_q(dX_i)b_{q+s}(dX_j) + a_q(dX_j)b_{q+s}(dX_i)],$$

for $r > 0$ and $s \geq 0$. Moreover, by using the multivariate version of Fourier-Fejér inversion formula, Malliavin and Mancino (2002) propose the Fourier spot co-volatility estimator as

$$\hat{\Sigma}_t^{(i,j)} = \sum_{k=0}^{N_S} \left(1 - \frac{k}{N_S}\right) \left[\hat{a}_k(\Sigma^{(i,j)}) \cos(kt) + \hat{b}_k(\Sigma^{(i,j)}) \sin(kt)\right], \qquad (4.45)$$

where $t \in (0, 2\pi)$, similar to (4.39) for the univariate case.

We close this section with a remark by Malliavin and Mancino (2002, 2009). Let $\widetilde{[X_i, X_j]}_T$ be some estimator of integrated covariation based on a synchronization scheme. The obvious way of estimating spot covariation from $\widetilde{[X_i, X_j]}_T$ is

$$\tilde{\Sigma}_t^{(i,j)} = \frac{\widetilde{[X_i, X_j]}_{t+h} - \widetilde{[X_i, X_j]}_t}{h}, \qquad (4.46)$$

which suffers serious numerical instability because $h > 0$ is typically small. The root cause of this problem is the implicit concept of "differentiating the process" in the synchronization procedure. Note that no synchronization is needed in the computation of the Fourier estimator because the methodology is "integration-based". Malliavin and Mancino (2002, 2009) attribute the success of Fourier methods for estimating the spot covariation to its use of integration rather than differentiation.

4.6   Other econometric models involving TAQ

The NYSE Trade and Quote (TAQ) database contains intraday transactions data (trades and quotes) for all securities listed on NYSE, AMEX and NAS-DAQ. In this section, by TAQ data we mean similar intraday transactions data from these and other exchanges, and we summarize several classes of models of these data, beyond (4.4) and (4.5) described above, in the high-frequency econometrics literature. We begin with models for the durations $t_i - t_{i-1}$. In particular, Hayashi and Yoshida (2005, 2008), discussed in the last paragraph of Section 4.4.4, assume the Poisson process model of the number of transactions $N(t)$ up to time t, or equivalently i.i.d. exponential durations. However, because equity markets only operate within the trading hours, transactions are more frequent when the markets open and close, violating the i.i.d. durations assumption. Section 4.6.1 considers *autoregressive conditional duration* (ACD) models for the intraday time series of durations and Section 4.6.2 uses a *self-exciting point process* (SEPP) model for $N(t)$. We then describe the use of marked point processes to decompose the transaction price dynamics into two components: (a) the point process model of the transactions and (b) the model of the price differences as the marks; specifically,

$$P_t - P_0 = \sum_{i=1}^{N(t)} D_i, \qquad (4.47)$$

where P_t is the transaction price at time t and D_i is the price difference of the ith transaction. Section 4.6.3 considers econometric modeling of D_i, and Sections 4.6.4 and 4.6.5 describe joint models of $N(t)$, D_i or other marks of the marked point process. In Section 4.6.6, we describe econometric models that use TAQ data to predict future volatilities for applications to portfolio optimization, derivatives pricing and risk management. Section 4.6.7 uses semimartingales with jumps to model the efficient price process.

4.6.1 ACD models of inter-transaction durations

For the transaction times $t_i \in [0, T]$ introduced in Section 4.2.2, Engle and Russell (1998) propose the following method to analyze and model the inter-transaction times $\delta_i = t_i - t_{i-1}$, particularly from intraday transactions data, for which times 0 and T represent the opening and closing times of the day, collected over m consecutive days. They propose to first estimate the intraday periodicity using a nonparametric regression model of the form

$$\log(\delta_i) = g(t_i) + \eta_i, \qquad (4.48)$$

in which g is a smoothing or regression spline (Lai and Xing, 2008, Sect. 7.2) and η_i are i.i.d. with common zero mean and finite variance. Defining the adjusted duration by $\delta_i^* = \delta_i / \exp\{\hat{g}(t_i)\}$, ACD assumes a model of the form $\delta_i^* = \psi_i \varepsilon_i$, where ε_i are i.i.d. positive random variable with mean 1. In particular, if ψ_i has the GARCH(r, s) form

$$\psi_i = \omega + \sum_{j=1}^{r} \gamma_j \delta_{i-j}^* + \sum_{j=1}^{s} \beta_j \psi_{i-j}, \qquad (4.49)$$

with non-negative parameters $\omega, \gamma_1, \ldots, \gamma_r, \beta_1, \ldots, \beta_s$, the ACD model is denoted by ACD(r, s). Specific choices for the distribution of ε_i are the exponential distribution with mean 1 and the standardized Weibull distribution with a shape parameter α (Lai and Xing, 2008, p. 293), yielding the exponential ACD (EACD) and the Weibull ACD (WACD) models, respectively. As in GARCH(r, s) models, ACD(r, s) can be written as an ARMA model with non-Gaussian martingale difference innovations. In particular, under the weak stationarity condition $\sum_{j=1}^{r} \gamma_j + \sum_{j=1}^{s} \beta_j < 1$,

$$E(\delta_i^*) = \frac{\omega}{1 - (\sum_{j=1}^{r} \gamma_j + \sum_{j=1}^{s} \beta_j)}.$$

Making use of the joint density function of the i.i.d. $\varepsilon_i = \delta_i^*/\psi_i$, Engle and Russell derive the likelihood function of $(\omega, \gamma_1, \ldots, \gamma_r, \beta_1, \ldots, \beta_s)$ in the case of EACD(r, s) and $(\alpha, \omega, \gamma_1, \ldots, \gamma_r, \beta_1, \ldots, \beta_s)$ in the case of WACD(r, s).

The asymptotic theory of the maximum (conditional) likelihood estimator can be derived under the weak stationarity condition and some other regularity assumptions; see Engle and Russell (1998) for details. For the EACD(1,1) model, let $u_i = d_i \sqrt{\delta_i^*}$, where d_i is independent of δ_i^* and is equal to -1 and 1 with probability 0.5. Note that the expected value of u_i is 0 and its variance is ψ_i. Regarding d_i as if it were standard normal would give the same likelihood function as that of the GARCH model. Since this likelihood function only involves u_i^2 (which is equal to the observed δ_i^*), GARCH software can be used to compute the quasi-maximum likelihood estimator (QML), which is still consistent and asymptotically normal; see Lai and Xing (2008, p. 294–296).

4.6.2   Self-exciting point process models

The ACD models implicitly assume that g in (4.48) is exogenous. An alternative approach that removes this assumption is to model directly the intensity of $N(t)$ (rather than the inter-transaction durations) by a self-exciting point process (SEPP). Hautsch (2012) and Bauwens and Hautsch (2009) give an overview of stochastic intensity models. The intensity process of $N(t)$ is defined by

$$\lambda(t) = \lim_{s \downarrow 0} \frac{1}{s} E\left[N(t + s) - N(t)|\mathcal{F}_t\right], \tag{4.50}$$

where $\mathcal{F}_t$ is the information set (σ-field) of the point process up to time t and $N(t) = \max\{i : t_i \leq t\}$. Appendix C shows that the intensity process contains the key information about $N(t)$; specifically, $N(t) - \Lambda(t)$ is a zero-mean locally square integrable martingale, where $\Lambda(t) = \int_0^t \lambda(s)\, ds$ is the integrated intensity function. The point process $N(t)$ is called *self-exciting* if its intensity process $\lambda(t)$ depends on the information set up to (but not including) the jump at time t and takes the form

$$\lambda(t) = \phi\left(\mu(t) + \int_{u<t} h(t - u)\, dN(u)\right), \tag{4.51}$$

where $\phi : \mathbb{R} \to \mathbb{R}_+$, and the kernel function h is a non-negative function on $\mathbb{R}_+$ that models the self-excitation of the intensity at time t by previous event times $t_i (< t)$.

A special case of (4.51) is $\phi(x) = x$ and $\mu(t) \equiv \mu$. Hawkes (1971) considers the exponential kernel $h(x) = \alpha e^{-\beta x}$, which leads to the intensity process

$$\lambda_\theta(t) = \mu + \int_{u<t} \alpha e^{-\beta(t-u)} \, dN(u) = \mu + \sum_{i:t_i<t} h_\theta(t - t_i) \qquad (4.52)$$

of the univariate *Hawkes process* $N(t)$, where $\theta = (\mu, \alpha, \beta)$ denotes the parameter vector that can be estimated by maximum likelihood based on successive transaction times $t_i \in [0, T]$. The R package `ppstat` and a recent enhancement `NHPoisson` can be used to fit Hawkes processes and more general SEPPs by maximum likelihood and to perform diagnostic checks when the intensity process in (4.50) has the parametric form $\mu(t) \equiv \mu$ and $h(s) = h_\theta(s)$, in which μ is a component of θ. Appendix C provides details of the likelihood function and the corresponding MLE. It also describes these software packages that provide additional functions to simulate these SEPPs. The simulation methods and software are summarized in Section C.4.

Whereas the preceding discussion has focused on the case $\mu_\theta(t) \equiv \mu$ because of the Hawkes process, Chen and Hall (2013) have pointed out the usefulness of time-varying background intensity $\mu_\theta(t)$ in modeling inter-transaction durations. They illustrate this with transactions data from the Australian Stock Exchange whose normal trading hours are 10:00–16:00 of Australian Eastern Standard Time. They model $\mu_\theta(t)$ by a B-spline (Lai and Xing, 2008, p. 169) with interior knots placed at 11:00, 12:00, 13:00, 14:00, and 15:00, and choose the kernel function $h_\theta(t) = \gamma_1 e^{-\gamma_2 t}$ for exponential decay and $h_\theta(t) = \gamma_1(1 + t)^{-(1+\gamma_2)}$ for polynomial decay. They show that for either choice, SEPPVB (in which VB stands for "time-varying background") fits the data better than the ACD and that SEPPVB with exponential decay performs slightly better than SEPPVB with polynomial decay. The R packages `ppstat` and `NHPoisson` can also be used for time-varying $\mu_\theta(t)$.

4.6.3 Decomposition of D_i and generalized linear models

Recall that $D_i = P_{t_i} - P_{t_{i-1}}$ is the transaction price difference and is discrete in nature, with a high probability that $D_i = 0$. Rydberg and Shephard (2003) therefore consider the decomposition

$$D_i = J_i A_i M_i, \qquad (4.53)$$

to model D_i, where $J_i = I_{\{D_i \neq 0\}}$ and conditional on $J_i = 1$, A_i is the sign of D_i and M_i is the magnitude of D_i in units of tick size. They express the conditional density of D_i given $\mathcal{F}_{i-1}$ as

$$P(D_i = y|\mathcal{F}_{i-1}) = P(J_i = 1|\mathcal{F}_{i-1}) \times P(A_i = 1|J_i = 1, \mathcal{F}_{i-1})$$
$$\times P(M_i = y|A_i = 1, J_i = 1, \mathcal{F}_{i-1}) \quad (4.54)$$

for $y > 0$, replacing $A_i = 1$ by $A_i = -1$ in (4.54) for $y < 0$. The binary nature of J_i and A_i suggests using logistic regression models to for the factors

$P(J_i = 1|\mathcal{F}_{i-1})$ and $P(A_i = 1|J_i = 1, \mathcal{F}_{i-1})$ in (4.54). Specifically, let

$$\text{logit}[P(J_i = 1|\mathcal{F}_{i-1})] = \boldsymbol{\beta}^\top \boldsymbol{x}_i, \tag{4.55}$$

$$\text{logit}[P(A_i = 1|J_i = 1, \mathcal{F}_{i-1})] = \boldsymbol{\gamma}^\top \boldsymbol{z}_i, \tag{4.56}$$

where $\text{logit}(P) = \log(P/(1-P))$, $\boldsymbol{x}_i$, $\boldsymbol{z}_i$ and $\boldsymbol{w}_i$ are finite-dimensional vectors consisting of elements of $\mathcal{F}_{i-1}$ and $\boldsymbol{\beta}$ and $\boldsymbol{\gamma}$ are parameter vectors; see Section 4.1.2 of Lai and Xing (2008) on logistic regression and generalized linear models. A simple probability model of $(M_i - 1)|(A_i, J_i = 1, \mathcal{F}_{i-1})$ is the geometric distribution with density function $\lambda_i(1 - \lambda_i)^k$, $k = 0, 1, \ldots$, whose parameter λ_i is related to $\boldsymbol{w}_i$ by

$$\text{logit}(\lambda_i) = \begin{cases} \boldsymbol{\theta}^\top \boldsymbol{w}_i & \text{if } A_i = 1 \\ \boldsymbol{\alpha}^\top \boldsymbol{w}_i & \text{if } A_i = -1. \end{cases} \tag{4.57}$$

The log-likelihood function in the Rydberg-Shephard model is therefore of the form $\sum_{i=1}^n f(D_i|\mathcal{F}_{i-1})$, in which the parameters $\boldsymbol{\beta}, \boldsymbol{\gamma}, \boldsymbol{\theta}$, and $\boldsymbol{\alpha}$ and the function f are defined by (4.55)–(4.57). Maximum likelihood can be used to estimate the unknown parameters. A more flexible alternative to the geometric distribution for $M_i - 1$ is the negative binomial distribution.

Rydberg and Shephard (2003) use $J_{i-1}, A_{i-1}, M_{i-1}, \ldots, J_{i-q}, A_{i-q}, M_{i-q}$ and the AIC[7] to determine the predictors $\boldsymbol{x}_i$, $\boldsymbol{z}_i$, and $\boldsymbol{w}_i$ in (4.55)–(4.57). Tsay (2010, p. 250–253) includes $J_{i-1}, A_{i-1}, M_{i-1}$, the inter-transaction duration δ_{i-1}, the volume (divided by 1000) at t_{i-1}, and the bid-ask spread right before t_i. Note that (4.54)–(4.57) is basically a generalized linear model with time-varying covariates.

4.6.4   McCulloch and Tsay's decomposition

While Rydberg and Shephard (2003) provide an appealing decomposition of D_i, the fitted model for J_i from Tsay (2010) shows that the inactive trades (that result in no price changes) are very likely to be followed by inactive trades. To model the clustering pattern of inactive trades more efficiently, McCulloch and Tsay (2001) propose the decomposition

$$P_t - P_0 = \sum_{i=1}^{\tilde{N}(t)} \tilde{D}_i, \tag{4.58}$$

where $\tilde{N}(t) = \sum_{i=1}^{N(t)} J_i$ is the number of active transactions up to time t and $\tilde{D}_i$ is the ith active transaction price change. Let $\tilde{t}_i$ be the transaction time of the ith active trade. Thus, $\tilde{\delta}_i = \tilde{t}_i - \tilde{t}_{i-1}$ is the duration between two consecutive active trades. Let $\tilde{Z}_i$ denote the number of inactive transactions

---

[7]AIC is the abbreviation of the Akaike Information Criterion; see Lai and Xing (2008, p. 122)

in the time interval $(\tilde{t}_{i-1}, \tilde{t}_i)$, hence $\tilde{Z}_i = \sum_{j=N(\tilde{t}_{i-1})+1}^{N(\tilde{t}_i)-1} (1 - J_j)$. Thus (4.58) represents the high-frequency data (δ_j, D_j), $j = 1, \ldots, N(T)$, as $(\tilde{\delta}_i, \tilde{Z}_i, \tilde{D}_i)$ for $j = 1, \ldots, \tilde{N}(T)$.

Since $P(\tilde{Z}_i = 0) > 0$ can be substantial, McCulloch and Tsay (2001) set $\tilde{Z}_i = \tilde{J}_i \tilde{U}_i$ where $\tilde{J}_i = I_{\{\tilde{Z}_i > 0\}}$ and $\tilde{U}_i = |\tilde{Z}_i|$. Moreover, because $P(\tilde{D}_i = 0) = 0$, $\tilde{D}_i$ can be encoded by the direction component $\tilde{A}_i = \mathrm{sign}(\tilde{D}_i)$ and the magnitude component $\tilde{M}_i = |\tilde{D}_i|$. Similar to Section 4.6.3, $\tilde{J}_i$ and $\tilde{A}_i$ can be modeled by logistic regression, while $(\tilde{M}_i - 1)|(\tilde{A}_i, \mathcal{F}_{i-1})$ and $(\tilde{U}_i - 1)|(\tilde{J}_i = 1, \mathcal{F}_{i-1})$ can be modeled by generalized linear models involving the geometric, or negative binomial, or Poisson distribution. Finally, unlike Section 4.6.3 that models the duration separately from the price movements, McCulloch and Tsay (2001) model $\tilde{\delta}_i|\mathcal{G}_{i-1}$ by an accelerated failure time model of the form

$$\log(\tilde{\delta}_i) = \beta_0 + \beta_1 \log(\tilde{\delta}_{i-1}) + \beta_2 \tilde{M}_{i-1} + \varepsilon_i, \qquad (4.59)$$

in which the ε_i are i.i.d. $N(0, \sigma^2)$. They call (4.59), in conjunction with (4.58) and the generalized linear models for $\tilde{J}_i$, $\tilde{A}_i$, $\tilde{U}_i$, and $\tilde{M}_i$, the *price and conditional duration model* because it offers an alternative way to model the joint dynamics of the durations and the price movements. Note that under this model, $\mathrm{Cov}(\tilde{A}_i, \tilde{A}_{i-1}) < 0$ refers to bid-ask bounce for consecutive *active* trades.

Another important issue considered by McCulloch and Tsay (2001) is how to fit gap-free time series models to high-frequency data across trading days. Note that Rydberg and Shephard (2003) only use the data after the market opens for 15 minutes and concatenate them with the data of the previous day, in order to remove the day gap effect under the assumption of constant parameters across days. Similar procedures have also been adopted by Engle and Russell (1998) in ACD estimation. In particular, ACD estimates the diurnal pattern g from (4.48) which is based on the assumption of constancy of g across days. Instead of assuming constant parameters, McCulloch and Tsay (2001) define $\boldsymbol{\Theta}_j$ as the vector of generalized linear model parameters on the jth trading day and assume that the $\boldsymbol{\Theta}_j$ are i.i.d. $N(\boldsymbol{\Theta}^*, \boldsymbol{\Sigma}^*)$, where $\boldsymbol{\Theta}^*$ and $\boldsymbol{\Sigma}^*$ are the mean vector and the covariance matrix of the prior distribution, respectively. They estimate the posterior distribution of the parameters by Markov Chain Monte Carlo (MCMC). In fact, their hierarchical model specification is equivalent to a system of generalized linear mixed models which can be alternatively estimated by a maximum likelihood procedure that approximates the likelihood function by Laplace's asymptotic quadrature formula and importance sampling; see Lai et al. (2006).

4.6.5 Joint modeling of point process and its marks

Engle (2000) proposes to model jointly the point process $N(t)$ and its marks (e.g., the price changes D_i) by specifying the joint conditional density of (δ_i, y_i) given $\mathcal{F}_{i-1}$, where $\delta_i = t_i - t_{i-1}$ and y_i is the mark (financial variable of inter-

est) for the ith transaction. Since $N(t)$ is basically a renewal process associated with the inter-arrival times δ_i, this specification also suffices for the joint
model in the title of this subsection. In Engle (2000), he discusses how various
market microstructure features can be modeled by defining y_i appropriately.
For example, y_i can represent P_{t_i}, D_i or the traded volume in the ith transaction, and the posted bid and ask prices of the transacted stock and any other
related stock at time t_i can be modeled jointly by a vector of marks $\boldsymbol{y}_i$. He also
considers using y_i to model categorical variables such as the counterparties
involved and the order mechanism used (for assets traded in a hybrid market)
in the ith transaction. He further points out that "economic hypotheses are
often associated with the distribution of the marks", in particular, the distribution of the marks at some fixed time in the future or after a certain number
of transactions have occurred, and that "the joint analysis of transaction times
and prices implements the standard time deformation models by obtaining a
direct measure of the arrival rate of transactions and then measuring exactly
how this influences the distribution of the other observables in the market."

As an application of the proposed joint modeling approach, Section 6 of
Engle (2000) introduces the UHF-GARCH model in which UHF stands for
"ultra-high frequency". Letting r_i be the return for the period from t_{i-1} to
t_i, the conditional variance per transaction is defined as

$$V_{i-1}(r_i|x_i) = h_i. \tag{4.60}$$

Since the conditional volatility σ_i of the ith transaction is frequently quoted in
annualized terms, $V_{i-1}(r_i/\sqrt{\delta_i}|\delta_i) = \sigma_i^2$. From this and (4.60) it follows that
$h_i = \delta_i\sigma_i^2$. Fitting an ARMA(1,1)-GARCH(1,1) model (Lai and Xing, 2008,
p. 156) to $r_i/\sqrt{\delta_i}$ yields

$$\begin{aligned}
\frac{r_i}{\sqrt{\delta_i}} &= \rho\frac{r_{i-1}}{\sqrt{\delta_{i-1}}} + e_i + \phi e_{i-1}, \\
\sigma_i^2 &= \omega + \alpha e_{i-1}^2 + \beta\sigma_{i-1}^2,
\end{aligned} \tag{4.61}$$

with innovations e_i. This is the simplest UHF-GARCH model, and richer
UHF-GARCH models that include additional terms associated with the EAD
model for δ_i are also considered by Engle (2000).

4.6.6   Realized GARCH and other predictive models relating low-frequency to high-frequency volatilities

In Section 2.6.3 we have referred to the R package `rugarch`, in which the
first r stands for "realized". The realized GARCH (RealGARCH) model was
introduced by Hansen et al. (2012) to relate realized measures of volatility in
high-frequency financial data to future low-frequency (e.g., daily) volatility.
They say:

> Standard GARCH models utilize daily returns (typically squared re-
> turns) to extract information about the current level of volatility, and

this information is used to form expectations about the next period's volatility. A single return only offers a weak signal about the current level of volatility. The implication is that GARCH models are poorly suited for situations where volatility changes rapidly to a new level. The reason is that a GARCH model is slow at 'catching up'.... High-frequency financial data are now widely available and the literature has recently introduced a number of realized measures of volatility.... Any of these measures is far more informative about the current level of volatility than is the squared return. This makes realized measures very useful for modeling and forecasting future volatility. Estimating a GARCH model that includes a realized measure in the GARCH equation (known as GARCH-X model) provides a good illustration of this point.

RealGARCH, however, differs from the GARCH-X models because it is a "complete" model (data generating mechanism) that also specifies stochastic dynamics of the high-frequency *realized measure* x_t. For example, the log-linear specification of RealGARCH(1,1) is characterized by $r_t = \sqrt{h_t} z_t$, with i.i.d. standard normal z_t and the GARCH and measurement equation

$$\log h_t = \omega + \beta \log h_{t-1} + \gamma \log x_{t-1}, \tag{4.62}$$
$$\log x_t = \xi + \varphi \log h_t + \tau(z_t) + u_t,$$

where u_t are i.i.d. with $E(u_t) = 0$ and $\mathrm{Var}(u_t) = \sigma_u^2$, and $\tau(z)$ is the *leverage function* which is designed to enable an asymmetric effect of z_t on volatility, while still satisfying $E[\tau(z)] = 0$. In particular, $\tau(z) = \tau_1 z + \tau_2(z^2 - 1)$ in Hansen et al. (2012), and this is used to induce an EGARCH-type structure (Lai and Xing, 2008, p. 152–153) in the GARCH equation. The uncertainties in the high-frequency risk measure x_t, as an approximation or estimate of some high-frequency trading entity such as $[X]_T$, is reflected in the latent variable u_t in (4.62), and assuming that $(z_t, u_t/\sigma_u)^\top \sim N(\mathbf{0}, \mathbf{I})$ completes the specification of RealGARCH.

An alternative to the log-linear specification (4.62) is the linear specification

$$h_t = \omega + \beta h_{t-1} + \gamma x_{t-1}, \qquad x_t = \xi + \varphi h_t + \tau(z_t) + u_t. \tag{4.63}$$

Earlier, Engle and Gallo (2006) introduced a similar model but with x_t replaced by $(h_{R,t}, h_{RV,t})$, in which R_t is the intraday range (high minus low) and RV_t is the realized variance

$$h_t = \omega + \alpha r_{t-1}^2 + \beta h_{t-1} + \delta r_{t-1} + \varphi R_{t-1}^2,$$
$$h_{R,t} = \omega_R + \alpha_R r_{t-1}^2 + \beta_R h_{t-1} + \delta_R r_{t-1} + \varphi R_{t-1}^2,$$
$$h_{RV,t} = \omega_{RV} + \alpha_{RV} RV_{t-1} + \beta_{RV} h_{RV,t-1} + \delta_{RV} r_{t-1}$$
$$+ \theta_{RV} RV_{t-1} I_{\{r_{t-1}<0\}} + \varphi_{RV} r_{t-1}^2,$$

in which $R_t^2 = h_{R,t} z_{R,t}^2$, $RV_t = h_{RV,t} z_{RV,t}^2$, and $(z_t, z_{R,t}, z_{RV,t})^\top \sim N(\mathbf{0}, \boldsymbol{I})$. This is called the multiplicative error model (MEM). Shephard and Sheppard (2010) subsequently proposed a more complicated model than (4.63), of the form

$$h_t = \omega + \alpha r_{t-1}^2 + \beta h_{t-1} + \gamma x_{t-1}, \qquad x_t = \mu_t z_{RK,t}^2,$$

$$\mu_t = \omega_R + \alpha_R x_{t-1} + \beta_R \mu_{t-1},$$

in which $(z_t, z_{RK,t})^\top \sim N(\mathbf{0}, \boldsymbol{I})$ and x_t is the realized kernel estimator of $[X]_T$. Shephard and Sheppard abbreviate their High-frEquency-bAsed VolatilitY model as HEAVY.

4.6.7   Jumps in efficient price process and power variation

While microstructure noise is used to quantify high-frequency deviations from a Brownian semimartingale efficient price process, caused by bid-ask bounce and other microstructure factors as in Section 4.2.2, the model of geometric Brownian motion (GBM) with jumps has been used to study large movements in asset prices since the 1960s when the analysis was mainly performed by using low-frequency data; see Press (1967), Merton (1976), and Ball and Torous (1983). In particular, the simplest form in this class of models is

$$dX_t = \mu dt + \sigma dB_t + J_t dN_t, \tag{4.64}$$

where X_t is the log price of an asset at time t, B_t is standard Brownian motion, μ and $\sigma > 0$ are the drift and volatility parameters, respectively, N_t is a Poisson process with arrival rate λ, and $J_t \sim N(\beta, \eta)$ is the logarithm of the jump size. The return under this model is a Poisson mixture of normal distributions, hence the fat-tailedness can be explained by (4.64). The non-zero mean β of the jump size J_t can also model the return's skewness.

Motivated by applications to option pricing and term structure modeling of interest rates, (4.64) has been generalized to

$$dX_t = \mu dt + \sigma dB_t + dL_t \tag{4.65}$$

in which L_t is a Lévy pure jump process, i.e., $L_0 = 0$ and L_t is a pure jump process with stationary independent increments; see Madan et al. (1998), Carr and Wu (2004), Eberlein et al. (1998), and Eberlein and Raible (1999). In particular, (4.64) with $dL_t = J_t dN_t$ corresponds to L_t being a compound Poisson process with normal jump sizes. More generally, if L_t is a compound Poisson process with i.i.d. random jump sizes, it is said to be of *finite activity* and the jumps are relatively sparse. All other Lévy pure jump processes have *infinite activity*, which means that jumps occur infinitely often on every open interval and most jump sizes are small. Aït-Sahalia (2004) outlines the following three broad research directions in financial econometrics associated with model (4.65):

(a) Estimation of the parameters and confidence (prediction) intervals.

(b) Testing from discrete-time data whether jumps are present.

(c) Analysis of statistical properties of quadratic variation and related quantities in the presence of jumps.

The research generated from these three directions has produced many fruitful applications of (4.65). To begin with, the quadratic variation of the jump diffusion process (4.64) is of the form

$$[X]_T = \sigma^2 T + \sum_{i=1}^{N_T} J_{s_i}^2, \tag{4.66}$$

where s_i is the ith arrival time of N_t. That is, $\sigma^2 T$ is the continuous component and $\sum_{i=1}^{N_T} J_{s_i}^2$ is the purely discontinuous component; see Appendix A. Instead of focusing only at quadratic variation, Lepingle (1976) has introduced the (absolute) *power variation* which is defined as

$$_p[X]_T = \operatorname*{p-lim}_{\operatorname{mesh}(\Pi) \to 0} \sum_{i=1}^{n} |X_{t(i)} - X_{t(i-1)}|^p \tag{4.67}$$

similar to (4.6), and has shown that under model (4.65), the contribution of the jump component to $_p[X]_T$ is 0 (after some normalization) for $p \in (0, 2)$, and can be ∞ if $p > 2$.

Aït-Sahalia and Jacod (2012) define the discrete-time analog of (4.67) as

$$B(p, u, \delta) = \sum_{i=1}^{n} |X_{i\delta} - X_{(i-1)\delta}|^p I_{\{|X_{i\delta} - X_{(i-1)\delta}| \le u\}},$$

where u is the truncation level and $\delta = T/n$ is the common width of the sampling intervals. Based on their analysis of the asymptotic behavior of $B(p, u, \delta)$ as $\delta \to 0$ and $u \to \infty$, they develop three strategies for the statistical tasks (a) and (b):

- Inflate the power $p > 2$ to magnify the jumps.

- Use the truncation level u to eliminate the big jumps.

- Compare power variations under different sampling frequencies.

In particular, they derive test statistics for (i) checking the possibility of finite activity if the null hypothesis assumes that "the jumps of X_t are infinitely active", and (ii) testing the presence/absence of continuous components. Aït-Sahalia and Jacod (2011, 2012, 2015) have also used these strategies to develop robust estimators in the presence of rounding errors and microstructure noise.

4.7   Supplementary notes and comments

1. **Microstructure noise model and $n^{-1/4}$-convergence**   The likelihood function $l(v_\varepsilon, \sigma^2)$ in (4.21) is derived from the normal microstructure noise model

$$Y_{i\delta} = X_{i\delta} + \varepsilon_i \qquad (4.68)$$

for $i = 1, \ldots, n$, where Y_t is the observed transaction price, $\delta = T/n$ is the length of the time-aggregation interval, X_t is the efficient price process under the working model $dX_t = \sigma dB_t$, and ε_i are i.i.d. $N(0, v_\varepsilon)$ and are independent of X_t. In spatial statistics, these models and the asymptotic behavior of the MLE have also been studied; see Stein (1990, 1999) and Chen et al. (2000). The case of an Ornstein-Uhlenbeck (OU) process $dX_t = -\alpha X_t dt + \sigma\sqrt{2\alpha}dB_t$, with $\alpha > 0$, has also been studied by Chen et al. (2000). The microstructure noise ε_i is called "measurement error" or "nugget effect" in spatial statistics. Irrespective of whether $dX_t = \sigma dB_t$ or $dX_t = -\alpha X_t dt + \sigma\sqrt{2\alpha}dB_t$, the results of Chen et al. (2000) and Xiu (2010) show that the QMLE of σ^2 is consistent with a rate of convergence $n^{1/4}$, i.e., $n^{1/4}(\hat{\sigma}^2 - \sigma^2) \xrightarrow{\mathcal{L}} N(0, \psi)$ as $n \to \infty$ and T is fixed, which we assume to be 1 without loss of generality (called "infill asymptotics" in spatial statistics, in which ψ differs for both models.) Note that the stationary distribution of the OU process is Gaussian with variance σ^2. The QMLE $\hat{v}_\varepsilon$ has the same asymptotic distribution $\sqrt{n}(\hat{v}_\varepsilon - v_\varepsilon) \xrightarrow{\mathcal{L}} N(0, 2v_\varepsilon^2)$ for both models and is asymptotically independent of $\hat{\sigma}^2$. Letting $\xi_i = (X_{i/n} - X_{(i-1)/n})/(\sigma/\sqrt{n})$ and recalling that $\Delta_i = Y_{i/n} - Y_{(i-1)/n}$, model (4.68) with $T = 1$ and $dX_t = \sigma dB_t$ can be expressed as

$$\Delta_i = \sqrt{v_\varepsilon}(e_i - e_{i-1}) + \sqrt{\frac{\sigma^2}{n}}\xi_i, \qquad (4.69)$$

where ξ_i and $e_i = \varepsilon_i/\sqrt{v_\varepsilon}$ are i.i.d. $N(0,1)$ random variables. When $n \to \infty$, Δ_i is dominated by the moving average term $\sqrt{v_\varepsilon}(e_i - e_{i-1})$. The information provided by the data for the estimation of σ^2 is much less than that for v_ε, hence the $n^{-1/4}$ rate for estimation of σ^2 instead of the usual $n^{-1/2}$ convergence rate for estimation of v_ε.

2. **Computational issues of QMLE**   Because of the dominance of $e_i - e_{i-1}$ in (4.69), the likelihood function $l(v_\varepsilon, \sigma^2)$ given in (4.21) is relatively flat along the direction of σ^2 but is already peaked in the direction of v_ε; see the top panel of Figure 4.1 based on a simulated dataset from (4.69) with $n = 1000$ and $(\sigma^2, v_\varepsilon) = (1, 1)$. Hence, the MLE obtained by using Newton-type iterations can be sensitive to the choice of starting values. The top panel of Figure 4.1 shows $\hat{v}_\varepsilon$ to be close to 1 but that a wide range of values can be chosen as the estimate of σ^2. To get around

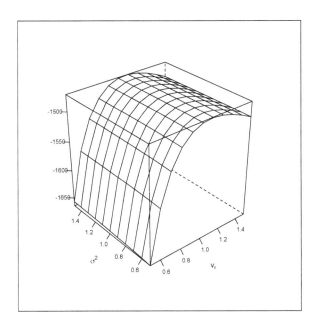

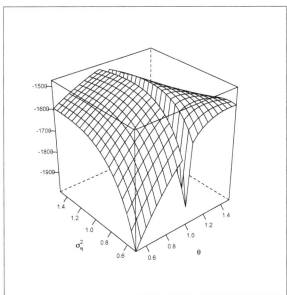

FIGURE 4.1: Log-likelihood function of a simulated data set of size $n = 1000$ from model (4.21), which is equivalent to (4.69), in the top panel, and using the MA(1) parameterization (4.70) in the bottom panel.

this difficulty, one can use an alternative parameterization of the MA(1) structure suggested by Hansen et al. (2008):

$$\Delta_i = \eta_i - \theta\eta_{i-1}, \tag{4.70}$$

where η_i are i.i.d. $N(0, \sigma_\eta^2)$. The log-likelihood function of (4.70) has the same form as $l(v_\varepsilon, \sigma^2)$ but with the diagonal and off-diagonal nonzero elements of the tri-diagonal matrix $\mathbf{\Omega}$ in (4.22) replaced by $\sigma_\eta^2(1+\theta^2)$ and $-\theta\sigma_\eta^2$, respectively, where

$$\sigma_\eta^2(1+\theta^2) = 2v_\varepsilon + \frac{\sigma^2}{n}, \qquad \theta\sigma_\eta^2 = v_\varepsilon. \tag{4.71}$$

By solving σ^2 from the preceding equations, Hansen et al. (2008) suggest estimating σ^2 by $n\hat\sigma_\eta^2(1+\hat\theta^2)$ where $(\hat\sigma_\eta^2, \hat\theta)$ is the MLE of the MA(1) model (4.70). From (4.71), it follows that

$$\theta = 1 + \frac{\sigma^2}{nv_\varepsilon} - \sqrt{\frac{\sigma^2}{4nv_\varepsilon^2}\left(1 + \frac{\sigma^2}{nv_\varepsilon}\right)},$$

$$\sigma_\eta^2 = \frac{v_\varepsilon}{1 + \frac{\sigma^2}{nv_\varepsilon} - \sqrt{\frac{\sigma^2}{4nv_\varepsilon^2}\left(1 + \frac{\sigma^2}{nv_\varepsilon}\right)}},$$

and therefore θ is close to 1, making the MA(1) model (4.70) nearly non-invertible, while σ_η^2 close to v_ε. The bottom panel of Figure 4.1 plots the log-likelihood function with parameterization (σ_η^2, θ) showing the maximum near the grove $\theta = 1$ that exhibits numerical instability of the MLE $(\hat\sigma_\eta^2, \hat\theta)$; see also Brockwell and Davis (2002, p. 196).

3. **Deconvolution and minimax estimator for model (4.68)**   Cai et al. (2010) address the preceding computational issue by developing a minimax estimator of $(v_\varepsilon, \sigma^2)$ which can be computed without using any hill climbing-type iterations and which has the same marginal asymptotic distribution as Xiu's QMLE[8]. This estimator uses deconvolution to the model (4.68), treating $X_{i\delta} = \sigma B_{i\delta}$ as the signal and the observed data $Y_{i\delta}$ as the convolution of the signal and the microstructure noise.

Singular value decomposition (SVD)

Cai et al. (2010) solve the inverse problem of (4.68) by using $\mathrm{Cov}(B_t, B_s) - s \wedge t$, that is, $\mathbf{Y} \sim N(\mathbf{0}, \sigma^2\mathbf{C} + v_\varepsilon\mathbf{I})$ in which the (i,j)th entry of $\mathbf{C}$ is $(i/n) \wedge (j/n)$. The SVD of $\mathbf{C}$ is $\mathbf{C} = \mathbf{U\Lambda U}^\top$, where $\mathbf{U} = (u_{i,j})_{1\leq i,j\leq n}$ is an orthogonal matrix defined by

$$u_{ij} = \sqrt{\frac{4}{2n+1}} \sin\left(\frac{(2j-1)i\pi}{2n+1}\right), \tag{4.72}$$

---

[8]While the QMLE of σ^2 and v_ε are asymptotically uncorrelated, Cai et al. (2010) have not established that this property holds for their deconvolution estimator.

and $\boldsymbol{\Lambda}$ is a diagonal matrix whose ith diagonal element is

$$\lambda_i = \left[4n\sin^2\left(\frac{(2i-1)\pi}{4n+2}\right)\right]^{-1}. \tag{4.73}$$

Hence, by using the transformation $\boldsymbol{Z} = \boldsymbol{U}^\top \boldsymbol{Y} \sim N(\boldsymbol{0}, \sigma^2\boldsymbol{\Lambda} + v_\varepsilon\boldsymbol{I})$, we have n independent variables

$$Z_i \sim N(0, \sigma^2\lambda_i + v_\varepsilon), \ i = 1,\ldots,n, \tag{4.74}$$

which suggests a linear estimator of σ^2 of the form

$$\hat{\sigma}^2 = \frac{1}{b_n}\sum_{i=k+1}^{m}\frac{\lambda_i}{(\tilde{\sigma}^2\lambda_i + \tilde{v}_\varepsilon)^2}(Z_i^2 - \tilde{v}_\varepsilon) \text{ with } 1 < k < m < n, \tag{4.75}$$

where $\tilde{v}_\varepsilon$ and $\tilde{\sigma}^2$ are preliminary estimators of v_ε and σ^2 given by

$$\tilde{v}_\varepsilon = \frac{1}{n-m}\sum_{i=m+1}^{n}Z_i^2, \ \tilde{\sigma}^2 = \frac{1}{k}\sum_{i=1}^{k}\frac{Z_i^2 - \tilde{v}_\varepsilon}{\lambda_i}. \tag{4.76}$$

Cai et al. (2010) show that as $n \to \infty$ and $m/n \to 0$ but $m/\sqrt{n} \to \infty$, $\sqrt{n}(\tilde{v}_\varepsilon - v_\varepsilon) \overset{\mathcal{L}}{\to} N(0, 2v_\varepsilon^2)$ and $\tilde{v}_\varepsilon$ is asymptotically minimax. They also show that by choosing $k = [n^{1/2-\gamma}]$ and $m = [n^{1/2+\gamma}]$, with $0 < \gamma < 1/20$, and

$$b_n = \sum_{i=k+1}^{m}\frac{\lambda_i^2}{(\tilde{\sigma}^2\lambda_i + \tilde{v}_\varepsilon)^2}, \tag{4.77}$$

$n^{1/4}(\tilde{\sigma}^2 - \sigma^2) \overset{\mathcal{L}}{\to} N(0, 8\sqrt{v_\varepsilon}\sigma^3)$ and $\hat{\sigma}^2$ is asymptotically minimax. The transformation of $\boldsymbol{Z} = \boldsymbol{U}^\top \boldsymbol{Y}$ requires only $O(n\log n)$ operations and the computation of (4.76), (4.77) and then (4.75) needs $O(n)$ steps, and requires no iteration.

This technique of transforming $\boldsymbol{Y}$ with correlated entries into $\boldsymbol{Z}$ with independent entries via $\boldsymbol{Z} = \boldsymbol{U}^\top \boldsymbol{Y}$ has been used by Munk et al. (2010) in estimating $\int_0^T \sigma_t^2 \, dt$ and $\int_0^T v_{\varepsilon,t} \, dt$ under the assumption that both σ_t^2 and $v_{\varepsilon,t}$ are deterministic smooth functions rather than constant parameters. Note that although the same notation of $dX_t = \sigma_t dB_t$ is used in Xiu (2010), σ_t^2 is instead a stochastic process (locally bounded positive semimartingale) in Xiu (2010). The assumption of deterministic and smooth σ_t^2 came from the deconvolution literature, which was used by Hoffmann et al. (2012) to model the spot volatility σ_t^2 via wavelet expansion and extended by Sabel et al. (2015) to handle more general forms, such as jumps and rounding errors, of microstructure noise.

4. **The QML principle and asymptotic theory** The preceding discussions on QMLE has already shed light on the quasi-maximum likelihood

principle that estimates a parameter vector $\boldsymbol{\theta}$ by maximizing the likelihood function under a "working parametric model" that relates the data to the parameters of interest, while recognizing that the actual data-generating mechanism is in fact much more complicated than the assumed model. A case in point is related to spatial statistics, and more generally spatio-temporal Markov random fields in image reconstruction for which the computation of the actual likelihood is formidable; see Lai and Lim (2015) who use QML that partitions the random field into blocks and assumes the disjoint blocks to be independent as their working model, and earlier works by Besag (1974, 1986) who uses a composite likelihood, which is a product of component likelihoods with each component being a conditional density of the value of the image at a site given the values of the site's neighbors according to some neighborhood system. Despite model oversimplification and potential misspecification, QMLE can still be consistent and asymptotically normal if it captures the essence of how the parameter vector is related to the data-generating mechanism; see White (1982), Lee and Hansen (1994), and Cox and Reid (2004).

5. **Market making and negative MA(1) autocorrelations**   Prior to Roll's model (4.1), the relationship between certain stylized facts of security price movements and the market makers was discussed by Mandelbrot (1963) who pointed out how the specialists could possibly eliminate the big price jumps predicted by the stable distribution model, and by Mandelbrot and Taylor (1967) who suggested using the subordinated stochastic process model to quantify specialists' activities. Niederhoffer and Osborne (1966) attributed the behavior in tick-by-tick transaction price differences (negative MA(1) feature) to the specialists' compensation received for their services. Roll (1984) used (4.1) to explain the MA(1) feature by modeling the bid-ask bounce.

6. **Nontrading explanation for negative MA(1) autocorrelations and nonsynchronous trading**   Miller et al. (1994) mention that stocks do not trade continuously. They may trade in consecutive intervals, but not at the close of each interval; this is called "nonsynchronous trading". They may also not trade in an interval, which is called "non-trading". This is in fact a special case of the general model (4.4) because there is no trading between t_i and t_{i+1} and because the t_i do not need to be evenly spaced, Lo and MacKinlay (1990) introduce a stochastic model of non trading and use it to explain negative MA(1) autocorrelations; details can be found in Lai and Xing (2008, p. 291–292).

7. **Round-the-clock volatility modeling by RealGARCH**   All realized estimators measure the volatility within the trading hours. RealGARCH is conceptually different from MEM and HEAVY, presented in Section 4.6.6, because it explicitly models the realized measure (denoted by x_t) with the contemporaneous round-the-clock volatility (given by h_t) via $x_t =$

$\xi + \varphi h_t + \tau(z_t) + u_t$ as in (4.63) (or via the log-linear version (4.62)). The round-the-clock volatility modeling provided by RealGARCH can be shown by the following schematic diagram:

$$h_t \left\{ \begin{array}{c} \text{Market close on day } t - 1 \\ \downarrow \\ \text{Market open on day } t \\ \downarrow\downarrow \\ \text{Market close on day } t \end{array} \right\} v_t$$

and $x_t = v_t + $ noise, where v_t is a measure of the conditional variance of high-frequency return within the trading hours of day t. Thus, v_t is modeled by $\xi + \varphi h_t$. Also, φ is expected to be less than 1 if r_t is a close-to-close return. The empirical results in Section 5.3 of Hansen et al. (2012) have shown the following pattern in this connection: (a) $\hat{\varphi} \approx 1$ from taking r_t as the open-to-close return and x_t as the realized kernel volatility estimate of an exchange-traded fund SPY, and (b) $\hat{\varphi} = 0.74 < 1$ when r_t is the close-to-close return.

8. **Causes of the Epps effect** The Epps effect is mainly caused by two features of high-frequency data. The first is the stickiness of inactive trades under high sampling frequency described in Section 4.6.4. In particular, when the width of the synchronization interval in Section 4.4.2 decreases, the synchronized price differences of both assets are mostly zeros and are small if non-zero (say, 1 tick for liquid stocks). Another cause of the Epps effect is asynchronicity, leading to situations where an unusually large movement occurs in the synchronized price differences of an asset but the other assets still have zero or small synchronized price differences. Zhang (2011) applies a Previous Tick approach for synchronization to address the asynchronicity effect, but because of the aforementioned reasons, the amount of useful information for covariance estimation diminishes when the synchronization interval is small. The Fourier method proposed by Malliavin and Mancino (2002, 2009) is more effective in addressing the issue of asynchronicity because the Fourier basis provides a wide spectrum of data-driven windows to capture the covariation without removing observations from analysis.

9. **Misplacement error and generalized synchronized method** Although the arbitrary selection mechanisms of the generalized synchronized method in Section 4.4.2 seems quite counterintuitive, Aït-Sahalia et al. (2010) claim that the random selection of $\tilde{t}_{i,j} \in (\tau_{j-1}, \tau_j]$, rather than the Previous Tick approach, is capable of handling the misplacement errors. They say: "In both cases, the previous ticks of the assets, if needed, are regarded as if they were observed at the sampling time τ_j's. By contrast, we advocate choosing an arbitrary tick for each asset within each interval. In practice, it may happen that the order of consecutive ticks is not recorded correctly. Because our synchronization method has no requirement on tick

selection, the estimator is robust to data misplacement error, as long as these misplaced data points are within the same sampling intervals." Informatics issues in aggregating transaction data across many exchanges and potential misplacement errors will be discussed in Section 8.3.1.

10. **Symmetric and positive definite Fourier estimator of spot co-volatility**   While the Fourier estimator of integrated covariance does not employ synchronization and is therefore exempt from the Epps effect, the Fourier estimator of spot co-volatility in (4.45) may lead to a spot volatility matrix with negative eigenvalues as pointed out by Liu and Ngo (2014). Malliavin and Mancino (2009) have proposed another estimate of spot co-volatility:

$$
\tilde{\Sigma}_t^{(j,k)} = \sum_{l=-N_S}^{N_S} \left(1 - \frac{|l|}{N_S}\right) e^{ilt} \left[\frac{2\pi}{2N_X + 1} \sum_{n=-N_X}^{N_X} \hat{c}_n(dp_j)\hat{c}_{l-n}(dp_k)\right],
$$
(4.78)

where $\hat{c}_n(dp_j) = \hat{a}_n(dp_j) - \sqrt{-1} \times \hat{b}_n(dp_j)$, $\hat{a}_n(dp_j)$ and $\hat{b}_n(dp_j)$ are defined in (4.33)–(4.35) by extending the same formula to negative n. The estimator (4.78) has been studied by Akahori et al. (2014) who express (4.78) as

$$
\tilde{\Sigma}_t^{(j,k)} = \sum_{r=2}^{n_j} \sum_{s=2}^{n_k} \mathcal{G}_{N_S}(t - t_{j,r}) D_{N_X}(t_{j,r} - t_{k,s}) R_{j,r} R_{k,s},
$$
(4.79)

where $R_{j,r} = X_{j,t_{j,r}} - X_{i,t_{j,r-1}}$, $R_{k,s} = X_{k,t_{k,s}} - X_{k,t_{k,s-1}}$, $\{t_{j,r} : r = 1, \ldots, n_j\}$ and $\{t_{k,s} : s = 1, \ldots, n_k\}$ are the transaction times for the log-prices $X_{i,t}$ and $X_{j,t}$, respectively. In (4.79), $D_N(x)$ is the rescaled Dirichlet kernel as in (4.42), and $\mathcal{G}_M(t)$ is the *Fejér kernel* of the form

$$
\mathcal{G}_M(t) = \sum_{|l| \leq M-1} \left(1 - \frac{|l|}{M}\right) e^{ilt} = \frac{1}{M}\left(\frac{\sin(Mt/2)}{\sin(t/2)}\right)^2.
$$

Akahori et al. (2014) point out that $\tilde{\Sigma}_t^{(j,k)} \neq \tilde{\Sigma}_t^{(k,j)}$ because $\mathcal{G}_{N_S}(t - t_{j,r})D_{N_X}(t_{j,r} - t_{k,s})$ is not a symmetric function in j and k. To address the issue of asymmetry and to ensure the positive definiteness of the estimator, they define a wider class of estimators by

$$
\check{\Sigma}_t^{(j,k)} = \sum_{l \in \mathbb{L}} f(l) e^{ilt} \left[\frac{2\pi}{|\mathbb{S}(l)|} \sum_{(m,n) \in \mathbb{S}(l)} \hat{c}_m(dp_j)\hat{c}_n(dp_k)\right],
$$
(4.80)

where $|\mathbb{S}(l)|$ is the cardinality of $\mathbb{S}(l)$. They show that $\check{\Sigma}_t^{(j,k)}$ reduces to $\tilde{\Sigma}_t^{(j,k)}$ if $\mathbb{L} = \{-N_S, \ldots, -1, 0, 1, \ldots, N_S\}$, $\mathbb{S}(l) = \{(m, l - m) : m = -N_X, \ldots, -1, 0, 1, \ldots, N_X\}$, and $f(l) = (1 - |l|/N_S)$ for any $l \in \mathbb{L}$. In particular, by choosing $\mathbb{L} = \{-2M, -2M+1, \ldots, -1, 0, 1, \ldots, 2M-1, 2M\}$ and for any $l \in \mathbb{L}$,

$$
\mathbb{S}(l) = \begin{cases} \{(-M+l+v, M-v) : v = 0, 1, \ldots, 2M-l\} & \text{if } l \geq 0 \\ \{(M+l-v, -M+v) : v = 0, 1, \ldots, 2M+l\} & \text{if } l < 0, \end{cases}
$$

and taking f to be a positive definite function on $\mathbb{L}$, Akahori et al. (2014) show that $\check{\Sigma}_t^{(j,k)}$ is indeed positive definite. They also recommend the following choices of f: (i) $f(l) = (1 - |l|/2M) \exp\{-\gamma|l|\}$, (ii) $f(l) = (1 - |l|/2M) \exp\{-2\pi^2 l^2/C\}$, and (iii) $f(l) = (1 - |l|/2M)^2$ with $2M+1$ being a prime number.

11. **Positive semi-definite multivariate realized kernel estimator**
The direct generalization of the univariate realized kernel estimator does not give a positive definite

$$
\tilde{K}(\boldsymbol{Y}_\tau) = \tilde{\boldsymbol{\Gamma}}_0 + \sum_{h=1}^{H} \kappa\left(\frac{h-1}{H}\right)(\tilde{\boldsymbol{\Gamma}}_h + \tilde{\boldsymbol{\Gamma}}_{-h}),
$$

where

$$
\tilde{\boldsymbol{\Gamma}}_h = \sum_j \left(\boldsymbol{Y}_{\tau_j} - \boldsymbol{Y}_{\tau_{j-1}}\right)\left(\boldsymbol{Y}_{\tau_{j-h}} - \boldsymbol{Y}_{\tau_{j-h-1}}\right)^\top;
$$

see Barndorff-Nielsen et al. (2011, Sect. 6.1). Conditions (a)–(d) in Section 4.4.4 are crucial for ensuring that the MRK estimator $\boldsymbol{K}(\boldsymbol{Y}_\tau)$ defined in (4.29) is positive definite.

4.8 Exercises

1. Consider the microstructure noise model (4.4). Assume $dX_t = \sigma dB_t$ where B_t is standard Brownian motion. Suppose the ε_i are i.i.d. uniformly distributed on $[-a, a]$ and are independent of the X_t. Let $\Delta_i = Y_{t_i} - Y_{t_{i-1}}$, with $t_i = i/n$.

   (a) Show that $E(\varepsilon_i^2) = a^2/3$ and $E(\varepsilon_i^4) = a^4/5$.

   (b) Let $\hat{a}$ be the QMLE of a. What is the asymptotic distribution of $\hat{a}$? (Hint: Use (4.23), (4.24) and the delta method.)

2. Consider Roll's model in Section 4.2.1. Suppose the efficient price process is of the form $P_t^* - P_0^* = \sigma B_t$, where B_t is standard Brownian motion. Let $\boldsymbol{D} = (D_1, \ldots, D_n)^\top$ denote the observed vector of transaction price changes, i.e., $D_1 = P_{t_1}$, $D_i = P_{t_i} - P_{t_{i-1}}$, and $t_i - t_{i-1} = 1/n$ for $i = 2, \ldots, n$. What is the asymptotic distribution of the QMLE of c?

3. Consider (4.32).

   (a) Show by using integration by parts that $a_k(dX)$ can be expressed as

   $$a_k(dX) = \frac{X_{2\pi} - X_0}{\pi} + \frac{k}{\pi} \int_0^{2\pi} X_t \sin(kt)\, dt. \qquad (4.81)$$

   (b) Derive (4.34) from the representation

   $$\hat{a}_k(dX) = \frac{X_{2\pi} - X_0}{\pi} + \frac{k}{\pi} \sum_{i=0}^{n} \int_{t_{i-1}}^{t_i} \tilde{X}_t \sin(kt)\, dt,$$

   where $\tilde{X}_s = X_{t_i}$ for all $s \in (t_i, t_{i+1}]$, i.e., $\tilde{X}_s$ is a piecewise constant approximation of X_t.

   (c) Prove (4.35) by using the same arguments as in (a) and (b).

   (d) Discuss whether a piecewise linear approximation can be used for X_t in each interval of $(t_{j-1}, t_j]$ to replace the piecewise constant approximation.

4. The Fourier coefficients of Σ_t in Section 4.5.1 can be more generally expressed as

   $$a_0(\Sigma) = \underset{N_x \to \infty}{\text{p-lim}} \frac{\pi}{N_X + 1 - n_0} \sum_{j=n_0}^{N_X} \left[a_j^2(dX) + b_j^2(dX) \right],$$

   $$a_k(\Sigma) = \underset{N_x \to \infty}{\text{p-lim}} \frac{2\pi}{N_X + 1 - n_0} \sum_{j=n_0}^{N_X} \left[a_j(dX) a_{j+k}(dX) \right], \qquad \text{for } k > 0,$$

   $$b_k(\Sigma) = \underset{N_x \to \infty}{\text{p-lim}} \frac{2\pi}{N_X + 1 - n_0} \sum_{j=n_0}^{N_X} \left[a_j(dX) b_{j+k}(dX) \right], \qquad \text{for } k \geq 0.$$

   Malliavin and Mancino (2002, 2009) suggest choosing $n_0 = 1$ to remove the effect of the drift term. Explain how it works. (Hint: Note that $a_0(dX) = \hat{a}_0(dX) = (X_{2\pi} - X_0)/(2\pi)$.)

5. Use the data in the file NASDAQ_ITCH.txt dataset to estimate the realized variance using the following subsampling methods:

   (a) sparse sampling as in Section 4.3.1, and

   (b) two time scales as in Section 4.3.3,

   with subsampling at 1-, 10-, and 30-second intervals and with $k = 1, 5, 10$. Also, estimate the realized kernel with $H = 10$ in Section 4.3.4.

6. The file hkex.txt contains the transacted price data of HKEx traded on Hong Kong Stock Exchange on March 11, 2014. Use the morning session of the dataset to estimate $[X]_t$ under model (4.4)–(4.5) by the following:

(a) QMLE in Section 4.3.6 with $\delta = 20, 40, 60$ seconds, and give also estimate of v_ε;

(b) the realized kernel as in (4.17) with the same set of $\delta = 20, 40, 60$ seconds and $H = 5$;

(c) the Fourier estimator by $\widehat{[X]}_t^{(F)} = 2\pi\hat{a}_0(\Sigma)$, with $\hat{a}_0(\Sigma)$ defined in (4.36) and $N_X = 20$.

7. The file `hsbc.txt` contains the transacted price data of HSBC traded on Hong Kong Stock Exchange on March 11, 2014. Combine this data set and the one in Exercise 6 to estimate the integrated covariance of HKEx and HSBC in the morning session by the following methods:

(a) QMLE in Section 4.4.3 with $\delta = 60$ seconds;

(b) the realized kernel (4.28), with $H = 5$;

(c) the Fourier estimator, with $N_X = 20$.

8. Use the HSBC data in Exercise 7 to fit the EACD and WACD models in Section 4.6.1. Check which model provides a better fit.

9. Use the HSBC data in Exercise 7 to fit the model in Section 4.6.4. Test whether $\text{Cov}(\tilde{A}_i, \tilde{A}_{i-1})$ is significantly negative.

5

Limit Order Book: Data Analytics and Dynamic Models

Whereas Chapter 4 has focused on high-frequency econometrics involving transaction prices and times, this chapter focuses on the *limit order book* (LOB) about which Section 4.1 has given a brief introduction and which is a "book" only in the virtual sense. In fact, LOB is a trading mechanism used by electronic exchanges globally to match buyer and sellers, by aggregating demands from both sides into an anonymous *trading book*. At any time, the LOB contains multiple layers, or levels, on the bid and ask sides, with each layer corresponding to a different price level, and with consecutive layers separated by the minimum price increment. For most US exchanges, this minimum increment is $0.01 for most stocks. For futures, this can vary with the underlying contract notional value. Market agents have several options when it comes to placing an order to buy or sell financial securities. *Limit order* and *market order* are the two most common order types. A *limit order* is an order to buy or sell a specific amount of shares of a stock at a specific price. When a limit order arrives at the exchange's order management system, it joins the bid or ask order queue at the price level specified by the order. The only change to the LOB that is visible to other market agents is an increase of queue size at that level : no information linked to the order originator is disseminated. A *market order* is an order to buy or sell a stock at the prevailing market price. In the most basic form, a market order to sell 1,000 shares will take out 1,000 lots of liquidity at the top level of the bid side of the order book. If the available liquidity is less than 1,000 at that level, the order will continue to execute at the next level of the bid side order book with a lower price. This continues until 1,000 lots have been filled. The advantage of a market order is that it is almost always guaranteed to be executed. However, there is a positive probability of executing at a significantly worse price than the best bid and ask prices if there is insufficient liquidity to fill the original order at those prices. For this reason, most exchanges offer various features to protect market orders from executing beyond some predetermined price band, or simply from executing beyond the top levels.

In Sections 5.1 and 5.2, we give an introduction to message decoding of order book information broadcast by the exchanges and construction of LOB, together with some stylized facts about LOB data. Section 5.3 introduces a bivariate Hawkes process model to fit LOB data by maximum likelihood and illustrates with actual data. Section 5.4 introduces some machine learning methods for LOB data analytics and Section 5.5 describes queueing models of LOB dynamics and their applications. Supplements and problems are given in Section 5.6.

5.1   From market data to limit order book (LOB)

Book event dissemination of market data

There are two types of order book information disseminated by exchanges: order-based and level-based. Some exchanges provide only level-based data feeds, e.g., CME, whereas others provide both level- and order-based feeds, e.g., Intercontinental Exchange (ICE).

For a level-based book, the exchange broadcasts a snapshot update of the state of the order book upon each market event pertaining to the specific symbol. These events could be a result of the addition (ADD) of an order to the book, a cancellation (CXL) or cancel-replace (CXR) of an existing order or an execution of an order[1]. From these snapshots, it is possible to build an order book that contains information such as the available price levels and sizes at each level. Upon a cancellation of an order, the exchange will send out a removal event message, together with the state of the new level.

From an order-based feed, which provides more information than that in a level-based book, it is possible to build an order book that contains the number of orders and more granular attributes of each individual order in the order book. From this, it is possible to infer the exact queue positions of the orders that belong to a customer. An order-based book is more resource-intensive to build and, depending on the implementation of the book building engine, there is a trade-off between gaining insight to the composition of the limit order book and incurring latency required to build a fine-grained order-based book.

Message decoding and order book construction

Upon receiving an exchange message, the book building engine of a high-frequency trading platform decodes the message, and modifies the internal book state based on the type of the order event received. For example, for ADD events, the engine updates the information at the appropriate price level or, in the case that the level does not yet exist, it creates a new level and inserts the order at the top of the book queue. Efficient implementation of the book construction process is important, because a sub-optimal book engine

---

[1] Note that for some exchanges such as the CME, executions and cancellations are both broadcast as order removal events. The only way to distinguish the two is by inferring from the subsequent execution messages.

can substantially increase the tick-to-trade time of a strategy in view of the large number of book events.

5.2 Stylized facts of LOB data

Figure 5.1 shows a series of snapshots illustrating the evolution of a limit order book spanning one minute of trading. Each slice in this plot represents a snapshot of the state of the order book right before and after a trade has taken place. The top solid dashed line traces the best ask price, and the bottom dashed line traces the best bid. The gap of the two lines is known as the bid-ask spread, and the solid black line in the middle is the market mid price, which is simply the average of the best bid and ask prices. The top half of the order book, just above the mid-price, consists of the sell orders, and just below the mid-price the buy orders, The shaded bars indicate the relative volume at each price level. For ease of visualization, the order queue at each time instance is scaled by the maximum size at that instance, across the whole book; gray represents the state of the book just before a trade has taken place and a lighter shade of gray for just after.

Observe that the shape of the queue profile on both sides of the book varies as the price changes. The profile of the order book signals potential supply-demand imbalance of the market, and changes in this profile convey information related to investors' reaction to price changes.

5.2.1 Book price adjustment

A typical order book price adjustment mechanism consists of the following key elements: market or marketable limit order[2] arrival, execution, cancellation, price improvement, and finally, short-term price equilibrium. For example, typical price movements in the LOB following the arrival of a large sell order can be outlined follows:

- arrival of sell order that takes out one or multiple levels of orders on the bid side of the order book (09:00:22.860),

- cancellation of the bid side order at the same level or from levels below (09:00:40.625),

- widening of the bid-ask spread (09:00:43.625),

---

[2]A marketable limit order is a buy order with a price limit at or above the lowest offer in the market or a sell order with a price limit at or below the highest bid in the market. Thus, a market buy order is a marketable limit buy order with price limit being infinity. Similarly, a market sell order is a marketable limit sell order with price limit being zero. Moreover, a *marketable order* is either a marketable limit order or a market order.

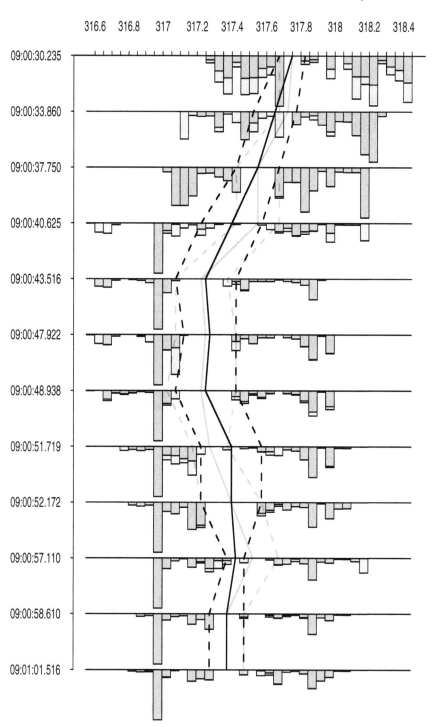

FIGURE 5.1: Snapshots showing the evolution of a ten-level deep limit order book just before a trade has taken place (gray lines) and just after (black lines) for British Petroleum PLC (BP). Dotted lines are for the best bid and ask prices. Solid line is the average or mid price. Bars are scaled by maximum queue size across the whole book and represented in two color tones

- arrival of sell orders, which closes the bid-ask spread (09:00:43.576),

- consolidation of the order book sizes on both sides, with the market reaching a short-term equilibrium (09:00:47.922).

Price movements are often induced by a large order or a series of small orders that consume a significant portion of the book. Once the rapid depletion of one side of the book is disseminated to the market, some market participants may interpret this as the arrival of informed traders and will likely adjust the fair value estimate accordingly, resulting in cancellation of outstanding orders. The combination of order consumption and cancellation leads to an initial widening of the bid-ask spread. This widening of the spread will in turn entice market-making strategies to submit price improving orders in order to capitalize on a wider bid-ask spread. The decision of whether the spread is too wide is a function of price dynamics (such as volatility) and also of the strategy (such as current inventory and utility maximization).

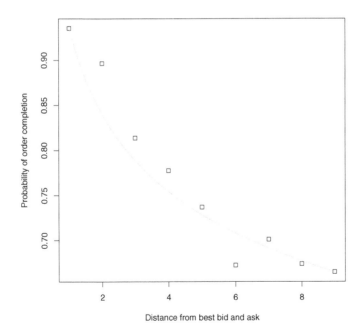

FIGURE 5.2: Probability of order completion within 5 seconds from submission for BP on June 25, 2010. Squares are relative frequencies based on empirical data and the solid curve is based on fitting a power-law function suggested by Bouchaud et al. (2002).

5.2.2   Volume imbalance and other indicators

Imbalance, when defined simply as the difference between total aggregate buy and sell orders, ignores the important fact that orders at different levels of the book have significantly different probabilities of being executed. An appropriately defined buy-sell imbalance measure can help extract information of the likely change in trade direction and intensity over the short run. From this, one can obtain the empirical cumulative distribution of order completion time, and hence deduce the probability of completion within a specific time period for a limit order submitted at a specific number of price increments away from the best bid and ask. Figure 5.2 shows the relative frequency (empirical probability) of a complete limit order fill within five seconds after submission, as a function of order-distance, assuming that the underlying true price process has zero drift–a reasonable assumption given the short time frame. For example, for a limit sell order, a distance of 0 corresponds to a sell order at the prevailing best bid (i.e., an effective market order), and a distance of 1 corresponds to a sell order at a price which is one price increment higher than the best bid (i.e., at the best ask price). The power-law function shown by the solid curve is a nonlinear regression function of the form $y = cx^{-\alpha}$.

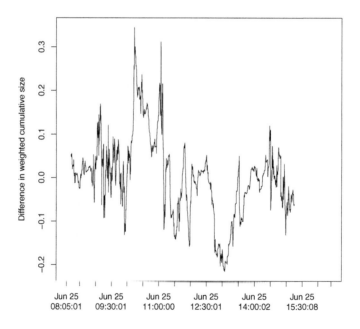

FIGURE 5.3: Time series for the difference $\bar{v}(t, \tau, L; 1) - \bar{v}(t, \tau, L; 2)$ in probability weighted cumulative volume for BP on June 25, 2010.

Probability weighted volume

To quantify market supply and demand, a probability weighted volume is defined as follows. Let $\pi_{l,i,\tau} = P(t_f < t + \tau | l, i)$ be the probability of an order of type $i \in \{1, 2\}$, 1 for bid side and 2 for ask side, submitted at book level l, to be completely filled (with t_f denoting that instant) within τ seconds from order submission at time t. Let $v_{t,l;i}$ be the queue size at time t, on side $i \in \{1, 2\}$ of the limit order book and at the book level l. Then the *probability weighted volume* is defined by

$$\bar{v}(t, \tau, L; i) := \frac{1}{\sum_{i,l} v_{t,l;i}} \sum_{l=0}^{L} v_{t,l;i} \pi_{l,i,\tau}. \tag{5.1}$$

The probability $\pi_{l,i,\tau}$ is typically estimated from the empirical density of order execution at specific levels over a defined window at each side of the LOB. Figure 5.3 shows the time series of the difference, $\bar{v}(t, \tau, L; 1) - \bar{v}(t, \tau, L; 2)$, between the bid and ask sides, of probability weighted cumulative volumes for $t \in \{t_i : i = 0, 1, \ldots, n\}$. This simple indicator gives a snapshot of the imbalance in the limit order book weighted by the probability of execution.

Putting aside the debate concerning EMH, there is uncertainty about the fair price or fair value of an asset when transactions take place at the discrete price levels of an exchange. For example, suppose BAC (Bank of America Corp.) has just been transacted at 15.89, but there still exists uncertainty as to the fair value of BAC when the market quote is 15.88 bid and 15.89 ask, although one might say that the true price might be closer to 15.89 than it is to 15.88. A way to estimate the fair value of an asset is to make use of the information embedded in the LOB as follows.

Inverse size weighted fair value (IWFV) price

Let p_0^B and v_0^B be the level-1 bid price and size, respectively, and similarly for the ask side price and size.[3] The *inverse size weighted fair value price* is defined by

$$\widehat{P}^{(IWFV)} = \frac{p_0^B v_0^A + p_0^A v_0^B}{v_0^B + v_0^A}.$$

The rationale is that the higher the ask side volume, the more is the selling pressure on the asset, hence the higher is the weight of the bid side price.

Threshold liquidity fair value (TLFV) price

For a given threshold θ, the bid and ask side depletion levels L_t^B and L_t^A at time t are defined by

$$L_t^B = \inf\left\{ L : \sum_{l=0}^{L} v_{l,t}^B \geq \theta \right\}, \qquad L_t^A = \inf\left\{ L : \sum_{l=0}^{L} v_{l,t}^A \geq \theta \right\},$$

---

[3]Level-1 data typically display the Best-Bid-Offer (BBO), i.e., the lowest ask and highest bid available at the time.

and the *threshold liquidity fair value price* is defined by

$$
\widehat{P}_t^{\,(\mathrm{TLFV})} = \frac{1}{2\theta} \left[\theta p_{L_t^B,t}^B + \sum_{l=0}^{L_t^B-1} v_{l,t}^B \left(p_{l,t}^B - p_{L_t^B,t}^B \right) \right.
$$

$$
\left. + \theta p_{L_t^A,t}^A + \sum_{l=0}^{L_t^A-1} v_{l,t}^A \left(p_{l,t}^A - p_{L_t^A,t}^A \right) \right].
$$

The economic intuition underlying this definition is related to the average buy and sell prices for θ units of the asset. Unlike $\widehat{P}^{(\mathrm{IWFV})}$ which always lies between the best bid and the best ask prices, $\widehat{P}_t^{\,(\mathrm{TLFV})}$ could be outside of the range of these level-1 prices.

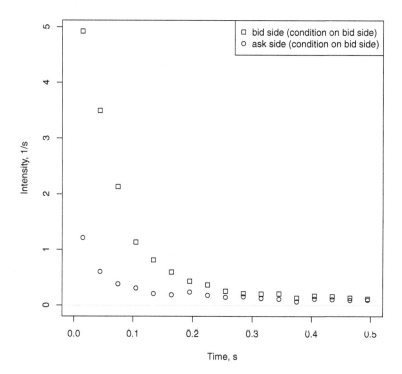

FIGURE 5.4: Conditional intensity of bid and ask side market orders following an order submitted on the bid side of the market, estimated with bin size ranging from 30 to 500 milliseconds, using BP tick data on June 25, 2010.

5.3 Fitting a multivariate point process to LOB data

In Section 4.6.2, we have introduced univariate point process models, and in particular the Hawkes self-exciting point process, to model the intraday trade intensity based on TAQ data. For LOB data of the type in Figure 5.1, we have multiple price levels on the bid and ask sides of the order book. Abergel and Jedidi (2015) introduce a multivariate extension of the Hawkes process to accommodate the multivariate data from different levels of the order book.

Multivariate Hawkes process

A doubly stochastic point process $(N_t^1, \ldots, N_t^K)^\top$ is called a K-*variate Hawkes process with exponential kernel* if the intensity processes satisfy

$$\lambda_t^{(k)} = \mu^{(k)} + \sum_{j=1}^{K} \alpha_{jk} \int_{u<t} e^{-\beta_{jk}(t-u)} \, dN_u^{(j)}, \qquad k = 1, \ldots, K. \tag{5.2}$$

The phrase "doubly stochastic" refers to the fact that both the point process and its intensity are stochastic; see Appendix C for further background on self-exciting multivariate point processes.

5.3.1 Marketable orders as a multivariate point process

The joint intensity model in (5.2) shows how the arrival of a marketable order on the bid or ask side of the order book can self-excite events, both on the same side and on the opposite side of the order book. The choice of exponential decay mimics real-world behavior, as illustrated in Figure 5.4 which shows the empirical conditional intensity given the buy and sell market order arrivals. For some asset classes, e.g., equities on US exchanges, transactions are in *round lots,* [4] which represent the normal size of trading for a security, generally 100 shares or more. Since the trade size is decided by the order originator, it is a potential source of information. Giving more weight to trades with the larger size makes them convey more information than the smaller *noise trades*. The baseline intensity $\mu^{(k)}$ in (5.2) incorporates the mean arrival rate of market orders. When there are more buy limit orders on the bid side of the order book than sell orders on the ask side, the likelihood of an uptick in price increases, and similarly it decreases when this imbalance is reversed. One way to model this in a point process framework is by using the probability weighted volume in (5.1) to scale the baseline intensity $\mu^{(k)}$ of the intensity process. We can

---

[4] *Odd lot* orders are orders with size less than the minimum round lot amount, and these orders have less favorable queue positions and may incur additional clearing fees at the exchanges. See rules posted by the exchanges for a comprehensive treatment of the regulation and requirements related to odd lot execution and other transaction related rules (e.g., Rule 124(c) of the NYSE Rules).

enrich the information set of the intensity process by incorporating these and other exogenous factors.

To illustrate the case for $K = 2$, let w_{1i} and w_{2j} be the trade sizes for the bid and ask market orders at times t_i and t_j, respectively. The intensity process can be modeled by

$$
\lambda_t^{(1)} = \mu^{(1)} \bar{v}_{2t} + \frac{1}{\bar{w}_1} \sum_{t_i < t} \alpha_{11} w_{1i} e^{-\beta_{11}(t - t_i)} + \frac{1}{\bar{w}_2} \sum_{t_j < t} \alpha_{12} w_{2j} e^{-\beta_{12}(t - t_j)},
$$

$$
\lambda_t^{(2)} = \mu^{(2)} \bar{v}_{1t} + \frac{1}{\bar{w}_2} \sum_{t_j < t} \alpha_{22} w_{2j} e^{-\beta_{22}(t - t_j)} + \frac{1}{\bar{w}_1} \sum_{t_i < t} \alpha_{21} w_{1i} e^{-\beta_{21}(t - t_i)},
$$

$$(5.3)$$

where $\bar{w}_1$ (or $\bar{w}_2$) is the average of the trade sizes w_{1i} (or w_{2i}) in the period $[0, t)$ and $\bar{v}_{1t}$ is the probability weighted volume for bid orders and $\bar{v}_{2t}$ is that for ask orders. The log-likelihood function for the intensity process can then be expressed as $\mathcal{L}(\theta_1) + \mathcal{L}(\theta_2)$, where $\theta_1 = (\mu^{(1)}, \alpha_{11}, \alpha_{12}, \beta_{11}, \beta_{12})$, $\theta_2 = (\mu^{(2)}, \alpha_{21}, \alpha_{22}, \beta_{21}, \beta_{22})$, and

$$
\mathcal{L}(\theta_1) = -\mu^{(1)} \sum_{i:t_i < T} \bar{v}_{2t_i} (t_i - t_{i-1}) - \frac{\alpha_{11}}{\beta_{11}} \sum_{i:t_i < T} \frac{w_{1i}}{\bar{w}_1} \left(1 - e^{-\beta_{11}(T - t_i)} \right)
$$

$$
- \frac{\alpha_{12}}{\beta_{12}} \sum_{j:t_j < T} \frac{w_{2j}}{\bar{w}_2} \left(1 - e^{-\beta_{12}(T - t_j)} \right)
$$

$$
+ \sum_{i:t_i < T} \log \left(\mu^{(1)} \bar{v}_{2t_i} + \alpha_{11} R_{11}(i) + \alpha_{12} R_{12}(i) \right), \quad (5.4)
$$

where R_{11} and R_{12} are given by the recursions

$$
R_{11}(i) = \frac{w_{1i}}{\bar{w}_1} e^{-\beta_{11}(t_i - t_{i-1})} [1 + R_{11}(i - 1)],
$$

$$
R_{12}(i) = \frac{w_{2i}}{\bar{w}_2} \left\{ e^{-\beta_{12}(t_i - t_{i-1})} [R_{12}(i - 1)] + \sum_{j':t_{i-1} \le t_{j'} < t_i} e^{-\beta_{12}(t_i - t_{j'})} \right\},
$$

and similarly for $\mathcal{L}(\theta_2)$; see Section C.3 in Appendix C for the background and derivation.

Abergel and Jedidi (2015) use K-dimensional vectors $\boldsymbol{a(t)} = (a_1(t), a_2(t), \ldots, a_K(t))$ and $\boldsymbol{b(t)} = (b_1(t), b_2(t), \ldots, b_K(t))$ to represent the ask and bid sides of the order book at time t, such that $a_i(t) \ge 0$ is the size available on the ask side i ticks away from the best bid, and similarly for negative integers $b_i(t)$ on the bid side. In other words, the model covers a $2K$-dimensional *moving-frame* price grid, with the minimum tick size as the mesh of the grid. The cumulative depth up to level i is given by $A_i(t) = \sum_{k=1}^{i} a_k(t)$, $B_i(t) = \sum_{k=1}^{i} |b_k(t)|$ so that their generalized inverse functions are

$$
A_t^{-1}(x) := \inf \left\{ p : \sum_{k=1}^{p} a_k(t) > x \right\}, \qquad B_t^{-1}(x) := \inf \left\{ p : \sum_{k=1}^{p} |b_k(t)| > x \right\},
$$

and the bid-ask spread is given by $A_t^{-1}(0) - B_t^{-1}(0)$. Abergel and Jedidi (2015, Sect. 5) use these generalized inverse functions to model the dynamics of the best ask and the best bid prices.

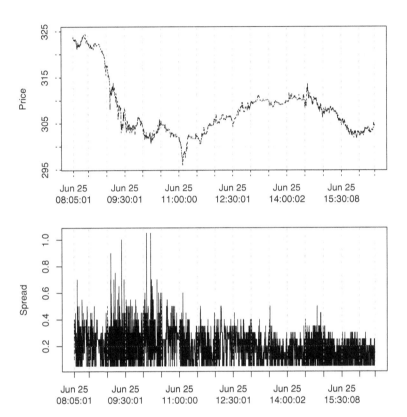

FIGURE 5.5: Time series plot of the best prevailing bid prices of the limit order book (top panel) and that of the bid-ask spread (bottom panel), for BP on June 25, 2010.

5.3.2   Empirical illustration

The data *Rebuild Order Book* (February 2011) dataset obtained from the London Stock Exchange (LSE) provides full market depth intraday order information and trading data, from which we reconstruct the complete order book. The dataset contains order detail, order deletion and trade detail information; records are time-stamped to the nearest millisecond. We use British Petroleum PLC (BP) for June 25, 2010 to illustrate the modeling framework and fitting process. Figure 5.5 shows the time series of bid prices and the bid-ask spread. The computer program that performs the order book reconstruction takes order details from the *order details record* file and then chronologically matches the trade and deletion information in the *order history record* file. The result of this reconstruction procedure is a time series of snapshots of the order book, at every trade event. A more comprehensive treatment of market data, order types, matching engines and LOB reconstruction is given in Chapter 8.

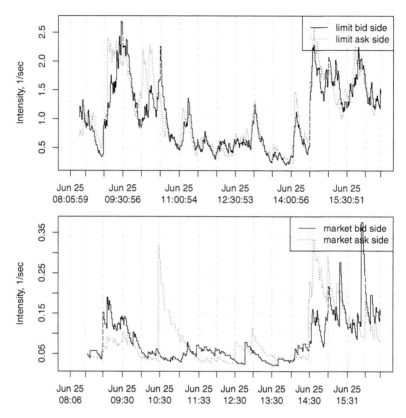

FIGURE 5.6: Arrival intensity of limit order (top panel) and market order (bottom panel) on bid side and ask side of the order book, estimated using overlapping windows of one-minute period, for BP on June 25, 2010.

The sample period is chosen to be the entire trading day on June 25, 2010, from 08:05:00.000 to 16:25:00.000. The first and last five minutes near the opening and closing of the market are discarded, so as to stay clear of the periods where there is often a lot of noise (e.g., incorrectly recorded orders or transactions) in the data; see Figure 5.1 for a 30-second picture of the evolution of the LOB for BP. There are four key order types: limit buy, limit sell, effective market buy and effective market sell. The effective market buy orders include market buy orders and limit orders that are submitted at or through the best ask price; and similarly for the effective market sell orders which include market sell orders and limit orders that are submitted at or through the best bid price. These are orders submitted with the intention to induce immediate fill. For the bid-side order intensity, an inter-arrival time is defined to be the time, measured in milliseconds, between two effective market orders at the bid side, and similarly for the ask side orders. Figure 5.6 shows the empirical intensity of bid and ask side orders. Here the intensity is calculated from the arrival rates of market orders within overlapping one-minute windows. Note the *U-shape* activity profile, which indicates that price discovery is concentrated near the opening and closing periods of the market.

TABLE 5.1: MLEs of parameters in four specifications of (5.3). Model 1: $\bar{v}_{1t} = \bar{v}_{2t} \equiv 1$ and $w_{1i} = w_{2i} \equiv 1$; Model 2: $w_{1i} = w_{2i} \equiv 1$; Model 3: $\bar{v}_{1t} = \bar{v}_{2t} \equiv 1$; Model 4: unrestricted $\bar{v}_{1t}, \bar{v}_{2t}, w_{1i}, w_{2i}$. Standard errors are given in parentheses; $\dagger$ denotes that the value is not significantly different from 0 at the 0.05 level.

	Model 1	Model 2	Model 3	Model 4
$\mu^{(1)}$	0.068	0.193	$0.000^{\dagger}$	$0.000^{\dagger}$
	(0.002)	(0.005)	(0.000)	(0.000)
$\mu^{(2)}$	0.005	0.015	0.005	0.013
	(0.001)	(0.002)	(0.001)	(0.002)
α_{11}	2.726	2.803	2.901	2.900
	(0.114)	(0.116)	(0.129)	(0.129)
α_{22}	2.624	2.631	2.424	2.430
	(0.126)	(0.126)	(0.126)	(0.126)
α_{12}	0.575	0.563	0.004	0.004
	(0.068)	(0.068)	(0.000)	(0.003)
α_{21}	0.002	0.002	0.001	0.001
	(0.000)	(0.000)	(0.000)	(0.000)
β_{11}	5.211	5.418	6.740	6.740
	(0.217)	(0.224)	(0.295)	(0.295)
β_{22}	6.990	7.016	7.659	7.684
	(0.347)	(0.349)	(0.400)	(0.400)
β_{12}	8.422	8.204	0.007	0.007
	(1.076)	(1.072)	(0.001)	(0.001)
β_{21}	0.004	0.004	0.002	0.002
	(0.001)	(0.000)	(0.000)	(0.000)
$\sup_\theta l(\theta)$	$-16,319$	$-16,208$	$-18,123$	$-18,120$

Maximum likelihood estimation

We fit the bivariate point process model (5.3) to the data, using four model specifications that are described in Table 5.1, which also gives the values of the MLEs of the parameters in each fitted model. Models 1 and 2 have substantially better fit than Models 3 and 4, based on the maximized log likelihood. For Model 1 or 2, $\hat{\mu}^{(1)}$ is substantially larger than $\hat{\mu}^{(2)}$. Moreover, the self-excitation parameters α_{11} and α_{22} are markedly larger than their cross-excitation counterparts α_{12} and α_{21}, which suggests that although submitted orders on both sides of the LOB would induce an overall increase in trading activity, they are more likely to induce more orders of the same type. The cross-excitation decay rate β_{12} is also markedly higher than β_{21}, which suggests that market orders traded on the ask side are more likely to induce orders traded on the bid side than on the ask side.

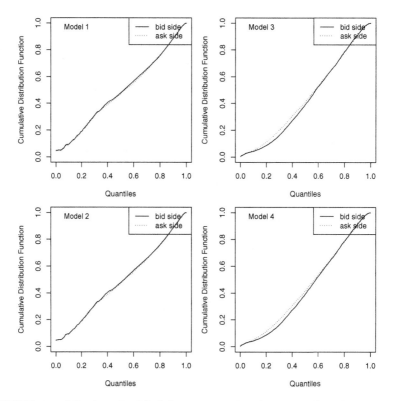

FIGURE 5.7: QQ-plots for Models 1, 2, 3, 4 fitted to the BP order book data on June 25, 2010.

QQ-plots and goodness of fit

To assess the goodness of fit of the proposed models, QQ-plots (Lai and Xing, 2008, p. 18) are used to visualize the relationship between empirical quantiles of the data and the theoretical quantiles under each assumed model. For a point process with intensity function λ_t, the *time-changed* sequence $\Lambda_{t_i} = \int_0^{t_i} \lambda_u \, du$, $i = 1, 2, \ldots$, is a Poisson point process, and therefore $\Lambda_{t_i} - \Lambda_{t_{i-1}}$

are i.i.d. exponential with mean 1. Figure 5.7 shows the QQ-plots for the four models fitted to the data on June 25, 2010. The plots show that including the order size information in Models 3 and 4 does not help improve fit, which could be a attributed to the presence of *slicing algorithms* that many trading systems adopt. These algorithms break large market orders into sizes comparable to the median trade size in order to minimize the signaling impact.

5.4   LOB data analytics via machine learning

Fletcher et al. (2010) discuss the usefulness of machine learning methods (in particular, support vector machines and multiple kernel learning) to analyze the high-dimensional data (from different levels of the LOB) at every order book update following a market event for making currency trading decisions. Zheng et al. (2013) give another application of machine learning methods (in particular, Lasso logistic regression) to predict price jumps in 40 largest French stocks in CAC40, a benchmark French stock market index. Kearns and Nevmyvaka (2013), abbreviated by KN in this section, give an overview of machine learning methods for LOB data analytics, saying :

> The inference of predictive models from historical data is obviously not new in quantitative finance: ubiquitous examples include coefficient estimation for the CAPM, Fama and French factors, and related approaches. The special challenges for machine learning presented by HFT generally arise from the very fine granularity of the data — often microstructure data at the resolution of individual orders, (partial) execution, hidden liquidity, and cancellations — and a lack of understanding of how such low-level data relates to actionable circumstances (such as profitably buying or selling shares, optimally executing a large order, etc.). In the language of machine learning, whereas models such as CAPM and its variants already prescribe what the relevant variables or "features" are for prediction or modeling (excess returns, book-to-market ratios, etc.), in many HFT problems one may have no prior intuitions about how (say) the distribution of liquidity in the order book relates to future price movements, if at all. Thus *feature selection* or *feature engineering* becomes an important process in machine learning for HFT, and is one of our central themes.

Section 4 of KN describes applications of machine learning to the problem of predicting near-term directional price movements from LOB data. In a case study of this problem, KN carries out data analytics to determine when (i.e., under what conditions in a given state space) and how (i.e., in which direction and with what orders) to trade. In particular, it mentions two main conceptual components of the analytics towards the "alpha generation purposes":

- the development of features that permit the reliable *prediction* of directional price movements from certain states (of the LOB), in which "reliable" means "(good) enough that profitable trades outweigh unprofitable ones" rather than being correct most of the time, and

- the development of learning algorithms for execution to capture this predictability or alpha at sufficiently low trading costs.

KN concludes from this case study and another one on machine learning for smart order routing (which we will discuss in Sections 7.5–7.7) that machine learning provides a powerful, scalable, and principled framework for data analytics and forecasting in high-frequency trading, but "no easy paths to profitability." As already noted in Section 2.8 and Chapter 3, profitable quantitative strategies involve not only efficient information collection and analysis for good prediction, but also the development of dynamic strategies that optimize prescribed performance criteria in the presence of uncertainty, as exemplified by reinforcement learning and AlphaGo in Supplement 7 of Section 3.4.

Feature engineering and machine learning

Feature engineering is an important technique to build prediction models using nonparametric regression methods. When time is taken into consideration, these methods are applied to moving windows of time. Machine learning is a subfield of computer science that evolved from computational learning theory and from pattern recognition and reinforcement learning in artificial intelligence. It overlaps with statistical learning in regression and classification (see Supplement 9 of Section 2.9). In the context of LOB data, classification aims at predicting the direction of price change, while regression aims at predicting the amount of price change and includes linear, logistic and nonlinear regression models together with forward/backward stepwise methods for variable selection described in Lai and Xing (2008, Chap. 1 and 4). Nonparametric regression (Chapter 7 of Lai and Xing, 2008) is mostly linear regression applied to a large number of basis functions. In machine learning, one has a large number p of features (predictors) and the issue is how to handle the case of p having the same or larger order of magnitude than n. Forward/backward stepwise feature selection can be used in conjunction with an appropriate selection criterion to choose a subset of features to build the model. Another solution is to add regularization terms to the loss function, and the regularization term serves as a constraint on the coefficients so that overfitting is prevented.

Linear and generalized linear regression and their nonlinear extensions (Lai and Xing, 2008, Chapters 1 and 4) are often used to build predictive models in machine learning and feature engineering. Another large class of predictive models consists of the *decision trees*. A decision tree is a tree whose branches are conditions of a certain feature (for example, $X_i > C$). To build such a tree, we start from the root node; if the condition is satisfied, go to the left node, otherwise go to the right node. The leaf nodes are the predictions of the sample. The fitted decision tree model usually has high variance. One way to reduce variance is to build many decision tree models using weighted training samples, where the weights are determined by the fitting results of previous trees. These trees then undergo a test phase that combines the trees via a boosting algorithm; see Chapter 10 of Hastie et al. (2009) and in particular the AdaBoost algorithm. Another popular decision tree method is the *random forest*, which trains many decision trees using random subsets of data and predicts with the average of all the trees. For the case of large p, both boosting decision trees and random forest provide feature importance scores, which can be used in feature selection.

Support Vector Machine (SVM) is another widely used machine learning algorithm for classification, and its main idea is to separate two classes by picking a hyperplane that maximizes the margin between the training points of the two classes. SVM can also be extended to nonlinear hypersurfaces using a kernel function to replace the inner product in SVM. This kernel method is faster in implementation than high-dimensional basis expansion; see Chapter 12 of Hastie et al. (2009). Finally, another popular choice of basis functions consists of *neural networks*, described in Hastie et al. (2009) and Lai and Xing (2008). An integrated approach to regression trees and neural networks is given by Lai et al. (2016b) and Deng et al. (2017). The latter also describes other applications of these advances of statistical/machine learning to financial technology.

5.5 Queueing models of LOB dynamics

Clearly the order book can be regarded as a queueing system, in which servers are from opposite sides of the multi-level bid and ask queues. Restricted to level-1 data, queueing models have been modified to model LOB dynamics and derive formulas for calibrating certain probabilities that predict future movements of the LOB. To begin with, it is natural to draw connections between LOBs and multi-class priority queues, especially between LOBs with cancellations and queues with reneging, and between market orders and services in priority queues in reducing queue lengths. Note also that similar to requests for service, limit orders increase queue lengths. Modeling LOB dynamics in a queuing framework can carry over the powerful diffusion and fluid limits for queue lengths to analyze the order book prices and waiting times. In Supplement 3 of Section 5.6, we describe Kruk's (2003) diffusion and fluid limits in an auction setting and his asymptotic theory that after rescaling, the lengths of the best bid and ask queues converge to a reflected two-dimensional Brownian motion in the first quadrant. This work paved the way for subsequent developments by Cont and de Larrard (2013) and Guo et al. (2015) for reduced-form queueing models of LOB level-1 data. An alternative model has been introduced by Huang et al. (2015) using additional levels of the LOB data, which will be described after we present the models of Guo et al. and of Cont and de Larrard.

5.5.1 Diffusion limits of the level-1 LOB reduced-form model

Motivated by empirical evidence (e.g., Biais et al., 1995) that the major component of the order flow occurs at the best bid and ask price levels, Cont and de Larrard (2013) propose a *reduced-form* model in which the state of the limit order book is represented by s_t^b, q_t^b, for the bid side price and queue size respectively, and similarly by s_t^a and q_t^a for the ask side. To simplify the model, they further assume that

(a) the bid-ask spread is a constant, say δ, and the bid and ask prices are multiples of δ;

(b) when the best bid queue is depleted, both the best bid price and the best ask price will move down one tick;

(c) when the best ask queue is depleted, the bid and ask prices will both move up one tick;

(d) once the best bid (respectively, ask) queue is depleted, a new best bid (respectively, ask) queue size is sampled from a density function $\tilde{f}(\cdot, y)$ (respectively, $f(x, \cdot)$) that depends on the queue history only through the current value y of the best ask queue size (respectively, x of the best bid queue size);

(e) limit orders, market orders, and cancellations are all of size 1 and arrive as independent Poisson processes with intensities λ, μ, and θ, respectively.

Assumption (d), called random re-initialization of queue sizes upon depletion, provides the key specification of the continuous-time Markov chain (s_t^b, q_t^b, q_t^a) on the discrete state space $\delta\mathbb{Z} \times \mathbb{N}^2$ via the sequence of depletion times $\tau_i = \tau_i^a \wedge \tau_i^b$, where $\tau_i^a = \inf\{t > \tau_{i-1} : q_t^a = 0\} - \tau_{i-1}$ and $\tau_i^b = \inf\{t > \tau_{i-1} : q_t^b = 0\} - \tau_{i-1}$. Under assumption (e), τ_i are i.i.d. Moreover, because of the renewal property of the model, the analysis of consecutive price movements can be reduced to the study of the first movement. Thus, for the ease of discussion, (τ_1^a, τ_1^b) and τ_1 are abbreviated as (τ_a, τ_b) and τ, respectively.

Various formulas arise from the analysis of queues via birth and death chains, the independence of (τ_a, q_0^a) and (τ_b, q_0^b) and certain symmetries between the bid side and the ask side in steady state. For example, by symmetry,

$$P(\tau_a > \tau_b | q_0^b = x, q_0^a = x) = P(\tau_b > \tau_a | q_0^b = x, q_0^a = x) = 1/2, \qquad (5.5)$$

which yields

$$
\begin{aligned}
P(\tau > t | q_0^b = x, q_0^a = x) &= P(\tau_a > t | q_0^b = x, q_0^a = x)/2 \\
&\quad + P(\tau_b > t | q_0^b = x, q_0^a = x)/2 \\
&= P(\tau_a > t | q_0^b = x, q_0^a = x) \\
&= P(\tau_b > t | q_0^b = x, q_0^a = x).
\end{aligned}
\tag{5.6}
$$

The first equality in (5.6) follows from (5.5), while $P(\tau_a > t | q_0^b = x, q_0^a = x) = P(\tau_a > t | q_0^a = x)$ follows from the independence between the bid and the ask queues. Thus, (5.6) can be simplified to

$$
P(\tau > t | q_0^b = x, q_0^a = x) = P(\tau_b > t | q_0^b = x) = P(\tau_a > t | q_0^a = x).
$$

Moreover, using the associated birth and death chain, an explicit formula is available for the Laplace transform $\mathcal{L}(s,x) = E(e^{-s\tau_b} | q_0^b = x)$ of the first passage time of state 0. Inverting the Laplace transform with this explicit formula then leads to the analytic formula (5.7); see Cont and de Larrard (2013, p. 9–10).

Conditional distribution of depletion time τ given (q_0^a, q_0^b)

$$
P[\tau > t | q_0^b = x, q_0^a = y] = \sqrt{\left(\frac{\mu+\theta}{\lambda}\right)^{x+y}} \, \xi_{x,\lambda,\theta+\mu}(t) \xi_{y,\lambda,\theta+\mu}(t),
\tag{5.7}
$$

where for any positive integer ν,

$$
\xi_{\nu,\lambda,\theta+\mu}(t) = \int_t^\infty \frac{\nu}{u} I_\nu\left(2\sqrt{\lambda(\theta+\mu)u}\right) \exp\left[-u(\lambda+\theta+\mu)\right] \, du,
$$

and I_ν is the modified Bessel function of the first kind, defined by

$$
I_\nu(x) = \sum_{m=0}^\infty \frac{1}{m!\Gamma(\nu+m+1)} \left(\frac{x}{2}\right)^{2m+\nu}.
$$

In the special case $\lambda = \mu + \theta$, which means in view of assumption (e) that the volume of limit orders is on average the same as that of market orders and cancellations combined, the LOB has *balanced order flow*, which is reminiscent of *heavy traffic* in classical queuing systems with service rate equal to the request inter-arrival rate.

Probability $\pi^{\mathrm{up}}(x,y)$ of price increase under $\lambda = \mu + \theta$

Conditional on having x orders on the bid side and y orders on the ask side, the probability that the next price move is an increase is

$$
\pi^{\mathrm{up}}(x,y) = \frac{1}{\pi} \int_0^\pi (2 - \cos(t) - \sqrt{(2-\cos(t))^2 - 1})^y \frac{\sin(xt)\cos\left(\frac{t}{2}\right)}{\sin\left(\frac{t}{2}\right)} \, dt. \tag{5.8}
$$

Assumption (e) on Poisson arrivals yielding i.i.d. τ_i can be relaxed (Cont and de Larrard, 2012, Sect. 3.2 and 3.3), but the corresponding formula is considerably more complicated, especially for the integrand. The derivation of (5.8) uses analytic formulas for the probability that a symmetric random walk (under the assumption of $\lambda = \mu + \theta$), starting in the upper quadrant of the plane, hits the x-axis before the y-axis.

Cumulative price change process s_t and its diffusion limit

Instead of the bid side price s_t^b, which also yields $s_t^a = s_t^b + \delta$ on the ask side by Assumption (a), we define the cumulative price change process s_t via the total number of price changes up to time t,

$$N_t = \max\{n \geq 1 : \tau_1 + \cdots + \tau_n \leq t\}, \qquad s_t = \sum_{i=1}^{N_t} X_i, \qquad (5.9)$$

where $X_1, X_2, \ldots$ denote the successive changes in prices, occurring at times τ_i, with $X_i = \delta$ if $q_{\tau_i^a} = 0$, and $X_i = -\delta$ if $q_{\tau_i^b} = 0$. Under the additional assumptions $\lambda = \mu + \theta$, $\tilde{f}(x,y) = f(y,x)$, and

(f) $f(x,y) = f(y,x)$ for all $(x,y), \in \mathbb{N}^2$,

(g) $\sum_{x=1}^{\infty} \sum_{y=1}^{\infty} xy f(x,y) < \infty$,

Cont and de Larrard (2013, Lemma 1 and Theorem 1) denote the infinite series in (g) by $D(f)$ and prove that as $n \to \infty$, $P(X_i = \delta) = P(X_i = -\delta) = 1/2$,

$$\frac{(\tau_1 + \cdots + \tau_n)}{n \log n} \xrightarrow{P} \frac{D(f)}{\pi \lambda}, \qquad (5.10)$$

$$\left\{ \frac{S_{\lfloor tn \log n \rfloor}}{\sqrt{n}}, t \geq 0 \right\} \text{ converges weakly to } \left\{ \delta \left(\frac{\pi \lambda}{D(f)} \right)^{1/2} B_t, t \geq 0 \right\}, \quad (5.11)$$

where B_t is standard Brownian motion. The weak convergence result (5.11) suggests that the order flow affects price volatility through the factor $1/\sqrt{D(f)}$, which is consistent with the previous work of Rosu (2009) using an equilibrium argument based on that "$\sqrt{D(f)}$, quoted in number of shares, represents an average of the size of the bid and ask queues after a price change."

To prove (5.10), Cont and de Larrard (2013) first make use of (5.7) to conclude that as $t \to \infty$,

$$P[\tau > t | q_0^a = x, q_0^b = y] \sim (xy/\pi\lambda) \, t^{-1}; \qquad (5.12)$$

see their equation (6). Multiplying both sides of (5.12) by $f(x,y)$ and summing over x, y gives $P(\tau > t) \sim D(f)/(\pi\lambda t)$ as $t \to \infty$. Making use of this, their Lemma 1 shows that the Laplace transform $\mathcal{L}(s) = E e^{-s\tau} = 1 - s \int_0^{\infty} e^{-st} P(\tau > t) \, dt$ satisfies

$$\lim_{n \to \infty} \left[\mathcal{L} \left(\frac{s}{n \log n} \right) \right]^n = \exp \left(\frac{-D(f)}{\pi \lambda} \right)$$

for every s. Since $[\mathcal{L}(s/(n\log n))]^n$ is the Laplace transform of $(\tau_1 + \ldots + \tau_n)/(n\log n)$, (5.10) follows. Alternatively, (5.10) also follows from the limiting stable distribution of the normalized sum $(\sum_{i=1}^{n}\tau_i - b_n)/a_n$ of i.i.d. random variables because τ is the domain of attraction of a stable law with index $\alpha = 1$, i.e., $P(\tau > t) \sim \text{constant} \times t^{-1}$; see Durrett (2010). The weak convergence result (5.11) is a consequence of (5.10) and Donsker's functional central limit theorem; see Appendix D.

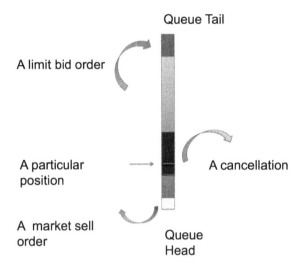

FIGURE 5.8: Orders in the best bid queue.

5.5.2   Fluid limit of order positions

Despite the intuitive appeal to connect LOB dynamics to multi-class queuing systems for which a rich toolbox of analytical techniques has been developed, there are fundamental differences between them. In particular, classical queuing theory tends to focus more on the stability of the entire system. In an LOB, however, it is important to analyze individual requests together with the overall state of the LOB. Indeed, because of the price-time priority (i.e., best-priced order first and First-In-First-Out) in most exchanges in accordance with regulatory guidelines, a better order position means less waiting time and higher probability of the order being executed. In fact, reducing low latency in trading and obtaining good order positions is one of the driving forces behind the technological arms race for high-frequency trading firms. Recent empirical studies by Moallemi and Yuan (2015) showed that values of order positions, if appropriately defined, have the same order of magnitude of half a spread. Indeed, analyzing order positions is one of the key components of algorithmic trading strategies. It is an important indicator of the likelihood of a limit order being executed. As shown in Figure 5.8, the position of an order in the best bid queue is affected by both the market orders and the cancellations, and its relative position in the queue is affected by limit orders as well.

Guo et al. (2015) therefore suggest modeling the dynamics of order positions together with the related queues. There are six types of orders in the best bid and ask queues: best bid orders (which we label as 1), market orders at the best bid (labeled 2), cancellation at the best bid (labeled 3), best ask (labeled 4), market orders at the best ask (labeled 5), and cancellation at the best ask (labeled 6). This basically amounts to using $(V_i^1, \ldots, V_i^6)$ to replace the bivariate vector (V_i^a, V_i^b) in the treatment of assumption (e) in Cont and de Larrard's page 7, where they define V_i^a (respectively, V_i^b) as the change in the ask (respectively, bid) queue size associated with the ith order on the ask (respectively, bid) side, yielding

$$P(V_i^a = 1) = P(V_i^b = 1) = \frac{\lambda}{\lambda + \mu + \theta},$$

$$P(V_i^a = -1) = P(V_i^b = -1) = \frac{\mu + \theta}{\lambda + \mu + \theta}.$$

Guo et al. likewise define V_i^j as the change in type-j ($j = 1, \ldots, 6$) queue size caused by the ith order. Note that only one component of $\boldsymbol{V}_i = (V_i^1, \ldots, V_i^6)$ is positive and all other components are 0. Letting T_i be the inter-arrival time between the ith and the $(i-1)$th orders, define

$$M_t = \max\left\{m : \sum_{i=1}^{m} T_i \leq t\right\}. \tag{5.13}$$

Whereas (5.9) is the renewal process of the inter-depletion durations τ_i, (5.13) is the renewal process of the inter-arrival times T_i.

Define the scaled net order flow process $\boldsymbol{C}_n(t) = [C_n^1(t), \ldots, C_n^6(t)]$ by

$$\boldsymbol{C}_n(t) = \frac{1}{n} \sum_{i=1}^{M_{[nt]}} \boldsymbol{V}_i, \tag{5.14}$$

and define the scaled order position $Z_n(t)$, the scaled queue length $Q_n^b(t)$ for the best bid queue, and $Q_n^a(t)$ for the best ask queue by

$$Q_n^b(t) = Q_n^b(0) + C_n^1(t) - C_n^2(t) - C_n^3(t),$$

$$Q_n^a(t) = Q_n^a(0) + C_n^4(t) - C_n^5(t) - C_n^6(t), \tag{5.15}$$

$$dZ_n(t) = -dC_n^2(t) - \frac{Z_n(t-)}{Q_n^b(t-)} dC_n^3(t).$$

As pointed out by Guo et al. (2015), it is straightforward to understand the definitions in (5.15) from the following considerations:

Bid/ask queue lengths increase with limit orders and decrease with market orders and cancellations according to their corresponding order processes; an order position will decrease and move towards zero with arrivals of cancellations and market orders; new limit orders arrivals will not change this particular order position. However, the arrival of limit orders may change the speed of the order position approaching zero (under the assumption that cancellations are uniformly distributed on every queue), hence the factor of $Z_n(t-)/Q_n^b(t-)$ (attached to $-dC_n^3(t)$ for order cancellation).

Fluid limits of scaled net order flow and bid-ask queues

The scaling in (5.15) follows from the law of large numbers for the inter-arrival times T_i and the changes V_i^j of queue size of type $j \in \{1, \dots, 6\}$, for which they assume:

(A1) $(T_1 + \cdots + T_n)/n \xrightarrow{P} \lambda^{-1}$;

(A2) $(V_1 + \cdots + V_n)/n \xrightarrow{P} V_*$ for some non-random vector V_*.

In fact, if the V_i are stationary ergodic, then $V_* = E(V_1)$. Moreover, since $M_{[nt]}/n \xrightarrow{P} \lambda t$ by (A1) and (5.14), combining (A2) with (5.15) yields $C_n(t) \xrightarrow{P} \lambda V_* t$ for every fixed t. This can be strengthened to the *functional weak law of large numbers*: As $n \to \infty$,

$$C_n(\cdot) \to \lambda V_* e(\cdot) \qquad \text{in } D^6[0, T], \tag{5.16}$$

where $e(\cdot)$ is the identity function. A more detailed discussion of the background and the assumptions underlying (5.16) is given in Supplement 4 of Section 5.6, which also provides some background queueing theory and the main argument to derive the fluid limits of (5.15). For simplicity of presentation they also assume:

(A3) Cancellations are uniformly distributed on every queue,

which they argue can be relaxed to include the possibility that the cancellation rate increases with queue length.[5]

---

[5] In practice, there are various motivations behind order cancellations. It has been reported that cancellations account for the majority of all orders in a liquid market (e.g., Cont, 2011). In fact, analyzing empirically the cancellation characteristics is important for both regulatory and trading purposes. Here is an extreme case reported by CNBC news that "a single mysterious computer program that placed orders — and then subsequently canceled them — made up 4 percent of all quote traffic in the US stock market for the week (of October 5, 2012). . . . The program placed orders in 25-millisecond bursts involving about 500 stocks and the algorithm never executed a single trade."

Fluid limits of (5.15) *Suppose there exist constants q^b, q^a, and z such that*
$(Q_n^b(0), Q_n^a(0), Z_n(0)) \Rightarrow (q^b, q^a, z)$. *Define*

$$v^b = -V_*^1 + V_*^2 + V_*^3, \qquad v^a = -V_*^4 + V_*^5 + V_*^6,$$

$$\alpha = \lambda V_*^2, \quad \beta = \frac{q^b}{\lambda V_*^3}, \quad c = -\frac{v^b}{V_*^3},$$

Then, as $n \to \infty$,

$$(Q_n^b, Q_n^a, Z_n) \Rightarrow (Q^b, Q^a, Z) \qquad \text{in } D^3[0, T], \qquad (5.17)$$

where $Q^b(\cdot)$, $Q^a(\cdot)$, and $Z(\cdot)$ are given by

$$Q^b(t) = q^b - \lambda v^b \min(t, \tau), \qquad Q^a(t) = q^a - \lambda v^a \min(t, \tau),$$

$$\frac{dZ(t)}{dt} = -\lambda \left(V_*^2 + V_*^3 \frac{Z(t-)}{Q^b(t-)} \right) \qquad \text{for } t < \tau, \ Z(0) = z. \qquad (5.18)$$

with

$$\tau = \min \left\{ \frac{q^a}{\lambda v^a}, \frac{q^b}{\lambda v^b}, \tau^z \right\},$$

$$\tau^z = \begin{cases} \left(\frac{(1+c)z}{\alpha} + \beta \right)^{c/(c+1)} \beta^{1/(c+1)} c^{-1} - \beta/c & c \notin \{-1, 0\}, \\ \beta(1 - e^{-\frac{z}{\alpha\beta}}) & c = -1, \\ \beta \log \left(\frac{z}{\alpha\beta} + 1 \right) & c = 0. \end{cases}$$

5.5.3   LOB-based queue-reactive model

The queueing model of Cont and de Larrard (2013) only uses level-1 data of
the LOB and makes a very strong *a priori* assumption (e) on the arrival of
limit/market/cancellation orders. We now describe a more data-driven model
introduced by Huang et al. (2015); this model uses a more flexible description
of a queueing system as a continuous-time Markov chain, some basic facts of
which are given in Appendix B. Let $\boldsymbol{q}_t = (q_t^{(-K)}, \ldots, q_t^{(-1)}, q_t^{(1)}, \ldots, q_t^{(K)}) \in$
$\mathbb{N}^{2K}$, with $q_t^{(-j)}$ and $q_t^{(j)}$ being the queue sizes at the jth best bid and jth best
ask prices, respectively, for $j = 1, \ldots, K$. It is assumed that $\boldsymbol{q}_t$ is a continuous-
time Markov jump process whose generator matrix $\boldsymbol{Q} = (Q_{\boldsymbol{x},\boldsymbol{y}})_{\boldsymbol{x},\boldsymbol{y} \in \mathbb{N}^{2K}}$ is
given by

$$Q_{\boldsymbol{x},\boldsymbol{y}} = \begin{cases} f_i(\boldsymbol{x}) & \text{if } \boldsymbol{y} = \boldsymbol{x} + \boldsymbol{e}_i, \\ g_i(\boldsymbol{x}) & \text{if } \boldsymbol{y} = \boldsymbol{x} - \boldsymbol{e}_i, \\ -f_i(\boldsymbol{x}) - g_i(\boldsymbol{x}) & \text{if } \boldsymbol{y} = \boldsymbol{x}, \\ 0 & \text{otherwise,} \end{cases} \qquad (5.19)$$

where $e_i \in \mathbb{N}^{2K}$ is the unit vector in which the ith entry is 1 $(i = \pm 1, \ldots, \pm K)$ and other entries are 0, and f_i and g_i satisfy the following conditions:

(i) There exists $\delta > 0$ such that $f_i(x) - g_i(x) < -\delta$ for all $i = \pm 1, \ldots, \pm K$ and all x whose entries all exceed some lower bound l.

(ii) There exists $H > 0$ such that $\sum_{i=-K}^{K} f_i \leq H$.

Assumptions (i) and (ii) ensure that the embedded chain has a stationary distribution and is ergodic so that the law of large numbers still applies; see Meyn and Tweedie (1993) and Appendix B of Lai and Xing (2008). In particular, assumption (ii) ensures that the order arrival speeds stay bounded and the system does not explode. A tractable model that captures the dependence of queue sizes in steady state (i.e., under the stationary distribution) among different price levels created by the price-time priority rule is developed via further specification for the functions $f_i(q)$ and $g_i(q)$.

Specification via intensities at individual queues

Important insights are provided by the simple (but unrealistic) case in which the $2K$ sequences of arrivals of limit (L), market (M), and cancellation (C) orders are independent so that

$$f_i(q) = \lambda_i^L(q^{(i)}) \quad \text{and} \quad g_i(q) = \lambda_i^M(q^{(i)}) + \lambda_i^C(q^{(i)}), \qquad (5.20)$$

where $\lambda_i^L(q^{(i)})$, $\lambda_i^M(q^{(i)})$, and $\lambda_i^C(q^{(i)})$ are the intensities of the arrivals of these orders for the ith queue, under the symmetry assumptions $\lambda_i^L = \lambda_{-i}^L$, $\lambda_i^M = \lambda_{-i}^M$, and $\lambda_i^C = \lambda_{-i}^C$.

To introduce dependence among these $2K$ queues created by the price-time priority rule, Huang et al. (2015) consider the following simple modification of $g_i(q)$ in (5.20):

$$g_i(q) = \begin{cases} \lambda_i^C(q^{(i)}) + \lambda_{\text{buy}}^M(q)I_{\text{bestask}(q)=i} & \text{if } i \geq 1, \\ \lambda_i^C(q^{(i)}) + \lambda_{\text{sell}}^M(q)I_{\text{bestbid}(q)=i} & \text{if } i \leq -1, \end{cases} \qquad (5.21)$$

where $\text{bestask}(q) = \min\{i > 0 : q^{(i)} > 0\}$, $\text{bestbid}(q)$ is defined similarly, and $\lambda_{\text{buy}}^M(q)$ and $\lambda_{\text{sell}}^M(q)$ are the steady-state intensities of the arrival of market buy and market sell orders, respectively. They point out that the model is a special case of quasi-birth-and-death (QBD) processes. A QBD process is a generalization of a birth-and-death process that moves up and down to the nearest neighbor one step at a time, but the time between these transitions has a more complicated distribution than classical birth-and-death processes; see Motyer and Taylor (2006) and Latouche (2010). The stationary distribution of a QBD process can be computed by using the matrix geometric method. This method is used in Sections 2.4.2, 2.4.3, and Appendix A.0.3 of Huang et al. (2015), who use the law of large numbers (via the ergodicity assumptions) to substitute the parameters of the QBD by the corresponding sample values and illustrate with data from the France Telecom stock.

Queue-reactive model with mean-reverting reference price

Sections 2.4.4 and 2.4.5 of Huang et al. (2015) describe another way to modify (5.20) at $i = \pm 1$ to induce dependence among the queues. Their empirical studies of (5.20) and its modifications suggest a new model, which they call *queue-reactive*, that also includes a mean-reverting reference price p_t^{ref} so that (5.20) holds during periods when p_t^{ref} does not change, but (5.20) is modified when p_t^{ref} undergoes changes, which can be triggered by one of the following events that modify the mid-price:

- The insertion of a buy limit order within the bid-ask spread while $q_t^{(1)} = 0$ at the time of this insertion, or the insertion of a sell limit order within the bid-ask spread while $q_t^{(-1)} = 0$ at the time of this insertion.

- A cancellation of the last limit order at one of the best offer queues.

- A market order that consumes the last limit order at one of the best offer queues.

Since the model is to be used for computing "many useful short term predictions" including "execution probabilities of passive orders, probabilities of price increase, ..." and for simulating performance of "complex trading tactics", Huang et al. prescribe estimable dynamics for p_t^{ref} and the afore-mentioned events that trigger its changes. Specifically, letting δ denote the tick size, they assume that p_t^{ref} changes with probability θ, and that when it changes at time t, whether $q_t^{(1)}$ or $q_t^{(-1)}$ is 0 (which corresponds to queue depletion in Section 5.5.1) is determined by whether $p_t^{\mathrm{ref}} - p_{t-}^{\mathrm{ref}}$ is δ or $-\delta$. They say:

> The value of $q_t^{(i)}$ switches immediately to the value of one of its neighbors (right if p_t^{ref} increases, left if it decreases) ... we keep records of the LOB up to the third limit ($i = \pm 3$) .... Note that the queue switching process must be handled very carefully: the average event sizes are not the same for different queues. So, when $q_{t-}^{(i)}$ becomes $q_t^{(j)}$, its new value should be renormalized by the ratio between the two average event sizes at price levels i and j (of the LOB).

They point out that the basic idea underlying the queue-reactive model is:

> The market does not evolve like a closed physical system, where the only source of randomness would be the endogenous interactions between participants. It is also subject to external information, such as news, which increases the volatility of the price. Hence, it will be necessary to introduce an exogenous component within the queue-reactive model.

To incorporate this exogenous component ("external information"), they assume that with probability $\tilde{\theta}$, the LOB state is drawn from its invariant distribution around the new reference price when p_t^{ref} changes. Section 3.1 of Huang et al. (2015) uses the France Telecom data to illustrate how θ and $\tilde{\theta}$ can be calibrated from the series of 10-minute standard deviations of the mid-price returns. The calibration uses the *mean reversion ratio* $\eta = N_c/(2N_a)$, where N_c (or N_a) is the number of consecutive p_t^{ref} moves of the same direction (or of the opposite direction) within the interval of interest. Including p_t^{ref} in the queue-reactive model means that the state of the Markov chain is now augmented to $(q_t, p_t^{\mathrm{ref}})$. Since p_t^{ref} plays such an important role in the queue-reactive model, Section 2.2.2 of Huang et al. (2015) gives the definition of p_t^{ref} used in the paper: if the bid-ask spread is an odd number, p_t^{ref} is defined as the mid-price; if the spread is an even number, p_t^{ref} equals mid-price $\pm$ half of the tick size, choosing the sign to make the reference price closest to p_{t-}^{ref}. Huang et al. (2015) also give references to other definitions, which are considerably more complicated, of the reference price.

5.6 Supplements and problems

1. Figure 5.9 shows one sample path of the intensity process λ_t simulated from a univariate Hawkes process with intensity

$$\lambda_t = \mu + \sum_{\tau_i < t} \alpha e^{-\beta(t-\tau_i)},$$

with $\mu = 0.3$, $\alpha = 0.6$, and $\beta = 1.2$. Figure 5.10 gives the sample paths of the intensity processes of a bivariate Hawkes process with intensity processes given by

$$\lambda_t^{(1)} = \mu^{(1)} + \sum_{\tau_i < t} \alpha_{11} e^{-\beta_{11}(t-\tau_i)} + \sum_{\tau j < t} \alpha_{12} e^{-\beta_{12}(t-\tau_j)},$$

$$\lambda_t^{(2)} = \mu^{(2)} + \sum_{\tau_j < t} \alpha_{22} e^{-\beta_{22}(t-\tau_j)} + \sum_{\tau i < t} \alpha_{21} e^{-\beta_{21}(t-\tau_i)},$$

where $\mu^{(1)} = \mu^{(2)} = 0.5$, $\alpha_{11} = \alpha_{22} = 0.8$, $\alpha_{12} = \alpha_{21} = 0.5$, $\beta_{11} = \beta_{12} = 1.5$, and $\beta_{22} = \beta_{21} = 1.5$. Note the cross-excitation dynamics between the two processes and the exponential decay after each jump, and read Section C.4 of Appendix C on how to simulate Hawkes processes.

2. Use the NASDAQ ITCH data (from the book's website) to calibrate a univariate point process for the bid-side trade intensity. Comment on the features of the fitted model.

3. **Basic ideas behind Kruk's model in the introductory paragraph of Section 5.5** Consider an auction, indexed by n, in which there

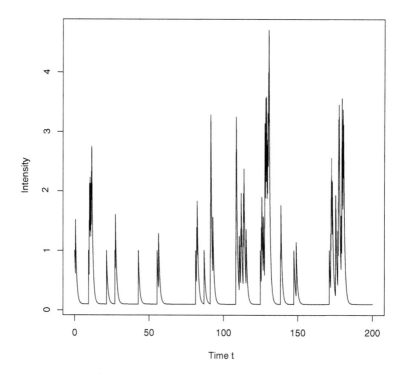

FIGURE 5.9: Simulated intensity of a univariate Hawkes process.

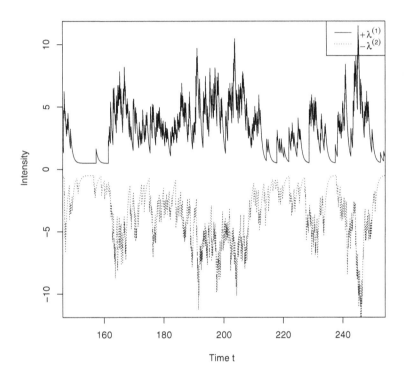

FIGURE 5.10: Simulated intensities of a bivariate Hawkes process, with $\lambda^{(1)}$ in the top panel and $\lambda^{(2)}$ inverted to aid visualization in the bottom panel.

are N price levels $P_1^{(n)}, \ldots, P_N^{(n)}$, and let $A_i^{(n)}(t)$ and $B_i^{(n)}(t)$ denote the outstanding ask and bid orders at price level P_i at time t. Clearly, $A_i^{(n)}(t)B_i^{(n)}(t) = 0$ as these bid and ask orders at the same price level are executed against each other. Kruk (2003) makes the following assumptions.

- Buyers arrive according to a renewal process with inter-arrival times $u_k^{(n)}$, $k = 1, 2, \ldots$, with mean $1/\lambda^{(n)}$ and standard deviation $\alpha^{(n)}$; sellers arrive also according to a renewal process with inter-arrival times $v_k^{(n)}$, $k = 1, 2, \ldots$, with mean $1/\mu^{(n)}$ and standard deviation $\beta^{(n)}$, and the two arrival processes are independent.

- The kth buyer (respectively, seller) is willing to buy (respectively, sell) $l_k^{(n)}$ (respectively, $m_k^{(n)}$) shares, where $l_k^{(n)}$ (respectively, $m_k^{(n)}$), $k = 1, 2, \ldots$, are positive i.i.d. random variables with mean $\bar{l}^{(n)}$ (respectively, $\bar{m}^{(n)}$) and standard deviation $\kappa^{(n)}$ (respectively, $\nu^{(n)}$).

- The kth buyer (respectively, seller) is willing to buy (respectively, sell) at the price $b_k^{(n)}$ (respectively, $a_k^{(n)}$), where $b_k^{(n)}$ (respectively, $a_k^{(n)}$), $k = 1, 2, \ldots$, are i.i.d. random variables with

$$P(b_k^{(n)} = i) = P(a_k^{(n)} = i) = p_i, \qquad i = 1, 2, \ldots, N.$$

Kruk's results that have motivated Sections 5.1.1 and 5.1.2

- For $N = 2$, under the scaling $\tilde{B}_k^{(n)}(t) = n^{-1/2}B_k^{(n)}(nt)$ and $\tilde{A}_k^{(n)}(t) = n^{-1/2}A_k^{(n)}(nt)$,

$$\left(\tilde{A}_1^{(n)}, \tilde{B}_1^{(n)}, \tilde{B}_2^{(n)}, \tilde{A}_2^{(n)}\right) \implies (0, 0, \boldsymbol{Z}),$$

in which $\boldsymbol{Z}$ is a reflected Brownian semimartingale.

- For $N \geq 3$, under the scaling $\tilde{B}_k^{(n)}(t) = n^{-1}B_k^{(n)}(nt)$ and $\tilde{A}_k^{(n)}(t) = n^{-1}A_k^{(n)}(nt)$,

$$\left(\tilde{A}^{(n)}, \tilde{B}^{(n)}\right) \implies (\bar{a}\boldsymbol{e}, \bar{b}\boldsymbol{e}),$$

where $\bar{a}$, $\bar{b}$ are two constants determined by the model parameters and $\boldsymbol{e}$ is the identity map $\boldsymbol{e}(t) = t$.

4. **Diffusion and fluid limits of queues** We provide here some background and additional references for the results in Sections 5.5.1 and 5.5.2. Cont and de Larrard's result (5.10) is a diffusion limit under $EX_i = 0$, and is a consequence of the functional central limit theorem. Blanchet and Chen (2013) also derive a pure jump limit for the price-per-trade process and a jump diffusion limit for the price-spread process. The results of Guo et al. in Section 5.5.2 are fluid limits and follow from the functional strong law of large numbers. The term diffusion limit occurs in heavy traffic, characterized by zero-mean increments that are associated with the

central limit theorem. The fluid model in queueing theory originates from modeling the fluid flow in a large tank connected to pipes that fill the tank and to pumps that remove the fluid. The random disturbances (stochastic inputs and outputs) are averaged out by the law of large numbers, resulting in a deterministic limiting distribution, as in (5.16); see Whitt (2002) for a survey of fluid and diffusion limits. It is more subtle to establish the weak limit for the stochastic system (5.17). In addition to the fluid limit, Guo et al. (2015) show that fluctuations of the order positions are Gaussian processes with mean reversion. The mean-reverting level is essentially the fluid limit of the order position relative to the queue length modified by the order book net flow, which is defined as the limit order minus the market order and the cancellation. The speed of the mean reversion is shown to be proportional to the order arrival intensity and the rate of cancellations.

For the reduced-form model in Section 5.5.1, Horst and Paulsen (2015) have derived the fluid limit for the entire LOB including both prices and volumes. Their analysis is extended in Horst and Kreher (2015) to include the order dynamics that can depend on the state of the LOB.

5. **An empirical study of LOB** Ahn et al. (2001) report an empirical study using the Trade Record and the Bid and Ask Record of 33 component stocks in the Hang Seng Index (HSI) of Hong Kong Stock Exchange between July 1996 and June 1997. The Trade Record consists of all transaction price-and-volume records with a time stamp recorded to the nearest second, while the Bid and Ask Record lists the size of outstanding limit orders in the best five bid-and-ask price levels in every 30-second interval. These 33 component stocks account for about 70% of the total market capitalization in their study period. Moreover, during that period, the trading sessions of each day are 10:00 to 12:30 and 14:30 to 15:55. Ahn et al. (2001) combine the two trading sessions and divide that into 16 fifteen-minute trading intervals. By defining the (short-term) volatility to be the sum of the squared transaction-to-transaction returns and the depth of a snapshot of the LOB as the total number of shares outstanding at the best five bid-and-ask prices, and averaging these two variables over all stocks, Figure 1 of Ahn et al. (2001) plots the volatility and the depth for each fifteen-minute interval. While the volatility shows the usual diurnal pattern of volatility, the average depth profile is inverted U-shaped. This depth pattern is also reported by Lee et al. (1993) for NYSE and Biais et al. (1995) for Paris Bourse. Ahn et al. (2001) formulate testing the following two hypotheses on LOBs as the primary objectives of the empirical study.

Hypothesis 1: An increase (a decrease) in short-term price volatility is followed by an increase (a decrease) in the placement of limit orders relative to market orders, hence the market depth will increase (decrease) subsequently.

Hypothesis 2: An increase (a decrease) in market depth is followed by a decrease (an increase) in short-term price volatility.

Let $R_{i,t}$ be the return of the ith transaction of the tth trading interval for $i = 1, \ldots, N_t$ where N_t is the number of transactions in that interval. They consider the following variables:

- $\text{RISK}_t = \sum_{i=1}^{N_t} R_{i,t}^2$ is the short-term volatility of the tth interval.

- $\text{RISK}_t^+ = \sum_{i=1}^{N_t} R_{i,t}^2 \, I_{\{R_{it}>0\}}$ is the upside volatility.

- $\text{RISK}_t^- = \sum_{i=1}^{N_t} R_{i,t}^2 \, I_{\{R_{it}<0\}}$ is the downside volatility.

- $\text{DEPTH}_t^{\text{bid}} =$ total outstanding number of shares at the best five bids at the end of the time interval t; $\text{DEPTH}_t^{\text{ask}}$ is similarly defined.

- $\text{DEPTH}_t = \text{DEPTH}_t^{\text{bid}} + \text{DEPTH}_t^{\text{ask}}$.

- $\text{DIFF}_t^{\text{buy}} =$ the difference between the number of shares of newly placed limit buy orders and market buy orders within the tth interval; $\text{DIFF}_t^{\text{sell}}$ is similarly defined.

- $\text{NBUY}_t =$ the number of shares of market buy orders during time interval t; NSELL_t is similarly defined.

- $\text{NTRADE}_t = N_t$.

- $\text{TIME}_{k,t} = 1$ if $t = k$; $\text{TIME}_{k,t} = 0$ otherwise.

To test Hypothesis 1, they first fit the cross-sectional regression model

$$\text{DEPTH}_t = \alpha_1 + \beta_1 \text{RISK}_{t-1} + \theta_1 \text{NTRADE}_t$$
$$+ \sum_{k=1}^{14} \gamma_{k,1} \text{TIME}_{k,t} + \rho_1 \text{DEPTH}_{t-1} + \varepsilon_{t,1}, \quad (5.22)$$

in which DEPTH_{t-1} and the binary variables $\text{TIME}_{k,t}$ are used to control for intraday variation and autocorrelation in the market depth. The model is fitted using the Generalized Method of Moments (GMM; see Section 9.7 of Lai and Xing, 2008) which gives t-statistics that are robust to heteroskedasticity, autocorrelation, and possible correlations between ε_t and the regressors. Two other definitions of DEPTH_t are also considered: only the best prices, and only the second to the fifth prices. For all three definitions, DEPTH_t is significantly negatively related to NTRADE_t, in agreement with the result reported by Lee et al. (1993), which reflects that transactions consume the available liquidity. Moreover, Table II of Ahn et al. (2001) shows that β_1 is significantly positive, which is consistent with Hypothesis 1.

To investigate how short-term volatility impacts the order flow, Ahn et al.

(2001) use GMM to fit another cross-sectional model:

$$\Delta\text{DEPTH}_t = \alpha_2 + \beta_2\text{RISK}_{t-1} + \rho_2\Delta\text{DEPTH}_{t-1}$$
$$+ \sum_{k=1}^{13} \gamma_{k,2}\text{TIME}_{k,t} + \varepsilon_{t,2}, \quad (5.23)$$

where $\Delta\text{DEPTH}_t = \text{DEPTH}_t - \text{DEPTH}_{t-1}$ is a measure of order flow composition because it is the difference between the size of the newly placed limit orders and that of the market orders and the cancellations submitted during time interval t. Their Table III shows that ρ_2 is significantly negative, which means order flow is self-adjusting in the sense that there are more trades when the order book is thick and more limit orders submitted when the book is thin. They also point out that the impact of short-term volatility on order flow is not strong. In fact, by using only the queue length of the best prices for the computation of DEPTH, β_2 is significantly positive for only seven stocks. To further explore how order flow composition is related to the liquidity-driven price volatility arising from the bid and ask sides, they also consider a third cross-sectional model:

$$\text{DIFF}_t^{\text{buy}} = \alpha_3 + \beta_3^+\text{RISK}_{t-1}^+ + \beta_3^-\text{RISK}_{t-1}^-$$
$$+ \rho_3\text{DIFF}_{t-1}^{\text{buy}} + \sum_{k=1}^{13} \gamma_{k,3}\text{TIME}_{k,t} + \varepsilon_{t,3}, \quad (5.24a)$$

$$\text{DIFF}_t^{\text{sell}} = \alpha_4 + \beta_4^+\text{RISK}_{t-1}^+ + \beta_4^-\text{RISK}_{t-1}^-$$
$$+ \rho_4\text{DIFF}_{t-1}^{\text{sell}} + \sum_{k=1}^{13} \gamma_{k,4}\text{TIME}_{k,t} + \varepsilon_{t,4}. \quad (5.24b)$$

Their Table IV shows that both ρ_3 and ρ_4 are significantly positive. Moreover, β_3^- and β_4^+ are significantly positive for 17 and 20 stocks, respectively, when the market depth is computed from the best prices. For the the third definition of market depth (the second through the fifth best prices), the numbers increase to 30 and 31, respectively. These results are consistent with liquidity providers' tendency to place limit orders on the side that gives higher profits for providing liquidity.

Ahn et al. (2001) then evaluate the subsequent effect of market depth onto the volatility by using GMM to fit the cross-sectional model

$$\text{Risk}_t = \alpha_5 + \beta_5\text{DEPTH}_{t-1} + \theta_5\text{NTRADE}_t$$
$$+ \sum_{k=1}^{14} \gamma_{k,5}\text{TIME}_{k,t} + \rho_5\text{RISK}_{t-1} + \varepsilon_{t,5}. \quad (5.25)$$

To further understand the direction of impacts, they also separate the depth into the bid and ask sides, and relate them to the downside and upside volatility via

$$\text{Risk}_t^+ = \alpha_6 + \beta_6^- \, \text{DEPTH}_{t-1}^{\text{bid}} + \beta_6^+ \text{DEPTH}_{t-1}^{\text{ask}} + \theta_6 \text{NBUY}_t$$
$$+ \sum_{k=1}^{14} \gamma_{k,6} \text{TIME}_{k,t} + \rho_6 \text{RISK}_{t-1} + \varepsilon_{t,6},$$

$$\text{Risk}_t^- = \alpha_7 + \beta_7^- \, \text{DEPTH}_{t-1}^{\text{bid}} + \beta_7^+ \text{DEPTH}_{t-1}^{\text{ask}} + \theta_7 \text{NSELL}_t$$
$$+ \sum_{k=1}^{14} \gamma_{k,7} \text{TIME}_{k,t} + \rho_7 \text{RISK}_{t-1} + \varepsilon_{t,7}.$$

Their Table VI shows that both the upside and the downside volatilities are significantly associated to the market depth in the best queue. Specifically, while the upside volatility is significantly negatively related to the ask depth but not to the bid depth, the downside volatility is significantly negatively related to the bid depth but not to the ask depth.

6. An important finding of Ahn et al. (2001) concerning LOB dynamics is that it provides empirical evidence for the cycle of alternating between influx of limit orders and high density of market orders. That is, when limit orders are scarce, temporary order imbalance triggers an increase in short-term volatility and attracts public investors to place limit orders instead of market orders. Such influx of limit orders continues until the short-term price volatility decreases, which in turn increases the density of market orders. Since the analysis is performed on the equities data of Hong Kong Stock Exchange in which no system for market makers is implemented, Ahn et al. (2001) conclude that an order-driven trading mechanism without the presence of market makers can be viable and self-sustaining.

7. *Exercise: Data analysis* The file `hsbc_L1.txt` contains the level-1 data of HSBC traded in Hong Kong Stock Exchange from March 17 to March 19, 2014. Use `hsbc_L1.txt` to fit models (5.22) and (5.25) similar to Ahn et al. (2001) by dividing the data into fifteen-minute time intervals. Note that the trading hours of Hong Kong Stock Exchange in 2014 are from 9:30 to 12:00 and 13:00 to 16:00. Examine whether the statistical results are sensitive to the length of the time intervals by inflating it to thirty minutes, as well as by deflating it to five minutes.

8. *Exercise on weak convergence* Let $\{X_i\}_{i \geq 1}$ be a sequence of identically

distributed random variables taking values in $\{-1, 1\}$ such that

$$P(X_1 = 1) = P(X_1 = -1) = \frac{1}{2},$$

$$P(X_{i+1} = 1 | X_i = 1) = P(X_{i+1} = -1 | X_i = -1) = \frac{3}{4} \quad \text{for } i > 1.$$

(a) Define $S_n(t) = \frac{1}{\sqrt{n}} \sum_{i=1}^{\lfloor nt \rfloor} X_i$. Use the invariance principle for ϕ-mixing stationary sequences to show that S_n converges weakly to the scaled Brownian motion $\sqrt{3}B(t)$, $t \geq 0$.

(b) Define a sequence of SDE's $dY_n(t) = Y_n(t)dS_n(t)$ with $Y_n(0) = 1$. Show that as $n \to \infty$, Y_n converges weakly to the limiting process $\exp\{\sqrt{3}B(t) - \frac{t}{2}\}$, $t \geq 0$.

Note: The solution to the SDE $dY(t) = Y(t)d(\sqrt{3}B(t))$ with $Y(0) = 1$ is given by $Y(t) = \exp\{\sqrt{3}B(t) - \frac{3t}{2}\}$.

9. **Multi-agent models for LOB: Game theory, econophysics, and computational learning theory** In Supplements 1–3 of Section 1.7 we have briefly reviewed the basic economic principles underlying the stock trading infrastructure that culminates in today's double auction market consisting of multiple buyers and sellers: heterogeneously informed market participants, market/mechanism design, adverse selection, game theory and Nash equilibrium.[6] Here, in the spirit of LOB models described in Sections 5.3 and 5.5, we continue the review and discuss contributions by other disciplines to this topic. First we want to point out that the models in Sections 5.3 and 5.5 are basically those introduced by the financial mathematics, operations research, and statistics communities. However, there have also been many important developments in other research communities. Section 3 of Gould et al. (2013) discusses the developments in (a) the economics literature, highlighting perfectly rational traders who attempt to maximize their expected utilities of payoffs from the trades in markets driven by information, and (b) the econophysics literature, using a "zero-intelligence" approach that assumes traders (or more precisely, their computers) blindly follow a set of rules that treat the order flows as "governed by specific stochastic processes." Their Section 3.10 points out statistical issues in the econophysics studies of LOB data, showing statistical regularities across different markets in a large empirical literature. However, Gould et al. (2013, p. 1717–1718) have pointed out that the empirical

---

[6]The game theory here pertains to that of a non-cooperative game with n players so that player i has strategy set S_i. Let $f_i(\mathbf{x})$, $\mathbf{x} = (x_1, \ldots, x_n)$, be the payoff for player i when the other players j use $x_j \in S_j$; $\mathbf{x}$ is called a strategy profile. Let $\mathbf{x}_{-i}$ be the strategy profile of the $n - 1$ players with player i removed. A strategy profile $\mathbf{x}^*$ is a Nash equilibrium if $f_i(x_i^*, \mathbf{x}_{-i}^*) \geq f_i(x_i, \mathbf{x}_{-i}^*)$ for all $1 \leq i \leq n$ and all $x_1 \in S_1, \ldots, x_n \in S_n$. Hence, in Nash equilibrium, no unilateral deviation in strategy by any player is profitable for that player. John Nash, who won the Nobel Prize in Economic Sciences in 1994, proved the existence of a Nash equilibrium for finite S_i if mixed strategies (by randomization over S_i) are allowed.

analysis "is fraught with difficulties because assumptions such as independence and stationarity, which are often required to ensure consistency of estimation (in statistical mechanics), are rarely satisfied by LOB data." Moreover, "suboptimal estimators have been employed... often produced estimates with large variance or bias," and they also point out that "different (empirical) studies often present conflicting conclusions" because "different trade-matching algorithms operate differently,... and different researchers have access to data of different quality."

One of the main findings of this empirical literature is the prevalence of long-memory processes. This poses statistical difficulties in estimating autocorrelations; see Section 3.10.2 of Gould et al. (2013). On a more conceptual level, this challenges the mean-reverting property that is often assumed in studying market impact and developing high-frequency trading strategies. As shown by Lamoureux and Lastrapes (1990), Hillebrand (2005), and Lai and Xing (2013), spurious long memory is bound to occur if one does not incorporate parameter changes in fitting time series data involving two time-scales, one being the short scale, in which the data are collected, and the other being the long time-scale that includes parameter jumps. Gould et al. (2013, Sect. 6) point out that understanding nonstationarity behavior is a key unresolved problem in LOB research, for which "almost all LOB models to date have focused on some form of equilibrium," contrary to empirical evidence strongly suggesting that LOBs are subject to frequent shocks in order flow that cause them to display nonstationary behavior, so they may never settle into equilibrium."

Section 5 of Gould et al. (2013) comments on the substantial progress with and interest in LOB modeling in the economics and physics communities, and in particular, on agent-based models (ABMs) in their Section 5.3. It says, however, that "work by the two communities has remained largely independent." The LOB models produced by economists have generally treated order flow as static, and "are trader-centric, using perfect-rationality frameworks to derive optimal trading strategies given certain market conditions." In contrast, models from physicists treat order flow as dynamic (Farmer et al., 2005). Concerning ABMs, in which "a large number of possibly heterogeneous agents interact in a specified way", they lie between the two extremes of zero-intelligence (in the physics community) and perfect-rationality models (in the economics community).

There is an ongoing debate between the economics and econophysics communities about which approach provides more useful models for multi-agent systems. On the one hand, Lo and Mueller (2010) argue against the "physics envy" in economics/financial modeling, saying: "The quantitative aspirations of economists and financial analysts have for many years been based on the belief that it should be possible to build models of economic systems—and financial markets in particular—that are as predictive as those in physics. While this perspective has led to a number of important breakthroughs in economics, physics envy has also created a false sense of mathematical precision in some cases." On the other hand, Hommes (2002) is another economist collaborating with statistical physicists and says: "In the past two decades economics has witnessed an important paradigmatic change: a shift from a rational representative agent, analytically tractable model of the economy to a boundedly rational, heterogeneous agents computationally oriented evolutionary framework. This change has at least three closely related aspects: (i) from representative agent to heterogeneous agent systems; (ii) from full rationality to bounded rationality; and (iii) from a mainly analytical to a more computational approach."

The more computational approach to multi-agent systems is also taken by information and computer scientists; see for example, Wooldridge (2009), Bloembergen et al. (2015), and Helbing and Kirman (2013) in the context of economic theory. This is what computational learning theory, a subfield of computer science devoted to studying computational complexity, machine learning, and artificial intelligence, in the title of this supplement refers to.

10. **Mean field approximations and Nash equilibrium in LOB models** Mean field theory originates in physics, dating back to the works of Pierre Curie and then of Pierre Weiss (1907). The theory addresses a single complex system that consists of a large number of individual components which interact with each other, and approximates a many-body problem with a one-body problem that aggregates the effect of the other individuals on any given individual by a single average effect, which is tantamount to replacing the interactions of all the other particles in the external field by an arbitrary particle; see Kadanoff (2009). Smith et al. (2003, Sect. 3.6 and 3.7) point out the usefulness of mean field theory in deriving predictions in the (multi-agent) double auction markets.

Lions and Lasry (2007) and Lasry and Lions (2007), and independently Huang et al. (2007), developed the mathematics of mean field games that involve strategic decision making by a large number of players. Consider an n-player differential game in which the state process of the ith player is

$$dX_t^i = b(t, X_t^i, \alpha_t^i, \bar{\mu}_t)dt + \sigma dB_t^i, \tag{5.26}$$

where B^i, $i = 1, \ldots, n$, are independent Brownian motions, α_i^t is an admissible policy and $\bar{\mu}_t = \sum_{i=1}^n \delta_{X_t^i}$ is the empirical distribution of $X_t^1, \ldots, X_t^n$. The ith player's goal is to maximize

$$J_t^i = E\left[\int_0^T f(t, X_t^i, \alpha_t^i, \bar{\mu}_t)\, dt + g(X_T^i)\right]. \tag{5.27}$$

Here b and f are measurable real-valued functions on $[0, T] \times \mathbb{R} \times \mathcal{A} \times \mathbb{R}$ and g is a measurable function from $\mathbb{R}$ to $\mathbb{R}$. Note that each player solves essentially the same control problem. Because of the symmetry of the game, each player should have the same optimal policy and in turn the same law for the optimal control process and state process. Note also that each player interacts with other players only through the aggregated term $\bar{\mu}_t$. Therefore, when the number of players is large, the impact of a single player on other players is almost negligible. In a Nash equilibrium, a representative player would choose the optimal strategy against a population of players with known optimal strategies. When the number of players in the game approaches infinity, the empirical distribution $\bar{\mu}_t$ converges to μ_t, which is the true distribution of a representative player's optimal state. The resulting model is called Mean Field Game (MFG), where a representative player solves the following optimal control problem:

$$\sup_{\alpha} E\left[\int_0^T f(t, X_t, \alpha_t, \mu_t)\, dt + g(X_T)\right] \tag{5.28}$$

$$\text{subject to } dX_t = b(t, X_t, \alpha_t, \mu_t)dt + \sigma dB_t.$$

Here μ_t is the limit of the interaction term in an n-player differential game, describing the external context which the representative player plays against in a MFG, and thus it is exogenous to the optimal control problem in (5.28). On the other hand, by the nature of Nash equilibrium and the symmetry in the n-player differential game, μ_t should match the law of the optimal state process X_t^μ for which $\bar{\mu}_t$ in (5.27) is replaced by μ_t. Hence μ_t is a fixed point that can be determined iteratively as follows: given a value of the distribution μ_t, plug it in the system in (5.28) and solve the optimal control problem for the optimal state process X_t^μ and the optimal control α_t^μ. Then update μ_t and iterate until $\mu_t = \mathcal{L}_{X_t^\mu}$. The MFG involves $\mu_t \in \mathbb{R}$. It can be extended to higher-dimensional $\mathbb{R}^d$; see Carmona and Delarue (2013).

Lions and Lasry (2007) and Lasry and Lions (2007) use the dynamic programming principle to derive the HJB equation for the value function corresponding to the optimal α in (5.28) when μ_t is given, and determine μ_t via the Fokker-Planck equation in statistical mechanics (also known as the Kolmogorov forward equation in stochastic processes). Thus, they derive a system of coupled PDEs that characterize the Nash equilibrium of the n-player differential game as $n \to \infty$. They have collaborated with Guéant, Buera and Moll in Guéant et al. (2011) and Achdou et al. (2014) to give applications of MFG to double auctions markets and macroeconomics. Carmona and Delarue (2013) and Carmona and Lacker (2015) develop an alternative approach by using a stochastic maximum principle, which leads to a system of forward-backward stochastic differential equations (FBSDE) for fixed μ, and with μ determined as the fixed point of $\mu_t = \mathcal{L}_{X_t^\mu}$. The idea of stochastic maximum principle dates back to Kushner (1972) and was subsequently developed by Peng (1990) and others into a powerful methodology to evaluate optimal controls via FBSDEs instead of HJBs. Ren (2016) has recently further developed this approach and provided an efficient numerical method to compute the Nash equilibrium when the approach is applied to agent-based LOB modeling.

6

Optimal Execution and Placement

At a high level, algorithmic trades are based on two different time scales: the daily or weekly scale and a much smaller time scale of seconds (up to the hundredth of a second). According to a recent empirical study of Kirilenko et al. (2015), these two time scales reflect two different types of traders: the fundamental traders and the high-frequency traders (HFT), respectively; they display statistically different behaviors and played different roles during the flash crash of May 2010. The two time scales also correspond to the two stages by which the traders slice and place orders. The first stage optimally slices big orders into smaller ones on a daily basis to minimize the price impact (part of the transaction costs discussed in Chapter 3), and the second stage optimally places the orders within seconds. The former is known in the algorithmic trading literature as *optimal execution* and the latter is either *optimal placement* within one exchange or *smart order routing* (SOR) across different exchanges.

In this chapter, we discuss the main modeling ideas and analytical approaches for the optimal execution and optimal placement problems within one exchange, leaving the SOR problem across exchanges to the next chapter. Specifically, Section 6.1 begins with the optimal execution problem for a single asset and introduces the seminal papers of Bertsimas and Lo (1998) and Almgren and Chriss (2001) in this area. In this connection we introduce the basic concept of additive market impact underlying their work and its subsequent generalizations. A feature of the optimal strategies derived is their deterministic character explained in Section 6.1.3. To arrive at a nondeterministic strategy, Section 6.2 considers a multiplicative price impact that leads to a singular stochastic control problem solved by Guo and Zervos (2015). Section 6.3 considers optimal execution when limit order book data are used to infer the dynamics of the supply and demand of the asset in the trading period and describes the seminal works of Obizhaeva and Wang (2013) and Alfonsi et al. (2010) in this area. Section 6.4 reviews optimal execution of a portfolio of assets. In Section 6.5 we describe a class of stochastic optimization problems associated with optimal placement. Section 6.6 provides supplements and problems, and in particular introduces the commonly used performance measures VWAP, TWAP, and POV for execution algorithms used by high-frequency traders.

6.1   Optimal execution with a single asset

The major modeling premise behind the optimal execution problem is that any trading strategy, especially one that involves a large amount of buying or selling within a short period of time, would have an impact on the stock price. Thus, optimal execution is to find an appropriate trading strategy that balances two opposite risks: the market risk and the price impact. As shown in Chapter 5, a sell order that is too large may depress the price and reduce the potential profit, while too many small transactions may take too long to complete and thereby increase the market risk. In essence, a price impact model for optimal executions differs from classical hedging and pricing models in that it focuses on controlling the speed/quantity of trading to minimize the price impact.

The study of the optimal execution problem was pioneered by Bertsimas and Lo (1998) and by Almgren and Chriss (1999, 2001). The price impact is assumed to be additive and the optimal solution is a volume-average type deterministic trading strategy. With the development of more complex models, the price impact can be categorized into three main types: permanent, temporary, and transient. Permanent impact models assume that each trade affects the future underlying price, temporary impact models assume that each trade only affects the execution price at the time of that particular order being placed, and transient impact models assume that the price impact for the underlying asset price decays over time. Empirical studies of the trade impact on the price have been carried out by Hasbrouck (1991) and Potters and Bouchaud (2003). These studies, based on publicly available data, show that the price impact of trades is an increasing and concave function of their sizes. However, this conclusion is not confirmed by other studies that use proprietary data; see, for example, Almgren et al. (2005) and Obizhaeva and Wang (2013).

Bertsimas and Lo's expected cost minimization under linear price impact

Suppose a trader places orders at times $1, \ldots, T$ to purchase a total of $\tilde{N}$ shares. Suppose the trader buys N_t shares at time t $(1 \leq t \leq T)$, paying price P_t per share. Under the assumption of linear price impact and a random walk model for stock prices when there is no price impact,

$$P_t = P_{t-1} + \epsilon_t + \theta N_t, \tag{6.1}$$

in which θ is a positive constant, the ϵ_t are i.i.d. and $E\left[\epsilon_t | N_t, P_{t-1}\right] = 0$; θN_t is known as the price-impact function. Bertsimas and Lo (1998) consider the problem of minimizing the total expected cost:

$$\min_{N_1, \ldots, N_T} E\left[\sum_{t=1}^{T} P_t N_t\right] \qquad \text{subject to} \sum_{t=1}^{T} N_t = \tilde{N}. \tag{6.2}$$

The solution, given in Section 6.1.1, is $N_1^* = \cdots = N_t^* = \cdots = N_T^* = \tilde{N}/T$.

Almgren and Chriss' framework with permanent and temporary price impact

They consider a liquidation problem of selling a fixed number of shares by the fixed time T. In this framework, the price impact consists of two types: the permanent one and the temporary one. Specifically, let δ be the duration, assumed to be constant, between two consecutive trade times and let $K = T/\delta$, assumed to be an integer. Let P_k denote the asset price at time $k = 0, \ldots, K$. The asset prices are assumed to follow

$$P_k = P_{k-1} + \sigma\sqrt{\delta}\epsilon_k - \delta g\left(N_k/\delta\right), \tag{6.3}$$

in which the ϵ_k are i.i.d. with mean 0 and variance $\sigma^2 > 0$, and g is a given function, with $g(N_k/\delta)$ representing the permanent price impact that affects the asset's future price when N_k shares are sold. In addition, selling N_k shares also has temporary price impact $h(N_k/\delta)$ that only affects the execution price $\tilde{P}_k$ at time k, which is given by

$$\tilde{P}_k = P_{k-1} - h\left(N_k/\delta\right). \tag{6.4}$$

Given any trading strategy $(N_1, N_2, \ldots N_K)$ to sell $\tilde{N} = N_1 + \cdots + N_K$ shares by time T, the total revenue of selling the $\tilde{N}$ shares is

$$R(N_1, \ldots, N_K) = \sum_{k=1}^{K} \tilde{P}_k N_k. \tag{6.5}$$

The difference, denoted as $C(N_1, \ldots, N_K)$, between the total revenue without the temporary price impact and the total liquidation revenue is called the cost of liquidation. The optimal execution problem is to choose $N_1, \ldots, N_K$ to minimize the total expected cost of liquidation plus the total variance of trading revenue multiplied by a risk aversion parameter λ; that is,

$$\min_{N_1, \ldots, N_K} \{E\left[C(N_1, \ldots, N_K)\right] + \lambda \text{Var}\left[R(N_1, \ldots, N_K)\right]\}$$

$$\text{subject to } \sum_{k=1}^{K} N_k = \tilde{N}, \ N_k \geq 0, \ 1 \leq k \leq K. \tag{6.6}$$

Since δ is typically small, we can approximate (6.6) by a continuous-time problem whose solution is given in Section 6.1.2 and is again deterministic.

6.1.1 Dynamic programming solution of problem (6.2)

Let Y_t denote the number of shares that remain to be purchased. Clearly

$$Y_t = Y_{t-1} - N_{t-1}, \tag{6.7}$$

with $Y_1 = \tilde{N}$ and $Y_{T+1} = 0$. We find the optimal solution $(N_1^*, \ldots, N_T^*)$ by backward induction as in Supplement 4 of Section 3.4. Denoting the conditional expectation given the information set (σ-field) $\mathcal{F}_t$ up to time t by E_t, note that N_t is $\mathcal{F}_t$-measurable. Let

$$V_t(P_{t-1}, Y_t) = \min_{N_t, \ldots, N_T} E_{t-1}\left[\sum_{k=t}^{T} P_k N_k\right] \tag{6.8}$$

subject to $N_t + \cdots + N_T = \tilde{N} - N_1^* - \cdots - N_{t-1}^*$, and note that $N_T^* = Y_T$, $V_T(P_{T-1}, Y_T) = Y_T(P_{T-1} + \theta Y_T)$. For $t = T - 1$,

$$\begin{aligned}
V_{T-1}&(P_{T-2}, Y_{T-1}) \\
&= \min_{N_{T-1}} E_{T-2}[(P_{T-2} + \theta N_{T-1} + \epsilon_{T-1})N_{T-1} \\
&\qquad + V_T(P_{T-2} + \theta N_{T-1} + \epsilon_{T-1}, Y_{T-1} - N_{T-1})] \\
&= \min_{N_{T-1}} [(P_{T-2} + \theta N_{T-1})N_{T-1} \\
&\qquad + [P_{T-2} + \theta N_{T-1} + \theta(Y_{T-1} - N_{T-1})](Y_{T-1} - N_{T-1})] \\
&= \min_{N_{T-1}} (\theta N_{T-1}^2 - \theta Y_{T-1}N_{T-1} + \theta Y_{T-1}^2 + P_{T-2}Y_{T-1}).
\end{aligned}$$

This is the minimum of a quadratic function in N_{T-1}, hence

$$N_{T-1}^* = \frac{Y_{T-1}}{2} \quad \text{and} \quad V_{T-1}(P_{T-2}, Y_{T-1}) = \frac{3}{4}\theta Y_{T-1}^2 + P_{T-2}Y_{T-1}.$$

Continuing the analysis in this way, we derive the Bellman equation for $t < T$:

$$V_t(P_{t-1}, Y_t) = \min_{N_t} E_{t-1}[P_t N_t + V_{t+1}(P_t, Y_t - N_t)] \tag{6.9}$$

since $Y_{t+1} = Y_t - N_t$. Thus the optimal value function $V_{T-k}(P_{T-k-1}, Y_{T-k})$ can be solved recursively:

$$N_{T-k}^* = \frac{Y_{T-k}}{k+1}, \tag{6.10}$$

$$V_{T-k}(P_{T-k-1}, Y_{T-k}) = Y_{T-k}\left(P_{T-k-1} + \frac{k+2}{2(k+1)}\theta Y_{T-k}\right). \tag{6.11}$$

In particular, for $k = T - 1$, we have

$$N_1^* = \frac{Y_1}{T} = \frac{\tilde{N}}{T}, \tag{6.12}$$

$$V_1(P_0, Y_1) = Y_1\left(P_0 + \frac{T+1}{2T}\theta Y_1\right) = P_0\tilde{N} + \frac{\theta\tilde{N}}{2}\left(1 + \frac{1}{T}\right). \tag{6.13}$$

Hence $N_1^* = \cdots = N_t^* = \cdots = N_T^* = Y_1/T = \tilde{N}/T$.

6.1.2 Continuous-time models and calculus of variations

There are natural continuous-time counterparts of these models, which can be solved directly by using the Calculus of Variations (CoV) technique instead of the continuous-time dynamic programming approach in Supplement 5 of Section 3.4.

Fundamental problem of CoV and the Euler-Lagrange equation

Given a function $f = f(x, y, y')$ with $y = y(x)$, where $y' = dy/dx$, let

$$I = \int_{x_1}^{x_2} f(x, y, y') \, dx. \tag{6.14}$$

The fundamental problem of CoV is to find a function $y(x)$ that maximizes or minimizes the integral I subject to the boundary condition

$$y(x_1) = y_1, y(x_2) = y_2. \tag{6.15}$$

A necessary condition for f to achieve some extreme value of (6.14) is that the function $y = y(x)$ satisfies the Euler-Lagrange equation

$$\frac{\partial f}{\partial y} - \frac{\partial}{\partial x}\left(\frac{\partial f}{\partial y'}\right) = 0. \tag{6.16}$$

In the special case where the function f is only a function of y and y', the Euler-Lagrange equation becomes the Beltrami identity

$$y'\frac{\partial f}{\partial y'} - f = C, \tag{6.17}$$

where C is a constant.

A well-known example of CoV concerns finding the shortest path between two points $A = (x_1, y_1)$ and $B = (x_2, y_2)$. Suppose that $x_1 \neq x_2$ and suppose that the path between A and B is parameterized by $(x, y(x))$ with $y_1 = y(x_1)$, $y_2 = y(x_2)$. Then the length depends on the particular trajectory $y(x)$. Denoting the length as $I(y)$ and applying Pythagorean theorem, we can write

$$I(y) = \int_{x_1}^{x_2} \sqrt{1 + (y')^2} \, dx,$$

and therefore $f(x, y, y') = \sqrt{1 + (y')^2}$. Since $\partial f/\partial y' = y'/\sqrt{1 + (y')^2}$, the shortest path $y(x)$ would satisfy the Euler-Lagrange equation $y''/[1 + (y')^2]^{3/2} = 0$. Hence $y'' = 0$ and therefore the shortest path is the straight line connecting y_1 and y_2. For general f, there may be multiple solutions to the Euler-Lagrange equation and further checks are needed to determine which one is the global solution and whether it is the maximum or the minimum. Moreover, the Euler-Lagrange equation applies only when the function y is twice differentiable but the optimizer of (6.14) may not be twice differentiable. A comprehensive treatment of CoV to handle these situations is given by Gelfand and Fomin (2000).

Almgren and Chriss' model

Letting $\delta \to 0$ in (6.4) with linear $h(x) = \eta x$ and $g \equiv 0$ converts it to the following continuous-time problem. First $\sigma\sqrt{\delta}\epsilon_k$ in (6.3) becomes the increment $\sigma\, dB_t$, in which B_t is the standard Brownian motion. Next the execution price $\tilde{P}_t$ is given by

$$\tilde{P}_t = P_t - \eta \dot{Y}_t = P_t + \eta U_t, \tag{6.18}$$

where Y_t is the number of shares at time t and $\dot{Y}_t = -U_t$ is the rate of selling stock. Hence the expected liquidation cost function is given by

$$C = E\left[\int_0^T \tilde{P}_t U_t\, dt\right] - E\left[\int_0^T P_t U_t\, dt\right] = \eta E\left[\int_0^T U_t^2\, dt\right]. \tag{6.19}$$

The variance of the trading cost is given by

$$\mathrm{Var}\left[\int_0^T Y_t\, dP_t\right] = \sigma^2 E\left[\int_0^T Y_t^2\, dt\right]. \tag{6.20}$$

Combining (6.19) with (6.20) gives the following continuous-time counterpart of (6.6) since $U_t = -\dot{Y}_t$:

$$\inf_{Y_t} E\left[\eta \int_0^T \dot{Y}_t^2\, dt + \lambda\sigma^2 \int_0^T Y_t^2\, dt\right]. \tag{6.21}$$

In the CoV framework, the Euler-Lagrange equation associated with (6.21) is

$$\eta \frac{d^2 Y_t}{dt^2} - \lambda\sigma^2 Y_t = 0. \tag{6.22}$$

The solution to this second-order ODE is a linear combination of the basic solution e^{At} with $A^2 = \lambda\sigma^2/\eta$. The boundary conditions $Y_0 = \tilde{N}$ and $Y_T = 0$ can then be used to determine the unique solution

$$Y_t^* = \tilde{N} \frac{\sinh A(T-t)}{\sinh AT}$$

for the optimal trading strategy Y_t^*, which turns out to be deterministic.

Generalization

An obvious generalization of (6.21) is

$$\inf_{Y_t} E\left[\eta \int_0^T G(Y_t, \dot{Y}_t)\, dt\right]. \tag{6.23}$$

In particular, let $G(y, u) = C_1(y, u) + \lambda C_2(y, u)$, with

$$C_1(Y_t, U_t) = Y_t g(U_t) + U_t h(U_t), \quad C_2(Y_t, U_t) = \sigma^2 Y_t^2 + U_t^2 l^2(U_t),$$

where g, h, and l are given functions and $U_t = -\dot{Y}_t$. The case $l = 0$ corresponds to the permanent and temporary price impact functions considered in (6.3) and (6.4). The corresponding Euler-Lagrange equation is

$$0 = f_y(y, -\dot{y}) + \frac{d}{dt} f_u(y, -\dot{y}) = f_y(y, -\dot{y}) + \dot{y} f_{yu}(y, -\dot{y}) - \ddot{y} f_{uu}(y, -\dot{y}), \quad (6.24)$$

with $f(y, u) = yg(u) + uh(u) + \lambda \sigma^2 y^2 + \lambda u^2 l^2(u)$. This second-order differential equation can be solved with appropriate boundary conditions Y_0 and Y_T.

Continuous-time analog of Bertsimas and Lo's model

Similarly, a continuous-time analog of Bertsimas and Lo's model can be provided by replacing the increments ε_t of the random walk in (6.1) by σdB_t and θN_t by $\theta \dot{Y}_t$. Hence the continuous-time analog of (6.2) is

$$\min_{Y_t \in \mathcal{A}} E\left[\theta \int_0^T U_t dB_t + U_t^2 \, dt \right], \quad (6.25)$$

where $U_t = \dot{Y}_t$ is the rate of buying stock and $\mathcal{A}$ is the set of admissible strategies (satisfying $Y_0 = 0$ and $Y_T = \tilde{N}$). An application of CoV confirms that the optimal execution strategy agrees with its discrete-time counterpart: the optimal trading strategy $Y_T^* = \tilde{N}/T$, $Y_t^* = \tilde{N}[1 - (t/T)]$, with $Y_0 = 0$, $Y_T = \tilde{N}$.

6.1.3 Myth: Optimality of deterministic strategies

Modeling stock prices by a random walk or Brownian motion in conjunction with additive impact of large stock sales has been a widely used framework to solve the optimal execution problem. An intriguing consequence of this modeling approach is that the optimal strategies turn out to be deterministic. Such strategies may lead to predictable trading patterns, which can give rise to market manipulation with techniques such as predatory trading; see the game-theoretic formulations studied by Schied and Schöneborn (2009) and Moallemi et al. (2009).

Mathematically, a stochastic control problem is usually solved by analyzing the associated HJB equation, and not by CoV, as shown in Supplement 5 of Section 3.4. Predoiu et al. (2011) have given the following insightful explanation, via integration by parts and the martingale structure of P_t, of why the optimal execution problem is an exception to this rule of thumb. Take the initial position $Y_0 = 0$ and final position $Y_T = \tilde{N}$. Suppose the cost is $\int_0^T P_t \, dY_t$, in which P_t is a martingale; see Appendix A for the background on continuous-time martingales. Then

$$E\left[\int_0^T P_t \, dY_t \right] = E\left[P_T Y_T - P_0 Y_0 - \int_0^T Y_t \, dP_t \right] = P_0 \tilde{N}.$$

Hence the source of randomness P_t disappears in the optimization problem.

The same argument still holds even if a random variable that is independent of the underlying price process is added to the objective function, as similar techniques can be used to derive deterministic optimal strategies under the discrete-time framework; see Section 6.3.1.

6.2 Multiplicative price impact model

Guo and Zervos (2015) have considered the optimal execution problem with a Geometric Brownian Motion (GBM) model and with a multiplicative price impact. This problem yields an optimal policy that is stochastic and Markovian, meaning the optimal policy depends on the state variable associated with the efficient price process. The control formulation is of a *singular* type. We describe below the basic ideas of their model and results; see Supplement 6 of Section 3.4 for related background on singular stochastic control and Becherer et al. (2015) for additional details on multiplicative impact models.

6.2.1 The model and stochastic control problem

Suppose that the efficient price process X_t^0, which is free of any trading frictions and impact, is a GBM

$$dX_t^0 = X_t^0(\mu dt + \sigma dB_t), \tag{6.26}$$

with a standard $\mathcal{F}_t$-adapted Brownian motion B_t on a probability space $(\Omega, \mathcal{F}, \mathcal{F}_t, P)$ with constant μ and $\sigma(> 0)$, where $\{\mathcal{F}_t, t \geq 0\}$ is a filtration. Let ξ_t^s (respectively, ξ_t^b) be the total number of shares that the investor has sold (respectively, bought) by time t, so that ξ^s and ξ^b are $\mathcal{F}_t$-adapted increasing càdlàg processes such that $\xi_0^s = \xi_0^b = 0$. Let Y_t be the total number of shares held by the investor at time t, with $Y_0 = y \geq 0$, so that

$$Y_t = y - \xi_t^s + \xi_t^b$$

is of a finite variation, a technical constraint to avoid infinite buying and selling activities. For a time horizon $T \in (0, \infty]$, the task of liquidating all shares by time T yields the constraint

$$Y_{T+} = 0 \text{ if } T < \infty, \quad \text{and} \quad \lim_{t \to \infty} Y_t = 0 \text{ if } T = \infty.$$

The transaction price process X_t is related to the efficient price process X_t^0 via

$$X_t = X_t^0 \exp(-\lambda \xi_t^s + \lambda \xi_t^b),$$

where $\lambda > 0$ is the price impact factor for selling and buying. Let $\delta \geq 0$ be the discount factor, $C_b, C_s \geq 0$ the spreads of buying and selling each share, and let

$$X_t \circ_s d\xi_t^s = X_t \, d(\xi^s)_t^c + X_t \int_0^{\Delta \xi_t^s} e^{-\lambda u} \, du, \tag{6.27}$$

$$X_t \circ_b d\xi_t^b = X_t \, d(\xi^b)_t^c + X_t \int_0^{\Delta \xi_t^b} e^{\lambda u} \, du, \tag{6.28}$$

where $d\xi_t^c$ is the continuous part of $d\xi_t$; see Appendix A for the background martingale theory. To explain the meaning of the terms in (6.27) and (6.28), assume for simplicity that a small transaction made by the investor affects the share price proportionally to its value. In particular, if the investor sells (respectively, buys) a small amount $\varepsilon > 0$ of shares at time t, then the share price exhibits a jump of size

$$\Delta X_t = X_{t+} - X_t = -\lambda \varepsilon X_t \text{ (respectively, } \Delta X_t = X_{t+} - X_t = \lambda \varepsilon X_t),$$

for some constant $\lambda > 0$, or equivalently,

$$X_{t+} = (1 - \lambda \varepsilon) X_t \simeq e^{-\lambda \varepsilon} X_t \ \left(\text{respectively, } X_{t+} = (1 + \lambda \varepsilon) X_t \simeq e^{\lambda \varepsilon} X_t \right).$$

Consider the sale of $\Delta \xi_t^s$ shares within the time period $[0, t]$ as being equivalent to the sale of N packets of shares of small size $\varepsilon = \Delta \xi_t^s / N$ over N evenly spaced times prior to t. Such a sale to reduce the price impact would result in a revenue of $\sum_{j=0}^{N-1} e^{-\lambda j \varepsilon} X_t \varepsilon \simeq \lambda^{-1} X_t (1 - e^{-\lambda \Delta \xi_t^s}) = \int_0^{\Delta \xi_t^s} X_t e^{-\lambda u} \, du$.

Note that compared to the additive price impact model with continuous rate of trading in a regular control setting, the controls ξ^s and ξ^b are of singular type. Moreover, $\lambda X_t d(\xi^s)_t^c$ is nonlinear. In fact, $X_t \, d(\xi^s)_t^c + X_t \int_0^{\Delta \xi_t^s} e^{-\lambda u} \, du$ is a nonlinear price impact factor, which consists of both temporary and permanent types, with exponential decay over trade sizes for the permanent impact. To maximize the investor's total expected payoff in selling and buying shares of the stock in the time interval $[0, T]$, the stochastic control problem associated with optimal execution has value function

$$v(T, x, y)$$
$$= \sup_{\xi^s, \xi^b} E \left[\int_{[0,T]} e^{-\delta t} \left(X_t \circ_s d\xi_t^s - X_t \circ_b d\xi_t^b - C_s \, d\xi_t^s - C_b \, d\xi_t^b \right) \right].$$

6.2.2 HJB equation for the finite-horizon case

As noted in Supplements 5 and 6 of Section 3.4, a standard approach to solving the preceding stochastic control problem is to associate its value function with an HJB equation. Since the equation may have multiple solutions and it is also possible that none of the solutions can identify the value function, this approach needs to go through a "verification step" to confirm that the solution found is the desired one. For the problem in the preceding section with $T < \infty$, the HJB equation has the form

$$\max\{-w_t(t,x,y) + \mathcal{L}w(t,x,y), \ -\lambda x w_x(t,x,y) - w_y(t,x,y) + x - C_{\mathrm{s}},$$
$$\lambda x w_x(t,x,y) + w_y(t,x,y) - x - C_{\mathrm{b}}\} = 0, \quad (6.29)$$

with boundary condition

$$w(0,x,y) = \frac{x}{\lambda}\left(1 - e^{-\lambda y}\right) - C_s y, \qquad (6.30)$$

where $\mathcal{L}w(t,x,y) = \sigma^2 x^2 w_{xx}(t,x,y)/2 + \mu x w_x(t,x,y) - \delta w(t,x,y)$.

To derive the HJB equation, consider the following. At a given time $t > 0$, with the share price $x > 0$ and $y > 0$ shares, there are three options: (i) wait for a short time Δt and then continue optimally; (ii) sell a small amount $\varepsilon > 0$ of shares, and then continue optimally; and (iii) buy a small amount $\varepsilon > 0$ of shares, and then continue optimally. The action of waiting, not necessarily optimal, implies the inequality

$$v(t,x,y) \geq E\left[e^{-\delta \Delta t} v(t - \Delta t, X_{\Delta t}, y)\right].$$

Applying Itô's formula and dividing by Δt before letting $\Delta t \downarrow 0$, we obtain

$$-v_t(t,x,y) + \frac{1}{2}\sigma^2 x^2 v_{xx}(t,x,y) + \mu x v_x(t,x,y) - \delta v(x,y) \leq 0. \qquad (6.31)$$

The second option of selling is associated with the inequality

$$v(t,x,y) \geq v(t, x - \lambda x \varepsilon, y - \varepsilon) + (x - C_{\mathrm{s}})\varepsilon.$$

Rearranging terms and letting $\varepsilon \downarrow 0$,

$$-\lambda x v_x(t,x,y) - v_y(t,x,y) + x - C_{\mathrm{s}} \leq 0. \qquad (6.32)$$

The third option of buying suggests the inequality

$$v(t,x,y) \geq v(t, x + \lambda x \varepsilon, y + \varepsilon) - (x + C_{\mathrm{b}})\varepsilon.$$

Passing to the limit $\varepsilon \downarrow 0$,

$$\lambda x v_x(t,x,y) + v_y(t,x,y) - x - C_{\mathrm{b}} \leq 0. \qquad (6.33)$$

Since one of the three possibilities should be optimal, equality should hold for one of (6.31)–(6.33), proving (6.29). The boundary condition (6.30) corresponds to the value of the y shares at time 0 because the model assumes permanent price impact. Guo and Zervos (2015, Prop. 4) carry out the verification that associates a smooth solution to the HJB equation (6.29)–(6.30) with the value function of the singular stochastic control problem. In this connection, they define the waiting, sell, and buy regions as

$$W = \left[(t, x, y) \in [0, T] \times \mathbb{R}^*_+ \times \mathbb{R}_+ : \mathcal{L}w(t, x, y) = 0 \right],$$
$$S = \left[(t, x, y) \in [0, T] \times \mathbb{R}^*_+ \times \mathbb{R}_+ : \lambda x w_x(t, x, y) + w_y(t, x, y) - x + C_s = 0 \right],$$
$$B = \left[(t, x, y) \in [0, T] \times \mathbb{R}^*_+ \times \mathbb{R}_+ : \lambda x w_x(t, x, y) + w_y(t, x, y) - x - C_b = 0 \right],$$

in which they use the notation $\mathbb{R}_+ = [0, \infty)$ and $\mathbb{R}^*_+ = (0, \infty)$. They show that if $\delta \geq \mu$, then the optimal strategy does not involve buying shares, i.e., the set B can be discarded. Moreover, in this case, it is optimal to sell all shares at time 0 if $C_s = 0$. If $\mu = \delta$ and $C_s > 0$, then selling all available shares at time T gives rise to a sequence of ϵ-optimal strategies. On the other hand, if $\delta < \mu$, then $v(T, x, y) = \infty$ and arbitrarily high payoffs can be achieved.

6.2.3 Infinite-horizon case $T = \infty$ with $0 \leq \mu < \delta$ and $C_s > 0$

As pointed out in the preceding paragraph, for $0 \leq \mu < \delta$, the optimal strategy does not involve buying shares, and it also does not entail selling all shares at time 0 since $C_s > 0$. Moreover, in the infinite-horizon case $T = \infty$, not all shares need to be liquidated in finite time and instead of (6.30), we now have the boundary condition

$$w(x, 0) = 0 \qquad \text{for all } x > 0, \tag{6.34}$$

and the HJB equation reduces to

$$\max \left\{ \mathcal{L}w(x, y), \ -\lambda x w_x(x, y) - w_y(x, y) + x - C_s \right\} = 0, \tag{6.35}$$

since the value function and optimal policy for the infinite-horizon case is time-invariant. In the wait region W, w should satisfy the ODE (in x)

$$\frac{1}{2}\sigma^2 x^2 w_{xx}(x, y) + \mu x w_x(x, y) - \delta w(x, y) = 0,$$

for which the only bounded solution as $x \downarrow 0$ is $w(x, y) = A(y)x^n$, where $n > 1$ for $\delta > \mu$ is the positive solution of the quadratic equation $\sigma^2 n(n - 1)/2 + \mu n - \delta = 0$. The boundary condition (6.34) implies that $A(0) = 0$.

The sell region S has boundary $x = F(y)$ so that $(x, y) \in S$ if $x \geq F(y)$ and $(x, y) \in W$ if $x < F(y)$. Since

$$-\lambda x w_x(x, y) - w_y(x, y) + x - C_s = 0 \qquad \text{for } x \geq F(y), \tag{6.36}$$

differentiating with respect to x yields

$$-\lambda x w_{xx}(x, y) - \lambda w_x(x, y) - w_{yx}(x, y) + 1 = 0 \qquad \text{for } x \geq F(y). \tag{6.37}$$

Guo and Zervos (2015) use the smooth fit principle (see Supplement 5 in

Section 3.4) that w is $C^{2.1}$ (i.e., twice continuously differentiable in x and continuously differentiable in y) to combine (6.36) and (6.37) with $w(x,y) = A(y)x^n$, yielding two simultaneous equations in $F(y)$, $A(y)$, and $\dot{A}(y)$, which have solutions

$$F(y) = \frac{nC_s}{n-1} := F_0, \qquad \dot{A}(y)F_0^n = -\lambda n A(y)F_0^n + F_0 - C_s. \qquad (6.38)$$

Since $A(0) = 0$, the ODE in (6.38) has solution

$$A(y) = \frac{1}{\lambda n^2}\left(\frac{n-1}{nC_s}\right)^{n-1}(1 - e^{-\lambda n y}).$$

In the sell region $\mathcal{S}$, there are two possible actions: either sell everything all at once, or sell something and then continue. Let $\mathbb{Y}(x) = \lambda^{-1}\ln(x/F_0)$. Then $F_0 - x = -x(1 - e^{-\lambda\mathbb{Y}(x)})$, $y - \mathbb{Y}(x) > 0 \Leftrightarrow x < F_0 e^{\lambda y}$. Partitioning the sell region into

$$\mathcal{S}_1 = \{(x,y) \in \mathbb{R}_+^* \times \mathbb{R}_+^* : x \geq F_0,\ y \leq \mathbb{Y}(x)\}$$
$$\text{and}\quad \mathcal{S}_2 = \{(x,y) \in \mathbb{R}_+^* \times \mathbb{R}_+^* : x \geq F_0,\ y > \mathbb{Y}(x)\},$$

Guo and Zervos (2015) show that $\mathcal{S}_1$ is the part of the sell region where it is optimal to sell all available shares at time 0, with value function

$$w(x,y) = \frac{1}{\lambda}x(1 - e^{-\lambda y}) - C_s y \qquad \text{for } (x,y) \in \mathcal{S}_1,$$

and that $\mathcal{S}_2$ is the part where it is optimal to sell $\mathbb{Y}(x)$ shares at time 0, with value function

$$w(x,y) = w(x, y - \mathbb{Y}(x)) + \frac{1}{\lambda}x(1 - e^{-\lambda\mathbb{Y}(x)}) - C_s\mathbb{Y}(x) \qquad \text{for } (x,y) \in \mathcal{S}_2. \quad (6.39)$$

As illustrated in Figure 6.1, if the stock price takes values in the waiting region $\mathcal{W}$, then it is optimal to take no action; if the stock price at time 0 is inside $\mathcal{S}_1$, then it is optimal to sell all available shares immediately; if the stock price at time 0 is inside $\mathcal{S}_2$, then it is optimal to liquidate an amount that would have sufficient price impact to drop the stock price to F_0. Guo and Zervos (2015, Prop. 5) carry out the verification that the value function is a $C^{2.1}$ solution to the HJB equation and is given by

$$w(x,y) = \begin{cases} 0 & \text{if } y = 0 \text{ and } x > 0 \\ A(y)x^n & \text{if } y > 0 \text{ and } x \leq F_0 \\ A(y - \mathbb{Y}(x))F_0^n + \frac{x - F_0}{\lambda} - C_s\mathbb{Y}(x) & \text{if } y > 0 \text{ and } F_0 < x < F_0 e^{\lambda y} \\ \frac{1}{\lambda}x(1 - e^{-\lambda y}) - C_s y & \text{if } y > 0 \text{ and } F_0 e^{\lambda y} \leq x. \end{cases}$$

Note that (6.39) can be expressed as $A(y - \mathbb{Y}(x))F_0^n + \lambda^{-1}(x - F_0) - C_s\mathbb{Y}(x)$ because of (6.36) and the solution to the ODE in (6.38).

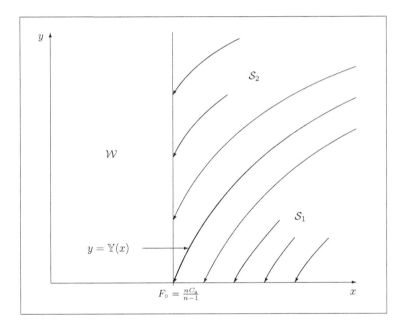

FIGURE 6.1: The "wait", "sell all", and "sell part" regions of the optimal strategy at time 0.

6.2.4   Price manipulation and transient price impact

The issue of price manipulation in optimal execution theory is concerned with whether profits can be generated by a round-trip trade, which is a trading strategy that involves 0 net buying or selling of shares over a given (finite) time horizon. Mathematically, an admissible round-trip trade with time horizon $T > 0$ is any pair $(\zeta^{\mathrm{s}}, \zeta^{\mathrm{b}})$ of $\mathcal{F}_t$-adapted increasing càdlàg processes such that $\zeta_0^{\mathrm{s}} = \zeta_0^{\mathrm{b}} = 0$, $\zeta_{T+}^{\mathrm{s}} = \zeta_{T+}^{\mathrm{b}}$ and $\sup_{t \in [0,T]}(\zeta_{t+}^{\mathrm{s}} - \zeta_{t+}^{\mathrm{b}}) \leq \Gamma$ for some constant $\Gamma > 0$, which may depend on the trading strategy itself. There are studies on the existence of price manipulation in additive price impact models. Huberman and Stanzl (2004) have shown that in the Almgren-Chriss model with $h = 0$, in order to ensure no price manipulation for all $T > 0$, $g(\cdot)$ has to be linear; see also Gatheral (2010) and Alfonsi and Blanc (2014). In the GBM model with multiplicative price impact, Guo and Zervos (2015) show that no price manipulation is possible if $\delta \geq \mu$, under which every round-trip trade has nonpositive expected payoff.

To extend their model to handle transient price impact, Guo and Zervos (2015, Remark 2) show that an additional state variable $Z_t = \lambda \int_0^t G(t-s)\, dY_s$ has to be introduced, where G is a kernel quantifying the decay of the price impact over time. In particular, for the case $G(u) = e^{-\gamma u}$, the additional state variable Z_t undergoes the stochastic dynamics $dZ_t = -\gamma Z_t\, dt - \lambda\, d\xi_t^{\mathrm{s}} + \lambda\, d\xi_t^{\mathrm{b}}$ and the value function is a function of $t, x, y,$ and z.

6.3   Optimal execution using the LOB shape

The optimal execution models discussed so far start with assumptions on the dynamics of the underlying asset price and the impact from the particular trader's order, and then introduce the objective functions. In particular, the price impact of a trade has been assumed to depend only on the size of the trade and does not indicate how the supply will respond over time to a sequence of trades. Obizhaeva and Wang (2013) argue that the intertemporal properties of the security's supply and demand should be considered for optimal execution. They propose an optimal execution model that incorporates the supply and demand information from the LOB. Their model, first proposed in a working paper in 2005, has attracted much attention and has inspired others to extend their work. In particular, Alfonsi et al. (2010) have built on the LOB-based market impact model of Obizhaeva and Wang (2013) and generalized it to allow for a general shape of the LOB. In view of Chapter 5 that shows the block shape assumed by Obizhaeva and Wang (2013) to be incompatible with the LOB data and the statistical models described therein, we follow the more general approach of Alfonsi et al. (2010) in this section and then use their result to derive a closed-form representation of the optimal execution strategy in the special case of a block-shaped LOB assumed by Obizhaeva and Wang.

Modeling time decay (resilience) of LOB response to large purchase order

Consider a trader who wants to purchase a large amount $x_0 > 0$ of shares within a certain time period $[0, T]$. Since the problem is about buy orders, we begin by concentrating on the upper part of the LOB, which consists of shares offered at various ask prices. The lowest ask price at which shares are offered is called the *best ask price* and is denoted by A_0 at time 0. Suppose first that the large trader is not active initially, so that the evolutions of the LOB are determined by the actions of noise traders only. We assume that the corresponding unaffected best ask price A_t^0 is a martingale and satisfies $A_0^0 = A_0$. This assumption includes in particular the case assumed by Obizhaeva and Wang, in which $A_t^0 = A_0 + \sigma B_t$ for a Brownian motion B_t. Alfonsi et al. (2010) emphasize that "we can take any martingale and hence use, e.g., a geometric Brownian motion", and that "we can allow for jumps in the dynamics of A_t^0 so as to model the trading activities of other large traders in the market." They also point out that "the optimal strategies will turn out to be deterministic, due to the described martingale assumption."

Above the unaffected best ask price A_t^0, they assume a continuous ask price distribution for available shares in the LOB: the number of shares offered at price $A_t^0 + x$ is given by $f(x)dx$ for a continuous function $f : \mathbb{R} \to (0, \infty)$, which is called the *shape function* of the LOB. Note that f is not a probability density function; its integral gives the number of available shares. The choice of a constant shape function corresponds to the block-shaped LOB model of Obizhaeva and Wang. The shape function determines the impact of a market order placed by a large trader. Suppose the large trader places a buy market order for $x_0 > 0$ shares at time $t = 0$. This market order will consume all shares located at prices between A_0 and $A_0 + D_{0+}^A$, where D_{0+}^A is determined by $\int_0^{D_{0+}^A} f(x)\, dx = x_0$. Consequently, the best ask price will be shifted up from A_0 to

$$A_{0+} = A_0 + D_{0+}^A.$$

Denote by A_t the actual best ask price at time t, i.e., the best ask price after taking the price impact of previous buy orders of the large trader into account. Let

$$D_t^A = A_t - A_t^0$$

denote the *extra spread* caused by the actions of the large trader. Another buy market order of $x_t > 0$ shares will consume all the shares offered at prices between A_t and $A_{t+} = A_t + D_{t+}^A - D_t^A$, hence

$$A_{t+} = A_t^0 + D_{t+}^A,$$

where D_{t+}^A is determined by the condition

$$\int_{D_t^A}^{D_{t+}^A} f(x) \, dx = x_t. \tag{6.40}$$

Thus, the process D_t^A captures the impact of market buy orders on the current best ask price. The price impact $D_{t+}^A - D_t^A$ is a nonlinear function of the order size x_t unless f is constant between D_t^A and D_{t+}^A. Hence the model includes the case of *nonlinear impact functions* considered by Almgren and Chriss that are described in Section 6.1. This is actually of the temporary type, and Alfonsi et al. (2010, p. 144) further explain why adding classical permanent impact "would be somewhat artificial in our model" and "would not change optimal strategies."

Another important quantity is the process of the number of shares "already eaten up" at time t:

$$E_t^A = \int_0^{D_t^A} f(x) \, dx. \tag{6.41}$$

It quantifies the impact of the large trader on the *volume* of the LOB. Letting

$$F(z) = \int_0^z f(x) \, dx, \tag{6.42}$$

we can express (6.41) as

$$E_t^A = F(D_t^A) \quad \text{and} \quad D_t^A = F^{-1}(E_t^A), \tag{6.43}$$

assuming that f is strictly positive to obtain the second identity. Note that (6.40) is equivalent to

$$E_{t+}^A = E_t^A + x_t. \tag{6.44}$$

It is an established empirical fact that order books exhibit certain resilience to the price impact of a large buy market order, i.e., after the initial impact the best ask price reverts back to its previous position; see Biais et al. (1995), Potters and Bouchaud (2003), Bouchaud et al. (2004), and Weber and Rosenow (2005). Accordingly Alfonsi et al. (2010) propose two models for the resilience of the market impact with *resilience speed* $\rho > 0$:

Model 1   The volume of the order book recovers exponentially, i.e., E evolves according to

$$E_{t+s}^A = e^{-\rho s} E_t^A \tag{6.45}$$

if the large investor is inactive during the time interval $[t, t + s)$.

Model 2   The extra spread D_t^A decays exponentially, i.e.,

$$D_{t+s}^A = e^{-\rho s} D_t^A \tag{6.46}$$

if the large investor is inactive during the time interval $[t, t + s)$.

The dynamics of E^A or D^A is completely specified by Model 1 or 2, respectively. Alfonsi et al. (2010) also use a similar idea to model the impact of a large sell order on the lower part of the LOB, which consists of a certain number of bids for shares at each price below *the best bid price*. As in the case of ask prices, they distinguish between an unaffected best bid price, B_t^0, and the actual best bid price, B_t, for which the price impact of previous sell orders of the large trader is taken into account. The distribution of bids below B_t^0 is modeled by the restriction of the shape function f to the domain $(-\infty, 0]$. More precisely, for $x < 0$, the number of bids at price $B_t^0 + x$ is equal to $f(x)\, dx$. The quantity $D_t^B := B_t - B_t^0$, which usually is negative, is called the *extra spread* in the bid price distribution. A sell market order of $x_t < 0$ shares placed at time t will consume all the shares offered at prices between B_t and

$$B_{t+} := B_t + D_{t+}^B - D_t^B = B_t^0 + D_{t+}^B,$$

where D_{t+}^B is determined by the condition

$$x_t = \int_{D_t^B}^{D_{t+}^B} f(x)\, dx = F(D_{t+}^B) - F(D_t^B) = E_{t+}^B - E_t^B,$$

in analogy with (6.43). Note that F is defined via (6.42) also for negative arguments. If the large trader is inactive during the time interval $[t, t+s)$, then the processes D^B and E^B behave just as their counterparts D^A and E^A, i.e., $E_{t+s}^B = e^{-\rho s} E_t^B$ in Model 1 and $D_{t+s}^B = e^{-\rho s} D_t^B$ in Model 2.

6.3.1 Cost minimization

When placing a single buy market order of size $x_t \geq 0$ at time t, the large trader will purchase $f(x)\, dx$ shares at price $A_t^0 + x$, with x ranging from D_t^A to D_{t+}^A . Hence, the total cost of the buy market order amounts to

$$\pi_t(x_t) = \int_{D_t^A}^{D_{t+}^A} (A_t^0 + x) f(x)\, dx = A_t^0 x_t + \int_{D_t^A}^{D_{t+}^A} x f(x)\, dx. \qquad (6.47)$$

In practice, large orders are often split into a number of consecutive market orders to reduce the overall price impact. Hence, the question at hand is to determine the size of the individual orders so as to minimize a cost criterion. Suppose that the large trader wants to buy a total of $x_0 > 0$ shares until time T and that trading can occur at $N + 1$ equidistant times $t_n = n\tau$ for $n = 0, \ldots, N$ and $\tau = T/N$. Alfonsi et al. (2010) define an *admissible strategy* to be a sequence $\boldsymbol{\xi} = (\xi_0, \xi_1, \ldots, \xi_N)$ of random variables such that

- $\sum_{n=0}^N \xi_n = x_0$,

- each ξ_n is measurable with respect to $\mathcal{F}_{t_n}$,

- each ξ_n is bounded from below,

in which ξ_n corresponds to the size of the order (number of shares) placed at time t_n, and can be negative (for an intermediate sell order) but is assumed to satisfy some lower bound.

The *average cost* $\mathcal{C}(\boldsymbol{\xi})$ of an admissible strategy $\boldsymbol{\xi}$ is defined as the expected value of the total costs incurred by the consecutive market orders:

$$\mathcal{C}(\boldsymbol{\xi}) = E\left[\sum_{n=0}^{N} \pi_{t_n}(\xi_n)\right]. \tag{6.48}$$

The goal is to find admissible strategies that minimize the average cost within the class of all admissible strategies. Note that the value of $\mathcal{C}(\boldsymbol{\xi})$ depends on whether Model 1 or Model 2 is chosen, and the quantitative features of the optimal strategies are model-dependent. For convenience, the function F is assumed to be unbounded in the sense that

$$\lim_{x \uparrow \infty} F(x) = \infty \quad \text{and} \quad \lim_{x \downarrow -\infty} F(x) = -\infty. \tag{6.49}$$

This implies unlimited order book depth, without impacting the development of an optimal strategy.

Simplified cost functional and reduction to deterministic strategies

Alfonsi et al. (2010, Appendix A) show how the minimization of the cost functional (6.48) can be reduced to minimization of certain cost functions $C^{(i)} : \mathbb{R}^{N+1} \to \mathbb{R}$, where $i = 1, 2$ refers to the model under consideration. They introduce simplified versions of the model by collapsing the bid-ask spread into a single value. More precisely, for any admissible strategy $\boldsymbol{\xi}$, they introduce a new pair of processes D and E that react on both sell and buy orders according to the following dynamics:

- $E_0 = D_0 = 0$ and

$$E_t = F(D_t) \quad \text{and} \quad D_t = F^{-1}(E_t).$$

- For $n = 0, \ldots, N$, regardless of the sign of ξ_n,

$$E_{t_n+} = E_{t_n} + \xi_n \quad \text{and} \quad D_{t_n+} = F^{-1}(\xi_n + F(D_{t_n})).$$

- For $k = 0, \ldots, N-1$, $E_{t_{k+1}} = e^{-\rho\tau} E_{t_k+}$ in Model 1, and $D_{t_{k+1}} = e^{-\rho\tau} D_{t_k+}$ in Model 2.

The values of E_t and D_t for $t \notin \{t_0, \ldots, t_N\}$ are not needed in this reduction. Note that $E = E^A$ and $D = D^A$ if $\boldsymbol{\xi}$ consists only of buy orders, while

$E = E^B$ and $D = D^B$ if $\boldsymbol{\xi}$ consists only of sell orders. They introduce the *simplified price* of ξ_n at time t_n by

$$\bar{\pi}_{t_k}(\xi_n) := A^0_{t_n} \xi_n + \int_{D_{t_n}}^{D_{t_n}+} x f(x)\, dx,$$

regardless of the sign of ξ_n, and show that

$$\bar{\pi}_{t_n}(\xi_n) \le \pi_{t_n}(\xi_n) \text{ with equality if } \xi_k \ge 0 \text{ for all } k \le n.$$

The *simplified price functional* is defined as

$$\bar{C}(\boldsymbol{\xi}) = E\left[\sum_{n=0}^{N} \bar{\pi}_{t_n}(\xi_n)\right].$$

They next reduce the minimization of $\bar{C}$ to the minimization of functionals $C^{(i)}$ defined on deterministic strategies. To this end, they use the notation

$$X_t := x_0 - \sum_{t_k < t} \xi_k \qquad \text{for } t \le T, \ X_{t_{N+1}} := 0.$$

The *accumulated simplified price* of an admissible strategy $\boldsymbol{\xi}$ is

$$\sum_{n=0}^{N} \bar{\pi}_{t_k}(\xi_n) = \sum_{n=0}^{N} A^0_{t_n} \xi_n + \sum_{n=0}^{N} \int_{D_{t_n}}^{D_{t_n}+} x f(x)\, dx.$$

Integrating by parts yields

$$\sum_{n=0}^{N} A^0_{t_n} \xi_n = -\sum_{n=0}^{N} A^0_{t_n}(X_{t_{n+1}} - X_{t_n}) = x_0 A_0 + \sum_{n=1}^{N} X_{t_n}(A^0_{t_n} - A^0_{t_{n-1}}). \quad (6.50)$$

Since $\boldsymbol{\xi}$ is admissible, X_t is a bounded predictable process. Hence, due to the martingale property of the unaffected best ask process A^0_t, the expectation of (6.50) is equal to $x_0 A_0$.

Next, observe that in Model $i = 1, 2$, the simplified extra spread process D evolves deterministically once the values $\xi_0, \xi_1, \dots, \xi_N$ are given. Hence, there exists a deterministic function $C^{(i)} : \mathbb{R}^{N+1} \to \mathbb{R}$ such that

$$\sum_{n=0}^{N} \int_{D_{t_n}}^{D_{t_n}+} x f(x)\, dx = C^{(i)}(\xi_0, \dots, \xi_N). \quad (6.51)$$

It follows that

$$\bar{C}(\boldsymbol{\xi}) = A_0 x_0 + E\left[C^{(i)}(\xi_0, \dots, \xi_N)\right].$$

In Sections 6.3.2 and 6.3.3 it will be shown that the function $C^{(i)}$ has a unique minimum for $i = 1, 2$ within the set

$$\Xi = \left\{ (z_0, \dots, z_N) \in \mathbb{R}^{N+1} : \sum_{n=0}^{N} z_n = x_0 \right\}.$$

The minimum gives the deterministic optimal strategies $\boldsymbol{\xi}^{(i)}$ for Model $i = 1, 2$.

6.3.2 Optimal strategy for Model 1

Minimum of the function $C^{(1)}$: *Suppose that the function $h_1 : \mathbb{R} \to \mathbb{R}_+$ defined by*

$$h_1(y) = F^{-1}(y) - e^{-\rho\tau} F^{-1}(e^{-\rho\tau} y)$$

is one-to-one. Then there exists a unique optimal strategy $\boldsymbol{\xi}^{(1)} = (\xi_0^{(1)}, \ldots, \xi_N^{(1)})$. The initial market order $\xi_0^{(1)}$ is the unique solution of the equation

$$F^{-1}\left(x_0 - N\xi_0^{(1)}(1 - e^{-\rho\tau})\right) = \frac{h_1(\xi_0^{(1)})}{1 - e^{-\rho\tau}}. \tag{6.52}$$

The intermediate orders are given by

$$\xi_1^{(1)} = \cdots = \xi_{N-1}^{(1)} = \xi_0^{(1)}(1 - e^{-\rho\tau}), \tag{6.53}$$

and the final order is determined by

$$\xi_N^{(1)} = x_0 - \xi_0^{(1)} - (N-1)\xi_0^{(1)}(1 - e^{-\rho\tau}).$$

In particular, the optimal strategy is deterministic, which is the minimum of the function $C^{(1)}$. Moreover, it consists only of nontrivial buy orders, i.e., $\xi_n^{(1)} > 0$ for all n.

Some remarks on this result are in order. First, the optimal strategy $\boldsymbol{\xi}^{(1)}$ consists only of buy orders and so the bid price remains unaffected; hence $E_t^B \equiv 0 \equiv D_t^B$. Moreover, once $\xi_0^{(1)}$ has been determined via (6.52), the optimal strategy consists of a sequence of market orders that consume exactly that amount of shares by which the LOB has recovered since the preceding market order, due to the resilience effect. At the terminal time $t_N = T$, all remaining shares are bought. Alfonsi et al. (2010, Appendix B) have given the following explicit formula of the function $C^{(1)}$. Let $a = e^{-\rho\tau}$. Define $\tilde{F}(z) = \int_0^z x f(x)\, dx$, $G(y) = \tilde{F}(F^{-1}(y))$. Then, for any $\boldsymbol{z} = (z_0, \ldots, z_N) \in \Xi$, $C^{(1)}(z_0, \ldots, z_N)$ is equal to

$$G(z_0) - G(0) + G(az_0 + z_1) - G(az_0) + G(a^2 z_0 + az_1 + z_2) - G(a^2 z_0 + az_1)$$

$$+ \cdots + G\left(\sum_{k=0}^{N} a^k z_{N-k}\right) - G\left(\sum_{k=1}^{N} a^k z_{N-k}\right),$$

and is twice continuously differentiable. Let

$$\hat{h}_1(y) = h_1(y) - (1-a)F^{-1}(x_0 - Ny(1-a)).$$

Alfonsi et al. (2010, Lemma B.2) show that if h_1 is one-to-one, then $\hat{h}_1$ has at most one zero. Combining this result with the fact that $C^{(1)}(\boldsymbol{z}) \to \infty$ as $\|\boldsymbol{z}\| \to \infty$ (and therefore $C^{(1)}$ has at least one minimizer in Ξ), they show that the minimizer is unique and is given by $\boldsymbol{\xi}^{(1)}$.

When is h_1 one-to-one?

The function h_1 is continuous with $h_1(0) = 0$ and $h_1(y) > 0$ for $y > 0$. Hence, h_1 is one-to-one if and only if h_1 is strictly increasing. To see when this is the case, note that the condition

$$h_1'(y) = \frac{1}{f(F^{-1}(y))} - \frac{e^{-2\rho\tau}}{f(F^{-1}(e^{-\rho\tau}y))} > 0$$

is equivalent to

$$l(y) := f(F^{-1}(e^{-\rho\tau}y)) - e^{-2\rho\tau}f(F^{-1}(y)) > 0.$$

Hence, h_1 will be one-to-one if the shape function f is decreasing for $y > 0$ and increasing for $y < 0$. In fact, it has been observed in empirical studies that typical shapes of LOBs have a maximum at or close to the best quotes and then decay as a function of the distance to the best quotes, which conform to this assumption.

6.3.3 Optimal strategy for Model 2

In a similar way, Alfonsi et al. (2010, Appendix C) derive the function $C^{(2)}$ and show that it has a unique minimizer given by the following.

Minimum of the function $C^{(2)}$: *Suppose that the function* $h_2 : \mathbb{R} \to \mathbb{R}$ *defined by*

$$h_2(x) = x\frac{f(x) - e^{-2\rho\tau}f(e^{-\rho\tau}x)}{f(x) - e^{-\rho\tau}f(e^{-\rho\tau}x)}$$

is one-to-one and that the shape function satisfies

$$\lim_{|x|\to\infty} x^2 \inf_{z\in[e^{-\rho\tau}x,x]} f(z) = \infty.$$

Then there exists a unique optimal strategy $\boldsymbol{\xi}^{(2)} = (\xi_0^{(2)}, \dots, \xi_N^{(2)})$. *The initial market order* $\xi_0^{(2)}$ *is the unique solution of the equation*

$$F^{-1}\left(x_0 - N\left[\xi_0^{(2)} - F\left(e^{-\rho\tau}F^{-1}(\xi_0^{(2)})\right)\right]\right) = h_2\left(F^{-1}(\xi_0^{(2)})\right),$$

and the intermediate orders are given by

$$\xi_1^{(2)} = \dots = \xi_{N-1}^{(2)} = \xi_0^{(2)} - F\left(e^{-\rho\tau}F^{-1}(\xi_0^{(2)})\right),$$

while the final order is determined by

$$\xi_N^{(2)} = x_0 - N\xi_0^{(2)} - (N-1)F\left(e^{-\rho\tau}F^{-1}(\xi_0^{(2)})\right).$$

In particular, the optimal strategy is deterministic, which is the minimum of the function $C^{(2)}$. Moreover, it consists only of nontrivial buy orders, i.e., $\xi_n^{(2)} > 0$ *for all n.*

6.3.4   Closed-form solution for block-shaped LOBs

Alfonsi et al. (2010, Sect. 6) illustrate their results with a block-shaped LOB corresponding to a constant shape function $f(x) \equiv q$ for some $q > 0$. They say: "In this case, there is no difference between Models 1 and 2" and the unique optimal strategy $\boldsymbol{\xi}^*$ is

$$\xi_0^* = \xi_N^* = \frac{x_0}{(N-1)(1 - e^{-\rho\tau}) + 2}$$

$$\text{and} \quad \xi_1^* = \cdots = \xi_{N-1}^* = \frac{x_0 - 2\xi_0^*}{N - 1}.$$

This result "provides an explicit closed-form solution of the recursive scheme obtained by Obizhaeva and Wang" who derive the recursions by dynamic programming under the following assumptions on the extra spread D_t^A:

$$D_t^A = \lambda \sum_{k:t_k < t} \xi_k + \sum_{k:t_k < t} (1/q - \lambda) e^{-\rho(t - t_k)} \xi_k,$$

in which $\lambda < 1/q$ quantifies the permanent impact and $1/q - \lambda$ the temporary impact.

6.4   Optimal execution for portfolios

Having analyzed the optimal execution problem for a single asset, it is natural to consider the same problem for portfolios. That is, if one has to purchase, within a fixed time period T, fixed blocks of shares in $m \geq 1$ stocks, how does one find the optimal sequence of trades that minimizes the expected execution costs? There is only limited literature on this problem, and there is almost no reference to the extensive econometrics literature on asynchronicity and synchronization described in Section 4.4.

Bertsimas and Lo's model with linear percentage price-impact

A multivariate generalization of Bertsimas and Lo's model has been carried out as in the single-asset case, with the notable difference being that the underlying price model is a GBM while the price impact still remains linear and additive. Let $\boldsymbol{N}_t = (N_{1t}, N_{2t}, \ldots, N_{mt})^\top$ so that N_{it} represents the number of shares of stock i bought at time t at price P_t^i, and let $\boldsymbol{P}_t = (P_t^1, P_t^2, \ldots, P_t^m)^\top$ for $t = 1, \ldots, T$. Let the execution price vector $\boldsymbol{P}_t$ be the sum of two components:

$$\boldsymbol{P}_t = \tilde{\boldsymbol{P}}_t + \boldsymbol{\delta}_t, \tag{6.54}$$

where $\tilde{\boldsymbol{P}}_t = (\tilde{P}_t^1, \ldots, \tilde{P}_t^m)^\top$ is the efficient price vector that would prevail in the absence of any market impact and is modeled as a multivariate GBM

$$\tilde{\boldsymbol{P}}_t = \text{diag}\left[\exp(z_t^1), \ldots, \exp(z_t^m)\right] \tilde{\boldsymbol{P}}_{t-1}, \tag{6.55}$$

where $z_t = (z_t^1, \ldots, z_t^m)^\top$ are i.i.d. $N(\boldsymbol{\mu}, \boldsymbol{\Sigma})$. The price impact term $\boldsymbol{\delta}_t$ is defined by

$$\boldsymbol{\delta}_t = \tilde{\boldsymbol{p}}_t(\boldsymbol{A}\boldsymbol{N}_t + \boldsymbol{B}\boldsymbol{X}_t), \tag{6.56}$$

where $\boldsymbol{A}$ is a positive definite $m \times m$ matrix, $\boldsymbol{X}_t$ is a vector of n information variables, $\boldsymbol{B}$ is an arbitrary $m \times n$ matrix, $\tilde{\boldsymbol{p}}_t = \mathrm{diag}(\tilde{P}_t^1, \ldots, \tilde{P}_t^m)$ and

$$\boldsymbol{X}_t = \boldsymbol{C}\boldsymbol{X}_{t-1} + \boldsymbol{\eta}_t, \tag{6.57}$$

in which $\boldsymbol{C}$ is an $n \times n$ matrix with eigenvalues all less than 1 in modulus and the $\boldsymbol{\eta}_t$ are i.i.d. with mean $\boldsymbol{0}$ and covariance matrix $\boldsymbol{V}$. Note that as a percentage of the efficient price $\tilde{P}_t$, the price impact is a linear function of the dollar value of the trade of the m stocks and other state variables $\boldsymbol{X}_t$. Let $\boldsymbol{Y}_t$ be the vector of shares remaining to be bought at time t. Then

$$\boldsymbol{Y}_1 = \tilde{\boldsymbol{N}}, \quad \boldsymbol{Y}_{T+1} = \boldsymbol{0}, \quad \text{and} \quad \boldsymbol{Y}_t = \boldsymbol{Y}_{t-1} - \boldsymbol{N}_{t-1}. \tag{6.58}$$

As in (6.2), the optimal execution problem is

$$\min_{\boldsymbol{N}_1, \ldots, \boldsymbol{N}_T} E\left[\sum_{t=1}^T \boldsymbol{P}_t^\top \boldsymbol{N}_t\right] \quad \text{subject to} \quad \sum_{t=1}^T \boldsymbol{N}_t = \tilde{\boldsymbol{N}}. \tag{6.59}$$

Bertsimas et al. (1999) carry out dynamic programming by backward induction and derive the following generalization of (6.10) and (6.11):

$$\boldsymbol{N}_{T-k}^* = \left(\boldsymbol{I} - \frac{1}{2}\boldsymbol{A}_{k-1}^{-1}\boldsymbol{A}^\top\right)\boldsymbol{Y}_{T-k} + \frac{1}{2}\boldsymbol{A}_{k-1}^{-1}\boldsymbol{B}_{k-1}^\top\boldsymbol{C}\boldsymbol{X}_{T-k}, \tag{6.60}$$

$$\begin{aligned}
V_{T-k} = {} & \boldsymbol{P}_{T-k-1}^\top\boldsymbol{Y}_{T-k} + \boldsymbol{Y}_{T-k}^\top\boldsymbol{A}_k\boldsymbol{Y}_{T-k} \\
& + \boldsymbol{X}_{T-k}^\top\boldsymbol{B}_k\boldsymbol{Y}_{T-k} + \boldsymbol{X}_{T-k}^\top\boldsymbol{C}_k\boldsymbol{X}_{T-k} + d_k,
\end{aligned} \tag{6.61}$$

for $k = 0, \ldots, T-1$, where

$$\begin{aligned}
\boldsymbol{A}_k &= \boldsymbol{A} - \frac{1}{4}\boldsymbol{A}\boldsymbol{A}_{k-1}^{-1}\boldsymbol{A}^\top, & \boldsymbol{A}_0 &= \boldsymbol{A}, \\
\boldsymbol{B}_k &= \frac{1}{2}\boldsymbol{C}^\top\boldsymbol{B}_{k-1}(\boldsymbol{A}_{k-1}^\top)^{-1}\boldsymbol{A}^\top + \boldsymbol{B}^\top, & \boldsymbol{B}_0 &= \boldsymbol{B}^\top, \\
\boldsymbol{C}_k &= \boldsymbol{C}^\top\boldsymbol{C}_{k-1}\boldsymbol{C} - \frac{1}{4}\boldsymbol{C}^\top\boldsymbol{B}_{k-1}(\boldsymbol{A}_{k-1}^\top)^{-1}\boldsymbol{B}_{k-1}\boldsymbol{C}, & \boldsymbol{C}_0 &= \boldsymbol{0}, \\
d_k &= d_{k-1} + E[\boldsymbol{\eta}_{T-k}^\top\boldsymbol{C}_{k-1}\boldsymbol{\eta}_{T-k}], & d_0 &= 0.
\end{aligned}$$

Portfolio extension of Obizhaeva and Wang's LOB-based model

Tsoukalas et al. (2014) point out two main strands in the literature on optimal execution for a single asset. The first "seeks to develop functional forms of price impact, grounded in empirical observations" such as Bertsimas and Lo, and Almgren and Chriss. The second, exemplified by Obizhaeva and Wang (2013), "focuses on market microstructure foundations", such as supply and demand of liquidity as reflected by the LOB. To extend the second strand to the portfolio setting, they develop a multi-asset order book model with correlated risks and coupled supply/demand dynamics, for which "an order executed in one direction (buy or sell) will affect both the currently available inventory of limit orders and also future incoming orders on either side." For tractability, they adopt Obizhaeva and Wang's assumptions of blocked-shaped LOBs and random walks for the efficient prices $\tilde{\boldsymbol{P}}_t$. Specifically, the shape functions of the LOBs of asset i at time t are

$$f_{i,t}^A(x) = q_i^A I_{\{x \geq a_{i,t}\}}, \qquad f_{i,t}^B(x) = q_i^B I_{\{x \leq b_{i,t}\}} \qquad (6.62)$$

on the ask and bid sides, and $\tilde{\boldsymbol{P}}_t - \tilde{\boldsymbol{P}}_{t-1}$ are assumed to be i.i.d. $N(\boldsymbol{0}, \tau\boldsymbol{\Sigma})$ with $\tau = T/N$ as in Section 6.3.1. Also, $a_{i,t}$ and $b_{i,t}$ are the best available ask and the best available bid of asset i at time t, respectively.

For asset i, the permanent price impact of an order to buy (sell) $N_{i,t}^+$ ($N_{i,t}^-$) shares at time t is assumed to be $\lambda_{ii}(N_{i,t}^+ - N_{i,t}^-)$. Noting that there is also cross-asset impact λ_{ij} on the price of asset j, Tsoukalas et al. (2014) define the matrix $\boldsymbol{\Lambda} = (\lambda_{ij})_{1 \leq i,j \leq m}$ and vectors $\boldsymbol{N}_t^+ = (N_{1,t}^+, \ldots, N_{m,t}^+)^\top$, $\boldsymbol{N}_t^- = (N_{1,t}^-, \ldots, N_{m,t}^-)^\top$, $\boldsymbol{N}_t = ((\boldsymbol{N}_t^+)^\top, (\boldsymbol{N}_t^-)^\top)^\top$. Since the trader's total net trades must sum up to $\boldsymbol{N}$, $\sum_{t=0}^T (\boldsymbol{N}_t^+ - \boldsymbol{N}_t^-) = \sum_{t=0}^T \boldsymbol{\delta}\boldsymbol{N}_t = \boldsymbol{N}$ where $\boldsymbol{\delta}$ is the $m \times (2m)$ matrix $(\boldsymbol{I}, -\boldsymbol{I})$ with $\boldsymbol{I}$ being the $m \times m$ identity matrix. Analogous to (6.58), define $\boldsymbol{Y}_0 = \boldsymbol{N}, \boldsymbol{Y}_t = \boldsymbol{Y}_{t-1} - \boldsymbol{\delta}\boldsymbol{N}_{t-1}$. Also define the temporary price impact matrices

$$\boldsymbol{Q}^A = \text{diag}\left(\frac{1}{2q_1^A}, \ldots, \frac{1}{2q_m^A}\right), \qquad \boldsymbol{Q}^B = \text{diag}\left(\frac{1}{2q_1^B}, \ldots, \frac{1}{2q_m^B}\right)$$

and the replenishment processes

$$\boldsymbol{d}_t^A = \text{diag}(e^{-\tau\rho_1^A}, \ldots, e^{-\tau\rho_m^A})\left\{\boldsymbol{d}_{t-1}^A + 2\boldsymbol{Q}^A\boldsymbol{N}_{t-1}^+ - \boldsymbol{\Lambda}\boldsymbol{\delta}\boldsymbol{N}_{t-1}\right\},$$

$$\boldsymbol{d}_t^B = \text{diag}(e^{-\tau\rho_1^B}, \ldots, e^{-\tau\rho_m^B})\left\{\boldsymbol{d}_{t-1}^B + 2\boldsymbol{Q}^B\boldsymbol{N}_{t-1}^- - \boldsymbol{\Lambda}\boldsymbol{\delta}\boldsymbol{N}_{t-1}\right\}$$

for the bid and ask sides of the LOB. Letting

$$\boldsymbol{a}_t = \tilde{\boldsymbol{P}}_t + \frac{1}{2}\boldsymbol{s}_0 + \boldsymbol{\Lambda}\sum_{k=0}^{t-1}\boldsymbol{N}_k + \boldsymbol{d}_t^A,$$

$$\boldsymbol{b}_t = \tilde{\boldsymbol{P}}_t - \frac{1}{2}\boldsymbol{s}_0 + \boldsymbol{\Lambda}\sum_{k=0}^{t-1}\boldsymbol{N}_k + \boldsymbol{d}_t^B,$$

where s_0 is the steady-state bid-ask spread, they show that the trader's reward at time t can be expressed as $\pi_t = (\boldsymbol{b}_t - \boldsymbol{Q}^B \boldsymbol{N}_t^-)^\top \boldsymbol{N}_t^- - (\boldsymbol{a}_t - \boldsymbol{Q}^A \boldsymbol{N}_t^+)^\top \boldsymbol{N}_t^+$, which is the difference between the revenues (from selling shares) and costs (from buying shares). Hence the trader's terminal wealth is $W_T = \sum_{t=0}^{T} \pi_t$. Tsoukalas et al. (2014) consider the problem of maximizing the expected exponential utility of terminal wealth, subject to the constraint $\sum_{t=0}^{T} \boldsymbol{\delta N}_t = \boldsymbol{N}$, and show that the optimal execution strategy $(\boldsymbol{N}_t^*, 0 \leq t \leq T)$ is again deterministic in the sense that it does not depend on $\tilde{\boldsymbol{P}}_t$.

6.5 Optimal placement

After big orders (parent orders) are sliced into orders of smaller sizes (child orders) to minimize the market impact subject to certain constraints, the next step is order placement. To carry out this step in a trading platform , traders must decide on (a) whether to use market orders, limit orders, or both, (b) the number of orders to place at different price levels, and (c) the optimal sequence of orders in a given time frame. In terms of cost when using limit orders, traders do not need to pay the spread and can even get a *rebate*, which comes with an execution risk as there is no guarantee of execution for limit orders. On the other hand, when using market orders, one has to pay both the spread between the limit and the market orders and the fee in exchange for a guaranteed immediate execution. Essentially, traders have to balance between paying the spread and fees when placing market orders against execution/inventory risks when placing limit orders. It is worth noting that the rebate structure varies from exchange to exchange and may lead to different optimization problems for different exchanges. In particular, in some exchanges, successful executions of limit orders receive a discount, which is a percentage of the execution price, whereas in other places the discount may be a fixed amount. In this section we focus on a single exchange with a particular fee structure.

Mathematically, the optimal placement problem can be stated as follows. Suppose N shares are to be bought by time $T > 0$ ($T \leq 10$ seconds). One may split the N shares into $(N_{0,t}, N_{1,t}, \dots)$, where $N_{0,t} \geq 0$ is the number of shares placed as a market order at time t, $N_{1,t}$ is the number of shares placed at the best bid at time t, $N_{2,t}$ is the number of shares placed at the second best bid price at time t, and so on, $t = 0, 1, \dots, T$. If the limit orders are not executed by time T, then one has to buy the non-executed orders at the market price at time T. When a share of limit order is executed, the market gives a rebate $r > 0$, and when a share of market order is submitted, a fee $f > 0$ is incurred. Although no intermediate selling is allowed at any time, one nevertheless can cancel any non-executed order and replace it with a new order at a later time. Given N and T, the optimal placement problem is to find the strategy $(N_{0,t}, N_{1,t}, \ldots)_{t=0,1,\ldots,T}$ that minimizes the overall total expected cost. One of the key quantities in the analysis of this problem is the probability that a limit order at a particular price level is executed. The evaluation of this quantity can be difficult, depending on the tractability of the underlying model. In the following, we present two models where the order placement has no price impact, hence the problem can reduced to the case $N = 1$.

6.5.1 Markov random walk model with mean reversion

Guo et al. (2013) study the optimal placement problem under the following assumptions:

(A1) The spread between the best bid price and the best ask price is 1 tick.

(A2) The best ask price increases or decreases 1 tick at each time step $t = 0, 1, \ldots, T$.

Let A_t be the best ask price at time t, expressed in ticks. Then

$$A_t = \sum_{i=1}^{t} X_i, \qquad A_0 = 0, \tag{6.63}$$

where $\{X_i\}_{i=1}^{T}$ is a Markov chain on $\{-1, 1\}$ with

$$P(X_1 = 1) = \bar{p} = 1 - P(X_1 = -1) \qquad \text{for some } \bar{p} \in [0, 1], \tag{6.64}$$
$$P(X_{i+1} = 1 | X_i = 1) = P(X_{i+1} = -1 | X_i = -1)$$
$$= p < 1/2 \qquad \text{for } i = 1, \ldots, T - 1. \tag{6.65}$$

Choosing $p < 1/2$ makes the price "mean revert", a phenomenon often observed in high-frequency trading. In fact, Cont and de Larrard (2013) have shown that the well-documented negative autocorrelation of price changes at lag 1 (see Sections 2.1.2 and 4.2) is equivalent to $p < 1/2$. They estimate $p = 0.35$ for liquid stocks such as CITI, and Chen and Hall (2013) have provided further statistical analysis of high-frequency data with mean reversion. Besides the price dynamics, Guo et al. (2013) also make the following assumptions on the conditional probability of a limit order being executed by time T given A_t.

- If $A_t \leq -k$ for some $t \leq T$, then a limit order at price $-k$ will be executed with probability 1.

- If $A_t > -k+1$ for all $t \leq T$, then a limit order at price $-k$ will be executed with probability 0.

- If $A_t = -k + 1$ and $A_{t+1} = -k + 2$, then a limit order at price $-k$ has probability q to be executed between t and $t + 1$.

Optimal placement strategy for the static case

Suppose the trader has to decide at what price level to place the buy order, only at time $t = 0$. If the order is placed as a limit order and is not executed at any time $t < T$, then a market order has to be placed at time $t = T$. Clearly, solving the optimal placement problem in this case amounts to comparing the expected costs, which depend on all the parameters $f, r, p, q, \bar{p}$, of placing an order at level $-k$ of the LOB (i.e., k ticks lower than the initial best ask price), for different values of k, where r is the rebate for limit orders and f the fee for market orders. We focus only on q and $\bar{p}$, and use $C(k, q, T, \bar{p})$ to denote the expected cost of a limit order placed at the price k ticks lower than the initial best ask price. Key results about this function are given in (R1)– (R3) below and their implication on computing the optimal placement strategy is also discussed. Note that $C(0, q, T, \bar{p})$ is the expected cost of a market order. Since all limit orders placed below $-T - 1$ will not be executed until time T because of (A2) and will have to be filled by market orders at time T, their expected costs are the same and are denoted by $C(T + 1, q, T, \bar{p})$. This leads to the optimization problem

$$\min_{0 \leq k \leq T+1} C(k, q, T, \bar{p}) = \min_{0 \leq k < \infty} C(k, q, T, \bar{p}).$$

For the optimal placement strategy, one can show the following by using the mean-reverting property of A_t and the monotonicity property of its running minimum (see Supplement 9 of Section 6.6):

(R1) *If $q = 0$, then $C(1, 0, T, \bar{p}) < C(2, 0, T, \bar{p}) < \cdots < C(T + 1, 0, T, \bar{p})$. That is, the best bid order is better than any other level of limit orders.*

(R2) *If $q \neq 0$, then $C(2, q, T, \bar{p}) < C(3, q, T, \bar{p}) < \cdots < C(T + 1, q, T, \bar{p})$. In particular, an optimal placement strategy depends only on comparing the expected costs at three levels: the market order, the best bid, and the second best bid.*

(R3) *If $\bar{p} \geq 1 - p$, then $C(1, q, T, \bar{p}) < C(2, q, T, \bar{p})$. That is, the optimal placement strategy involves only the market order and the best bid order.*

These results significantly reduce the computational cost of the optimal placement problem. Instead of comparing the expected cost at each single level of LOB, an optimal placement strategy usually involves comparing the expected costs at only the top three levels: $C(0, q, T, \bar{p})$ for the market order, $C(1, q, T, \bar{p})$ for the best bid, and $C(2, q, T, \bar{p})$ for the second best bid.

Optimal placement in a dynamic setting

Suppose the trader has to buy one share of stock by time T, but is allowed to place an order at any level at any time t and has the option of subsequently modifying the order by either canceling or changing of limit order to market order. At time T, the previously unexecuted limit order will automatically be replaced by a market order. In this dynamic setting, only two price levels need to be considered at any given time t: the best bid or the market order, or no order at all. This is because at each time period the price movement is at most one tick, placing an order at the level below the best bid is equivalent to placing no order at all, as this order will not be executed by the next time period. In view of the Markov structure of (A_t, X_t), the optimal placement problem is a Markov decision problem where the expected cost for taking each action at each step can be solved recursively. In fact, one can explicitly derive the optimal solution by further exploring the homogeneity of the value function. The optimal strategy turns out to be of a threshold type.

Optimal placement at time $t \le T - 1$

There exist integers t_1^* and t_2^* with $t_1^* < t_2^*$ such that it is optimal to

- place no order for $t < t_1^*$ and use the best bid for $t \ge t_1^*$ if $X_t = 1$,

- use the best bid for $t < t_2^*$ and the market order for $t \ge t_2^*$ if $X_t = -1$.

This is illustrated in Figures 6.2 and 6.3.

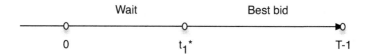

FIGURE 6.2: Optimal placement strategy for $1 < t \le T - 1$ when $X_t = 1$.

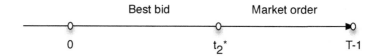

FIGURE 6.3: Optimal placement strategy for $1 < t \le T - 1$ when $X_t = -1$.

The following are explicit formulas for t_1^*, t_2^* and $V_t^*(1)$ and $V_t^*(-1)$, the value function evaluated at $X_t = \pm 1$:

$$t_1^* = T - \left\lceil \frac{1}{\log(p - pq)} \right. $$
$$\left. \cdot \log \frac{q(1 - 2p)}{((1 - p + pq)(f + 2 + r) - 1)(1 - q)(p^2 q + 1 + p^2 - 2p)} \right\rceil - 1;$$

$$t_2^* = T - \left\lceil \frac{1}{\log(p - pq)} \cdot \log \frac{(f + r + 1)(1 - p + pq) - (1 - p)(1 - q)}{(1 - p)(1 - q)((f + 2 + r)(1 - p + pq) - 1)} \right\rceil;$$

$$V_t^*(-1) = \begin{cases} f & \text{for } t \geq t_2^*, \\ (p + (1-p)q)(-1-r) + (1-p)(1-q)(1 + V_{t+1}^*(1)) & \text{for } t < t_2^*; \end{cases}$$

$$V_t^*(1) =$$

$$\begin{cases} (p - pq)^{T-t}\left(f + 2 + r - \dfrac{1}{1-p+pq}\right) - 2 - r + \dfrac{1}{1-p+pq} \\ \qquad\qquad\qquad\qquad\qquad \text{for } t_1^* \leq t \leq T, \\[2ex] A\left(\dfrac{a_1 + \sqrt{a_1^2 + 4a_2}}{2}\right)^{t_1^*-t+1} + B\left(\dfrac{a_1 - \sqrt{a_1^2 + 4a_2}}{2}\right)^{t_1^*-t+1} + a_4 \\ \qquad\qquad\qquad\qquad\qquad \text{for } t < t_1^*, \end{cases}$$

where $a_1 = p$, $a_2 = (1-p)(1-p-q+pq)$, $a_3 = 2p - 1 - (1+r)(1-p)(p + q - pq) + (1-p)(1-p-q+pq)$, $a_4 = a_3/(1 - a_1 - a_2)$, and the coefficients A and B are given by the following linear equations:

$$A + B = V_{t_1^*+1}^*(1) - a_4,$$

$$A\frac{a_1 + \sqrt{a_1^2 + 4a_2}}{2} + B\frac{a_1 - \sqrt{a_1^2 + 4a_2}}{2} = V_{t_1^*}^*(1) - a_4.$$

Note that the optimal placement policy is sensitive to both the remaining trading time and the market momentum because of price mean reversion. The policy becomes more conservative as t gets closer to T. If $X_t = -1$, then the price is more likely to go up in the next time period, hence one will use best bid or the market order, depending on $T - t$, and cannot afford placing no orders. If $X_t = 1$, then the price is more likely to go down in the next time period, hence one will either place no order or use the best bid order, depending on $T - t$. Note that both t_1^* and t_2^* could be negative in their respective formulas.

6.5.2 Continuous-time Markov chain model

For the optimal placement problem, Hult and Kiessling (2010) propose a continuous-time Markov chain model for the LOB. They consider the infinite-horizon setting (i.e., $T = \infty$). Using the Markov chain representation for the queue length of the LOB, they apply the potential theory of Markov chains (see Appendix B) to provide a method for computing the probability of an order ever being executed. They also prove the existence of the optimal placement policy and develop a value-iteration procedure to compute it.

Hult-Kiessling model

Let $\pi^1 < \pi^2 < \cdots < \pi^d$ denote all possible price levels in the LOB, and let $X_t = (X_t^1, \ldots, X_t^d)$ represent the volume of bid orders (negative values) and ask orders (positive values) at time $t > 0$, with $X_t^j \in \mathbb{Z}$. Let $\mathbb{S} \subset \mathbb{Z}^d$ be the state space of the Markov chain. For any $x \in \mathbb{S}$, define $j_B(x) = \max\{j : x^j < 0\}$, which is the highest bid level, and $j_A(x) = \min\{j : x^j > 0\}$, which is the lowest ask level. Then $x^j = 0$ for $j_B < j < j_A$, and $j_A - j_B$ is the spread. Let $e^j = (0, 0, \ldots, 1, 0, \ldots, 0)$ denote a vector in $\mathbb{Z}^d$ with 1 in the jth position. Denoting $j_B(x)$ simply by j_B and doing the same for j_A, the possible state changes associated with different orders are listed as follows.

- A limit bid order of size k at level j: $x \mapsto x - ke^j$, $j \leq j_A$.

- A limit ask order of size k at level j: $x \mapsto x + ke^j$, $j \geq j_B$.

- A market buy order of size k : $x \mapsto x - ke^{j_A}$, noting that ask orders are knocked out at j_A if $k > x^{j_A}$.

- A market sell order of size k : $x \mapsto x + ke^{j_B}$, noting that bid orders are knocked out at j_B if $k > |x^{j_B}|$.

- A cancellation of bid order of size k at level j: $x \mapsto x + ke^j$, $j \leq j_B$ and $1 \leq k \leq |x^j|$.

- A cancellation of ask order of size k at level j: $x \mapsto x - ke^j$, $j \geq j_A$ and $1 \leq k \leq x^j$.

It is difficult to implement this Markov chain model for the LOB because of the large transition probability matrix. Hult and Kiessling use a tractable parameterization that can be calibrated to data under the following conditions:

(C1) Limit bid (respectively, ask) orders at a distance of i levels from the best ask (bid) level arrive with intensity $\lambda_L^B(i)$ (respectively, $\lambda_L^A(i)$).

(C2) Market bid (respectively, ask) orders arrive with intensity λ_M^B (respectively, λ_M^A).

(C3) The distributions $(p_k)_{k \geq 1}$ and $(q_k)_{k \geq 1}$ of limit and market order sizes are given by

$$p_k = (e^\alpha - 1)e^{-\alpha k}, \qquad q_k = (e^\beta - 1)e^{-\beta k}.$$

(C4) The size of cancellation orders is 1, and a unit-size bid (respectively, ask) order located at a distance of i levels from the best ask (respectively, bid) level is canceled at a rate $\lambda_C^B(i)$ (respectively, $\lambda_C^A(i)$). Collectively, cancellations of bid (ask) orders at a distance of i levels from best ask (bid) level arrive with a rate $\lambda_C^B(i)|x^{j_A - i}|$ (respectively, $\lambda_C^A(i)|x^{j_B + i}|$) proportional to the volume at the level.

It follows from these conditions that the generator matrix Q is given by the following.

- Limit order:

$$x \mapsto x + e^j k \text{ for } j > j_B, \text{ with rate } p_k \lambda_L^A(j - j_B),$$
$$x \mapsto x - e^j k \text{ for } j < j_A, \text{ with rate } p_k \lambda_L^B(j_A - j).$$

- Market order of size with $k \geq 2$:

$$x \mapsto x + k e^{j_B}, \text{ with rate } q_k \lambda_M^A,$$
$$x \mapsto x - k e^{j_A}, \text{ with rate } q_k \lambda_M^B.$$

- Cancellation:

$$x \mapsto x - e^j \text{ for } j > j_B, \text{ with rate } \lambda_C^A(j - j_B)|x^j|,$$
$$x \mapsto x + e^j \text{ for } j < j_A, \text{ with rate } \lambda_C^B(j - j_A)|x^j|.$$

From the generator matrix $\boldsymbol{Q} = (Q_{xy})$ of X_t, in which Q_{xy} is the transition intensity from state $x = (x^1, \ldots, x^d)$ to state $y = (y^1, \ldots, y^d)$, we obtain the transition matrix $\boldsymbol{P} = (P_{xy})$ by $P_{xx} = 0$ and $P_{xy} = Q_{xy}/\sum_{z \neq x} Q_{xz}$ for $y \neq x$; see Appendix B. Hult and Kiessling use the embedded chain to compute the order execution probability and the optimal strategy as described below.

Order execution probability

To study the probability of the execution of a particular order that is placed at time 0 and price π^j, one needs to keep track of the queueing position of that order, in addition to the state of the entire LOB. Thus, one needs to combine the Markov chain of the LOB with the order position. Take the embedded chain X_n for the LOB. Let Y_n denote the number of bid orders ahead of and including the trader's order at time n and level j_0; for convenience a negative sign is attached to this number for bid orders. Suppose the trader puts a limit buy order at level j_0. The initial condition is $Y_0 = x_0^{j_0} - 1$ and the order can only move up toward 0 if an order ahead of the trader's order is executed or canceled. Note that $(X_n, Y_n)_{n \geq 0}$ is also a Markov chain with state space $\Omega = \mathbb{S} \times \{0, -1, \ldots\}$ and transition matrix $\boldsymbol{\Pi}$, where $\Pi_{(x,y),(x',y')} = P_{xx'} p_{yy'}$ and $(p_{yy'})_{y,y' \in \{0,-1,\ldots\}}$ is the transition matrix for the Markov chain $(Y_n)_{n \geq 0}$.

To derive the probability that an order is executed before the best ask level moves to J_1 when the order will be canceled, let $\tau = \inf\{n : (X_n, Y_n) \in \partial D\}$, where $\partial D = \{(x, y) \in \Omega : y = 0 \text{ or } j_A(x) = J_1\}$ is the boundary of $D = \{(x, y) \in \Omega : y < 0 \text{ and } j_A(x) < J_1\}$ and define the matrices $\boldsymbol{V}$ and $\boldsymbol{\Phi}$, whose (x, y) entries are

$$V(x, y) = \begin{cases} 1 & \text{if } y = 0 \\ 0 & \text{otherwise}, \end{cases}$$

$$\Phi(x, y) = E\left[V(X_\tau, Y_\tau) I_{\{\tau < \infty\}} | (X_0, Y_0) = (x, y)\right],$$

for $(x, y) \in \Omega$. Then according to potential theory (see Appendix B), Φ satisfies a system of linear equations

$$\Phi = \begin{cases} \Pi\Phi & \text{in } D, \\ V & \text{in } \partial D. \end{cases}$$

Optimal strategy for buying one share in infinite-horizon setting

In the infinite-horizon setting $T = \infty$, the optimal strategy is stationary. Suppose the initial state of the order book is x_0. A trader wants to buy one share, and places an order at price π^j. After n transitions of the order book, $j_n = j$ represents the level of the limit order, and Y_n represents the outstanding orders ahead of and including the trader's order at level j_n. This defines the discrete Markov chain $(\tilde{X}_n, Y_n, j_n)$. Hult and Kiessling (2010) assume that the trader has decided upon a best price level J_0 and a worst price level J_1 with $J_0 < j_A(x_0) < J_1$. The trader does not place bid orders at levels lower than J_0. The level J_1 is the worst-case buy price or stop-loss. If $j_A(\tilde{X}_n) = J_1$, the trader cancels the limit buy order immediately and places a market order at level J_1. The state space in this case is $\Omega = \mathbb{S} \times \{\ldots, -2, -1, 0\} \times \{J_0, \ldots, J_1\}$. The set of possible actions depend on the current state $\boldsymbol{s} = (x, y, j)$, at which the trader has three options if $y < 0$:

1. Do nothing and wait for a market transition.

2. Cancel the limit order and place a market buy order at $j_A(x)$.

3. Cancel the existing limit buy order and place a new limit buy order at any level j' with $J_0 \leq j' < j_A(x)$. This action results in the transition to $j_n = j'$, $\tilde{X}_n = x + e^j - e^{j'}$ and $Y_n = x^{j'} - 1$.

In a given state $\boldsymbol{s} = (x, y, j)$ with $y < 0$ and $j_A(x) < J$, the set of continuation actions is $C(x, y, j) = \{0, J_0, \ldots, j_A(x) - 1\}$. Here $a = 0$ represents the trader awaiting the next market transition, and the action $a = j'$ with $J_0 \leq j' < j_A(x)$ corresponds to canceling the outstanding order and submitting a new limit buy order at level j'. The cost of continuation is 0. If $y = 0$ or $j_A(x) = J_1$, then $C(\boldsymbol{s}) = \varnothing$ and only termination is possible. The set of termination actions is denoted by $\boldsymbol{T} = \{-2, -1\}$. If $y < 0$, the only termination action available is -1, representing cancellation of the limit order and submission of a market order at the ask price. If $y = 0$, the Markov chain always terminates since the limit order has been executed. This action is represented by -2. The Bellman equation (see Appendix B) in this case reduces to

$$V_\infty(\boldsymbol{s}) = \min \left(\min_{a \in C(\boldsymbol{s})} \sum_{s' \in \Omega} P_{\boldsymbol{s}\boldsymbol{s}'}(a) V_\infty(\boldsymbol{s}'), \min_{a \in \boldsymbol{T}(\boldsymbol{s})} v_{\boldsymbol{T}}(\boldsymbol{s}, a) \right) \qquad (6.66)$$

for $j_A(x) < J_1$, $y < 0$, $v_T(s, -2) = \pi^{j_0}$, $v_T(s, -1) = \pi^{j_A(x)}$, and

$$V_\infty(x, y, j) = \begin{cases} \pi^{J_1} & \text{for } j_A(x) = J_1, \ y < 0, \\ \pi^j & \text{for } y = 0. \end{cases} \qquad (6.67)$$

In (6.66), $P_{ss'}(a)$ denotes the transition probability from state s to state s' when action a is taken (just prior to the transition); see Section B.3 on controlled Markov chains.

6.6 Supplements and problems

1. **VWAP and TWAP** Volume Weighted Average Price (VWAP) and Time Weighted Average Price (TWAP) are two commonly used performance measures to evaluate execution algorithms. Using the same notations as those in Section 6.1, they are defined as

$$\text{VWAP} = \frac{\sum_{t=1}^T P_t N_t}{\sum_{t=1}^T N_t}, \qquad \text{TWAP} = \frac{1}{T}\sum_{t=1}^T P_t,$$

for discrete time. Their continuous-time definitions are

$$\text{VWAP} = \frac{\int_0^T P_t\, dY_t}{\int_0^T dY_t}, \qquad \text{TWAP} = \frac{1}{T}\int_0^T P_t\, dt.$$

Consider the random walk model (6.1) with linear and additive price impact. As discussed in Section 6.1, Bertsimas and Lo (1998) have shown that the optimal solution for execution problem (6.2) is uniform execution over time, i.e., $N_t^* = \tilde{N}/T$, which makes both measures equal.

$$\text{VWAP}_{N^*} = \frac{\sum_{t=1}^T P_t}{T} = \text{TWAP}_{N^*}.$$

Also, by (6.13),

$$E(\text{VWAP}_{N^*}) = \frac{V_1(P_0, Y_1)}{\tilde{N}} = P_0 + \frac{\theta}{2}\left(1 + \frac{1}{T}\right),$$

which increases linearly with the sensitivity-to-impact parameter θ. Moreover, if θ is small, $E(\text{VWAP}_{N^*})$ is essentially the initial price P_0.

Exercise Compare the optimal execution rule N^* of Bertsimas and Lo with that of Obizhaeva and Wang (OW) under the model (6.1) by using the performance measures VWAP and TWAP. The discrete-time OW rule can be obtained by replacing $(N - 1)(1 - e^{-\rho\tau})$ in the continuous-time solution in Section 6.3.4 by $\rho(T - 2)$. Show that

$$\text{OW}_1^* = \text{OW}_T^* = \frac{\tilde{N}}{\rho(T - 2) + 2} \quad \text{and} \quad \text{OW}_t^* = \frac{\rho\tilde{N}}{\rho(T - 2) + 2}$$

for $t = 2, \ldots, T - 1$. Show also that $\text{VWAP}_{\text{OW}} \neq \text{TWAP}_{\text{OW}}$ in general. Moreover, investigate the relationship between $E(\text{VWAP}_{\text{OW}})$ and $E(\text{VWAP}_{N^*})$ under different values of ρ.

2. Another commonly used performance measure is Percentage of Volume (PoV), which uses the actual volume traded in a period as the benchmark together with a participation rate β in the market during the trading period. For example, if Q shares are still to be purchased, PoV computes the volume V traded in the period $(t - \Delta t, t)$ and executes $\min(Q, \beta V)$.

3. *Exercise on the optimal execution problem of Bertsimas and Lo*   Solve the optimization problem (6.2) directly by rewriting the objective function in a quadratic form of $N_1, N_2, \ldots, N_T$.

4. A variant of the basic random walk model (6.1) with price impact incorporates market condition/private information X_t that undergoes AR(1) dynamics

$$P_t = P_{t-1} + \theta N_t + \gamma X_t + \epsilon_t,$$
$$X_t = \rho X_{t-1} + \eta_t,$$

where $-1 < \rho < 1$, $\{\epsilon_t\}_{t \geq 1}$ and $\{\eta_t\}_{t \geq 1}$ are two independent sequences of i.i.d. random variables with mean 0 and finite variance. Show that the optimal trading strategy is again deterministic in the sense that it is independent of P_t. Provide details to derive the optimal solution $N_{T-k}^* = \delta_{Y,k} Y_{T-k} + \delta_{X,k} X_{T-k}$, where

$$\delta_{Y,k} = \frac{1}{1 + k}, \quad \delta_{X,k} = \frac{\rho b_{k-1}}{2a_{k-1}}, \quad a_k = \frac{\theta}{2}\left(1 + \frac{1}{k + 1}\right),$$

$$b_k = \gamma + \frac{\theta\rho b_{k-1}}{2a_{k-1}} \ (> 0, \text{ if } \theta, \rho, \gamma > 0), \quad b_0 = \gamma.$$

5. The performance criteria used in Section 6.1 are the expected cost of Bertsimas and Lo (1998) and the mean variance criterion in Almgren and Chriss (2001). Other performance criteria that have been used in the literature include the expected utility of Schied and Schöneborn (2009) and the mean quadratic variation of Forsyth et al. (2012) for a GBM model. Gatheral and Schied (2013) argue for using a risk-neutral expected revenue or cost optimization criterion, especially in settings where market conditions should be independent of investor preferences. The equivalence between minimizing costs and minimizing the price impact of trading strategies has been established by Schied et al. (2010) for a class of models using the Euler-Lagrange equation.

6. *Exercise on Almgren and Chriss' model* Suppose one replaces Almgren and Chriss' variance term in Section 6.1.2 with a different risk measure. In particular, consider all admissible trading strategies to find

$$\inf_{Y_t \in \mathcal{A}} E\left[\eta \int_0^T \dot{Y}^2(t) \, dt + \lambda \sigma^2 \int_0^T Y(t) \, dt\right].$$

Show that the optimal liquidation strategy is of the form

$$Y^*(t) = (\tilde{N} - Ct)\left(1 - \frac{t}{T}\right),$$

where $C = \lambda \sigma^2 T / (4\eta)$.

7. *Exercise on Gatheral and Schied's model* Following Gatheral and Schied (2011), consider a GBM model for the optimal execution problem with a time-averaged Value-at-Risk (VaR). Specifically, consider

$$C(T, \tilde{N}, P_0) = \inf_{Y \in \mathcal{A}} E\left[\int_0^T \dot{Y}^2(t) \, dt + \lambda \int_0^T P(t)Y(t) \, dt\right],$$

subject to $Y(0) = \tilde{N}$, $Y(T) = 0$, where $dP(t) = \sigma P(t) \, dW(t)$. Derive the following HJB equation for $C(t, Y, P)$:

$$C_t = \frac{1}{2}\sigma^2 P^2 C_{PP} + \lambda PY + \inf_{\dot{Y}}(\dot{Y}^2 - \dot{Y}C_Y).$$

By considering $\tilde{N} = 0$ and letting $T \to 0$, derive the boundary condition for this HJB. Hence show that the unique optimal execution strategy is given by

$$Y_t^* = \frac{T-t}{T}\left[\tilde{N} - \frac{\lambda T}{4}\int_0^t P_u \, du\right].$$

8. For the multiplicative price impact model in Section 6.2.2, Guo and Zervos (2015, Lemma 2) have derived the following optimality properties explicitly without appealing to the HJB equation.

   - If $\mu \leq \delta$ and $C_s = 0$, then selling all available shares at time 0 is optimal.

   - If $\mu \leq \delta$, then buying shares is never part of an optimal strategy.

   - If $\mu = \delta$, then selling all available shares at time T is optimal.

9. The analysis of the optimal placement problem for the mean-reverting model in Section 6.4 uses certain results on the distribution of the random walk A_t and its running minimum $M_t = \min_{0 \leq u \leq t} A_u$, $0 \leq t \leq T$. In particular, Guo et al. (2013) have proved the following extensions of classical results on a random walk with i.i.d. increments and on its running minimum.

(a) If $(T - k)/2 \in \mathbb{Z}$ and $k > 0$, then $P(M_T = -k) = P(A_T = -k)$. This is called the *partial reflection principle* for Markov random walks.

(b) $P(A_T = k)$

$$
= \begin{cases}
\sum_{i=1}^{T-|k|} p^{T-i-1}(1-p)^i L_{T,k}^i & \text{for } \frac{T+k}{2} \in \mathbb{Z} \text{ and } |k| \leq T \\
0 & \text{otherwise,}
\end{cases}
$$

where

$$
L_{T,k}^i = (1 - \bar{p}) \binom{\frac{T+k}{2} - 1}{\lfloor \frac{i+1}{2} \rfloor - 1} \binom{\frac{T-k}{2} - 1}{\lfloor \frac{i+2}{2} \rfloor - 1}
$$
$$
+ \bar{p} \binom{\frac{T+k}{2} - 1}{\lfloor \frac{i+2}{2} \rfloor - 1} \binom{\frac{T-k}{2} - 1}{\lfloor \frac{i+1}{2} \rfloor - 1}.
$$

(c) $P(M_T = -k)$ is a decreasing function of k for $k = 1, \ldots, T$.

10. **Remarks on price manipulation**   Although absence of price manipulation is a desirable property of a model involving a short time-scale (such as seconds or minutes), it could be viewed as restrictive for models involving long time-scales (such as days or weeks). Indeed, in any model incorporating no price impact, the strategy that buys (respectively, short sells) one share of stock at time 0 and then sells it (respectively, buys it back) at time 1 is a price manipulation if the stock price is a submartingale (respectively, supermartingale), such as a geometric Brownian motion with a strictly positive (respectively, negative) drift; see Guo and Zervos (2015) and Gatheral and Schied (2011) and the references therein for detailed discussions on price manipulation.

Exercise   Compare the notion of no arbitrage for derivative pricing with that of no price manipulation, which is less widely known in quantitative finance and depends on specific performance criteria for the optimization problem.

11. Besides the optimal placement strategy, Hult and Kiessling (2010) also discuss a number of heuristic placement strategies. In particular, they consider the following naïve strategy that places a unit-size buy order at level J_0, and cancels the limit order at level J_1. They say: "The probability that the limit buy order is executed is all that is needed to compute the expected buy price", because the order will have to pay π^{J_0} (if executed) or π^{J_1} (if canceled and replaced by a market order at that price level). They show that potential theory (see Appendix B) conveniently reduces computation of the execution probability to solving a system of linear equations.

12. **Combining optimal execution and placement**   Guéant et al. (2012)

combine optimal trade scheduling (a.k.a. optimal execution) with optimally placing limit orders to liquidate a large number $\tilde{N}$ of shares of a given stock by time T. The inventory is $Y_t = \tilde{N} - \xi_t^a$, in which ξ^a is the jump process counting the number of shares sold. It is assumed that jumps are of size 1 and occur with intensity $\lambda(\delta^a)$ that depends on the price $X_t^a = X_t^0 + \delta_t^a$ quoted by the trader, in which X_t^0 is the efficient price process that is free of trading frictions and is assumed to be Brownian motion with drift: $dX_t^0 = \mu dt + \sigma dB_t$. This is, therefore, similar to the model in Section 6.2.1 which uses ξ^s to denote ξ^a and also does not model the process ξ^a. Guéant et al. (2012) assume that $\lambda(\delta^a) = A\exp(-\kappa\delta^a)$ and consider maximizing the expected exponential utility

$$\sup_{\delta^a \in \mathcal{A}} E\left(-\exp\left\{-\gamma\left[X_T + Y_{T-}(S_T - b)\right]\right\}\right), \tag{6.68}$$

where γ is the risk aversion parameter in the exponential utility $U(z) = 1 - e^{-\gamma z}$, $\mathcal{A}$ is the set of permissible trading strategies (predictable processes on $[0, T]$ that are bounded from below), b is the cost to liquidate all outstanding shares at time T, and $dX_t = X_t^a d\xi_t^a$ gives the amount of cash X_t the trader has at time t by selling the shares. They apply the dynamic programming principle to derive the HJB equation for the value function $u(s, x, y, t)$, which is (6.68) when E is replaced by the conditional expectation given $(X_t^0, X_t, Y_t) = (s, x, y)$. The HJB equation is

$$u_t + \mu u_s + \tfrac{1}{2}\sigma^2 u_{ss} + \max_{\delta^a} \lambda^a(\delta^a)\left[u(s, x + s + \delta^a, y - 1, t) - u(s, x, y, t)\right] = 0,$$

with the boundary conditions

$$u(s, x, 0, t) = -\exp(-\gamma x), \quad u(s, x, y, T) = -\exp\left\{-\gamma[x + y(s - b)]\right\}.$$

They solve this HJB equation by using a change of variables to transform the PDE into a system of linear ODEs. The basic idea is to "guess" the form of the solution based on the boundary condition. Once the correct form of the solution is derived, they use the usual verification approach to verify its optimality. The solution is of the form $u(s, x, y, t) = -\exp[-\gamma(x + ys)][v_y(t)]^{-\gamma/\kappa}$, in which v_y satisfies the linear ODE

$$\frac{dv_y(t)}{dt} = (\alpha y^2 - \beta y)v_y(t) - \eta v_{y-1}(t), \tag{6.69}$$

where $v_y(T) = \exp(-\kappa yb)$, $v_0(t) = 1$, $\alpha = \gamma\sigma^2/2$, $\eta = A[1 + (\gamma/\kappa)]^{-(1+\gamma/\kappa)}$. The optimal δ_*^a is given by

$$\delta_*^a(t, y) = \frac{1}{\kappa}\ln\left(\frac{v_y(t)}{v_{y-1}(t)}\right) + \frac{1}{\gamma}\ln\left(1 + \frac{\gamma}{\kappa}\right). \tag{6.70}$$

7

Market Making and Smart Order Routing

Market makers are as old as financial exchanges. They can be traced back to the late 1700s and were known as jobbers in London. Other than this almost forgotten name, they are more commonly known as dealers in NASDAQ and specialists in NYSE. Their primary business model is simple: buy securities from sellers and aim to resell them at a higher price to buyers in a relatively short time. There are minimum obligations imposed on market makers by the exchanges, for example, minimum spread width, minimum quoted volume, minimum percentage of the day the market maker must quote, and minimum time in force for market maker quotes; see Weaver (2012) for details.

From an economic perspective, a market maker is similar to a retail convenience store that operates 24 hours a day and is near many households. Without such stores, consumers who want milk in the middle of the night may need to wait until the next delivery from a farm. Also, the farmer who delivers 100 gallons of milk once a week is able to get paid from the store and does not need to wait until the 100 gallons of milk have been bought by individual consumers. Of course, if there are many nearby farms providing milk at various frequencies and with a full range of container sizes, the need for convenience stores that carry milk is diminished. Otherwise, convenience stores resolve the immediacy issue for both the farmer and the consumer. If liquidity is defined as the ease of converting assets into money and vice versa, the convenience store clearly provides liquidity to both the farmer and the consumer. Likewise, in the stock market, market making is designed to provide liquidity to both buyers and sellers of securities in the market. The market maker can switch from one security to another if the profit generated from the former is too low. Also, when the price of a security becomes too volatile, market makers of that security can stay away from it and resume trading when the price volatility returns to some tolerable level. Moreover, market makers operate under budget constraints and uncertainties in their forecasts of future supply and demand, and have therefore to manage their inventories well. In addition, they also need to face possible competition from other market makers because some exchanges, such as the London Stock Exchange and NASDAQ, adopt competing dealer systems. Compensation from the exchanges to market makers can be in various forms. For example, NYSE offers rebates and/or lower fees while Deutsche Börse and the Toronto Stock Exchange give information advantages and time priority, respectively.

This chapter first gives an overview of stochastic optimization problems associated with market making. Section 7.1 describes the optimization model of Ho and Stoll (1981) and the recent advances by Avellaneda and Stoikov (2008). Section 7.2 describes the solution by Guéant et al. (2013) who include an inventory constraint and subsequent extensions. In Section 7.3 we present another generalization by Guilbaud and Pham (2013) who consider market making with both limit and market orders. Their work uses impulse control, which is an area of stochastic control that has not been treated in previous chapters and is introduced here. Section 7.4 gives an overview of smart order routing (SOR) in fragmented primary markets and dark pools. Sections 7.5–7.7 consider the problem of optimal order splitting among exchanges and dark pools. Solution of these stochastic optimization problems involves approximate dynamic programming and machine learning which have been introduced in Supplement 7 of Section 3.4 and which we develop further here to address the problems at hand. Supplements and comments are given in Section 7.8, and exercises are presented in Section 7.9.

7.1 Ho and Stoll's model and the Avellanedo-Stoikov policy

Assume that the market maker's reference price process S_t is a Brownian motion with volatility σ. Assuming the CARA utility[1] function $1 - e^{-\gamma z}$, the agent's expected utility of simply holding q shares of stock until the terminal time T is $1 - u(x, s, q, t)$, where x is the initial wealth, s is the initial price per share, and

$$u(x, s, q, t) = E_t \{[-\exp(-\gamma(x + qS_T))]\}$$
$$= -\exp(-\gamma x)\exp(-\gamma qs)\exp\left[\gamma^2 q^2 \sigma^2 (T - t)/2\right]. \qquad (7.1)$$

Note that the agent's reservation (indifference) bid price r_t^b satisfies $u(x - r_t^b(s, q), s, q+1, t) = u(x, s, q, t)$, and is therefore given by $r_t^b(s, q) = s + (-1 - 2q)\gamma\sigma^2(T - t)/2$. Similarly, solving $u(x + r_t^a(s, q), s, q-1, t) = u(x, s, q, t)$ yields the reservation ask price $r_t^a(s, q) = s + (1 - 2q)\gamma\sigma^2(T - t)/2$.

The agent's market making policy is formulated as a stochastic control problem with two controls δ_t^a and δ_t^b defined below and three state variables: S_t, cash, and inventory. The agent quotes the bid price s_t^b and the ask price s_t^a, and the market buy/sell orders against the agent's ask (respectively, bid) orders arrive at Poisson rate $\lambda^a(\delta_t^a)$ (respectively, $\lambda^b(\delta_t^b)$), where $\delta_t^b = S_t - s_t^b$, $\delta_t^a = s_t^a - S_t$ are the spreads, and λ^a and λ^b are decreasing functions. Let N_t^a denote the amount of shares sold and N_t^b denote the amount of shares bought

---

[1] CARA stands for "constant absolute risk aversion"; Footnote 1 in Section 3.3.1 has introduced absolute risk aversion. The exponential utility function $U(z) = 1 - e^{-\gamma z}$ has absolute risk aversion coefficient γ, which does not change with z and is therefore "constant".

by the agent up to time t. Then they are Poisson processes with respective intensities $\lambda^a(\delta^a)$ and $\lambda^b(\delta^b)$. The cash X_t held by the agent satisfies the dynamics

$$dX_t = s_t^a dN_t^a - s_t^b dN_t^b. \tag{7.2}$$

The agent's inventory (number of shares held) at time t is $q_t = N_t^b - N_t^a$. The terminal wealth of the agent, therefore, is $X_T + q_T S_T$. The objective is to choose non-anticipative processes $\delta_a = \{\delta_t^a : 1 \le t \le T\}$ and $\delta_b = \{\delta_t^b : 1 \le t \le T\}$ to maximize the exponential utility so that the value function is equivalent (after subtracting 1) to

$$V(s, x, q, t) = \max_{\delta^a, \delta^b} E\left\{ -\exp\left[-\gamma(X_T + q_T S_T)\right] \,|\, (S_t, X_t, q_t) = (s, x, q)\right\}. \tag{7.3}$$

This problem was formulated by Ho and Stoll (1981).

Avellaneda and Stoikov (2008) suggest choosing $\lambda_a(\delta) = \lambda_b(\delta) = Ae^{-\kappa\delta}$, in which the positive constants A and κ characterize statistically the liquidity of the asset[2]. The CARA utility function reduces the value function to the form

$$V(s, x, q, t) = -\exp\left\{-\gamma[x + v(t, s, q)]\right\}, \tag{7.4}$$

and the HJB equation for v is

$$\max_{\delta^a} \gamma^{-1} A e^{-\kappa\delta^a} \left(1 - \exp\left\{-\gamma[s + \delta^a + v(t, s, q-1) - v(t, s, q)]\right\}\right)$$

$$+ \max_{\delta^b} \gamma^{-1} A e^{-\kappa\delta^b} \left(1 - \exp\left\{-\gamma\left[-s + \delta^b + v(t, s, q+1) - v(t, s, q)\right]\right\}\right)$$

$$+ v_t + \sigma^2 \frac{(v_{ss} - \gamma v_s^2)}{2} = 0, \tag{7.5}$$

with terminal condition $v(T, s, q) = qs$. The maxima in (7.5) are attained at

$$\begin{aligned}
\delta_*^a &= \gamma^{-1}\log(1 + \gamma/\kappa) - s - v(t, s, q-1) + v(t, s, q), \\
\delta_*^b &= \gamma^{-1}\log(1 + \gamma/\kappa) + s - v(t, s, q+1) + v(t, s, q).
\end{aligned} \tag{7.6}$$

Despite this simplification, the PDE (7.5)–(7.6) for v, with terminal condition $v(T, s, q) = qs$, is highly nonlinear and involves continuous variables t and s, and a discrete variable q. By using an asymptotic expansion to approximate v, Avellaneda and Stoikov (2008) develop an approximation to the optimal policy.

7.2 Solution to the HJB equation and subsequent extensions

Guéant et al. (2013) transform (7.5) and (7.6) into a system of ODEs; they

---

[2]Note that Guéant et al. (2012) in Supplement 12 of Section 6.6 use the same choice, which has the advantage of tractability when it is coupled with the exponential utility.

have already used a similar technique in another problem that Supplement 12 of Section 6.6 has considered. Including an additional constraint $|q| \leq Q$ on the inventory limits, the system is finite and can be solved explicitly. The solution can be expressed as

$$\exp\left\{-\gamma v(t, s, q)\right\} = \exp\{-\gamma qs\}\left(v_q(t)\right)^{-\gamma/\kappa}, \tag{7.7}$$

in which $(v_{-Q}(t), \ldots, v_{-1}(T), v_0(T), v_1(T), \ldots, v_Q(t))^\top$ is given explicitly by $e^{-(T-t)M}(1, \ldots, 1)^\top$, where letting $\beta = \kappa\gamma\sigma^2/2$ and $\eta = A(1 + \gamma/\kappa)^{-(1+\kappa/\gamma)}$,

$$M = \begin{pmatrix} \beta Q^2 & -\eta & 0 & \cdots & \cdots & \cdots & 0 \\ -\eta & \beta(Q-1)^2 & -\eta & 0 & \ddots & \ddots & \vdots \\ 0 & \ddots & \ddots & \ddots & \ddots & \ddots & \vdots \\ \vdots & \ddots & \ddots & \ddots & \ddots & \ddots & \vdots \\ \vdots & \ddots & \ddots & \ddots & \ddots & \ddots & 0 \\ \vdots & \ddots & \ddots & 0 & -\eta & \beta(Q-1)^2 & -\eta \\ 0 & \cdots & \cdots & \cdots & 0 & -\eta & \beta Q^2 \end{pmatrix},$$

where $e^A = \sum_{n=0}^{\infty} A^n/n!$ is the exponential of a square matrix A.

Extensions to other reference price models

The preceding HJB equation and its solution can be readily extended to Brownian motion with drift for the reference price, i.e., $dS_t = \mu dt + \sigma dB_t$. Extensions to the Ornstein-Uhlenbeck (OU) process

$$dS_t = \alpha(\mu - S_t)dt + \sigma dB_t \tag{7.8}$$

have been considered by Fodra and Labadie (2012) and Ahuja et al. (2016). Because the OU process is mean-reverting, Fodra and Labadie analyze the case when the reference price differs substantially from its steady-state mean and consider one or just part of a mean-reverting cycle, whereas Ahuja et al. are motivated by considering a long time horizon T covering multiple mean reversion cycles. Including the additional term $\alpha(\mu - S_t)$ in the reference price leads to an obvious modification of the HJB equation (7.5), which now has an additional summand $\alpha(\mu - s)$ on the right-hand side. The terminal condition still remains $v(T, q, s) = qs$. The highly nonlinear PDE (7.5) that includes an additional term $\alpha(\mu - s)$ with $\alpha > 0$, together with (7.6), can no longer be reduced to a finite system of ODEs. Ahuja et al. (2016) describe two numerical methods to solve it and compute the optimal policy. One is an implicit finite difference scheme and the other is a split-step scheme that splits the PDE into two PDEs, one of which can be further reduced to a finite system of ODEs under a boundedness constraint on the inventory limits.

7.3 Impulse control involving limit and market orders

In today's financial markets, limit order traders set the price of their orders subject to the discrete (tick) constraint of price quotes, and the market determines how fast their orders are executed. The stochastic control approach to market making described in the preceding section focuses on a limit order trader who optimizes the expected utility of terminal wealth generated by the market making activity. This model assumes a doubly stochastic point process of market order arrivals, with the intensity of the form $A \exp(-\kappa \delta_t)$, where δ_t is the spread between the limit order price and the observed underlying reference price. The stochastic control problem for market making thus formulated, however, does not use the information from the LOB about the actual arrival patterns and even ignores the discrete price levels of the LOB.

Guilbaud and Pham (2013) formulate a more flexible stochastic control problem for market making using both limit and market orders as the controls and LOB data as the information set. Thus, the agent can place buy and sell orders on the best bid and ask levels as well as one tick above the best bid and one tick below the best ask. Moreover, they assume the following continuous-time Markov chain model for the mid-price P_t and bid-ask spread D_t.

Model of the mid-price and bid-ask spread dynamics

- The mid-price P_t of the stock is assumed to be an exogenous Markov process on a given state space with a given infinitesimal generator.

- The bid-ask spread D_t of the LOB is modeled by a continuous-time Markov chain on a finite state space $\mathbb{S}$ consisting of multiples of the tick size δ, and subordinated by the Poisson process of the tick-time clock so that $D_t = \Delta_{N_t}$.

- The tick-time clock corresponds to a non-homogeneous Poisson process N_t with deterministic intensity $\lambda(t)$ generating the random times when the buy/sell orders in the market affect the spreads between the current best bid and best ask prices.

- Δ_n, $n = 0, 1, 2, \ldots$, is a Markov chain on $\mathbb{S}$ with states $i\delta$, $i = 1, \ldots, M$, and transition probabilities $\rho_{ij} = P[\Delta_{n+1} = j\delta | \Delta_n = i\delta]$ such that $\rho_{ii} = 0$.

- The processes P_t and D_t, $t \geq 0$ are independent.

7.3.1 Impulse control for the market maker

Limit order strategies

The market maker can submit at any time limit buy/sell orders at the best bid/ask prices as well as place orders at a marginally higher/lower price than the current best bid/ask. These four prices are defined, respectively, by $P_t^b = P_t - D_t/2$, $P_t^a = P_t + D_t/2$, $P_t^b - l_t$, $P_t^a + l_t$, in which $l_t = 0$ or δ represents the agent's choice. In addition, the number of shares q_t^b for a buy order and q_t^a for a sell order is also to be chosen. Hence, the market maker's control at time t is a predictable process

$$L_t = (q_t^b, q_t^a, l_t), \qquad t \geq 0. \tag{7.9}$$

For the limit order strategy L_t, the cash holdings X_t and the number of shares Y_t are given by

$$\begin{aligned} dY_t &= q_t^b dN_t^b - q_t^a dN_t^a, \\ dX_t &= -(P_t^b - l_t)q_t^b dN_t^b + (P_t^a + l_t)q_t^a dN_t^a. \end{aligned} \tag{7.10}$$

Market order strategy

The market order strategy is modeled by an impulse control that occurs at stopping times τ_i and consists of the number of shares $\zeta_i > 0$ (for buy orders) or $\zeta_i < 0$ (for sell orders) such that ζ_i is $\mathcal{F}_{\tau_i}$-measurable. Specifically, the market order strategy can be represented by the set of impulse controls

$$\mathcal{C} = \{(\tau_i, \zeta_i) : i = 1, 2, \dots\}.$$

Submissions of market orders take liquidity and are usually subject to fees, which Guilbaud and Pham (2013) assume to be fixed at $\phi > 0$. Therefore, the cash holding and the inventory jump at time τ_n are given by

$$\begin{aligned} Y_{\tau_n} &= Y_{\tau_n-} + \zeta_n, \\ X_{\tau_n} &= X_{\tau_n-} - c(\zeta_n, P_{\tau_n}, \Delta_{\tau_n}), \end{aligned} \tag{7.11}$$

where $c(\zeta, p, \Delta) = \zeta p + |\zeta|\Delta/2 + \phi$ represents the fixed fee ϕ plus a linear price impact for the market order.

7.3.2   Control formulation for the combined limit order and market order strategy to achieve optimality

The market maker has to liquidate the inventory by terminal date T. This leads to the liquidation function

$$L(x, y, p, \Delta) = x - c(-y, p, \Delta) = x + yp - |y|\Delta/2 - \phi,$$

which corresponds to the value an investor would obtain by liquidating all inventory at time T if $(X_T, Y_T, P_T, \Delta_T) = (x, y, p, \Delta)$. Let U be an increasing reward function and g be a non-negative convex function. Guilbaud and Pham (2013) formulate the optimal market making problem as the optimal choice of the regular controls L_t and the impulse control set $\mathcal{C}$, $0 \leq t \leq T$, in the

Markov control problem with value function

$$v(t, x, y, p, \Delta)$$
$$= \sup_{L, \mathcal{C}} E_{t,x,y,p,\Delta} \left\{ U\left[L(X_T, Y_T, P_T, \Delta_T)\right] - \gamma \int_t^T g(Y_u)\, du \right\}, \quad (7.12)$$

in which $E_{t,x,y,p,\Delta}$ stands for the conditional expectation given $(X_t, Y_t, P_t, \Delta_t)$ $= (t, x, y, p, \Delta)$.

Quasi-variational-inequality (QVI) for the stochastic control problem (7.12)

Guilbaud and Pham use the dynamic programming principle to show that the value function (7.12) is the unique viscosity solution[3] to the following extension of the HJB equation (7.5):

$$\min\left\{ -\frac{\partial v}{\partial t} - \sup_{q^b, q^a, l} \mathcal{L}^L v + \gamma g, v - \mathcal{M}v \right\} = 0, \quad (7.13)$$

together with the terminal condition $v(T, x, y, p, s) = U[L(x, y, p, \Delta)]$. In (7.13), $\mathcal{L}^L$ is the infinitesimal generator associated with the controlled Markov process with the regular control $L = (q^a, q^b, l)$ for placing limit orders, and $\sup_{q^b, q^a, l}$ refers to maximizing over $l \in \{0, 1\}$, the bid-quote size q^b and the ask-quote size q^a that have an upper bound Q as in Section 7.2. In addition to $\mathcal{L}^L$, (7.13) also has the *impulse operator* $\mathcal{M}$ associated with the impulse controls for placing market orders and is defined by

$$\mathcal{M}\psi(t, x, y, p, \Delta) = \sup_{|\zeta| \le Q} \psi\left[t, x - c(\zeta, p, \Delta), y + \Delta, p, \Delta\right]. \quad (7.14)$$

Pham (2005) gives a survey of viscosity solutions to HJB equations (or more general variational inequalities and quasi-variational inequalities). He first points out that the HJB equation is the infinitesimal version of the dynamic programming principle (DPP) for the value function v of the stochastic control problem $\sup_{a_s \in \mathcal{A}} E[\int_t^T f(s, X_s^{t,x}, a_s)\, ds + g(X_T^{t,x})]$:

$$-\frac{\partial v}{\partial t}(t, x) - \sup_{a \in \mathcal{A}} \{\mathcal{L}^a v(t, x) + f(t, x, a)\} = 0 \quad \text{on } [0, T) \times \mathbb{R}^n,$$

where $\mathcal{L}^a$ is the second-order infinitesimal generator associated to the diffusion X with constant control a. The function $H(t, x, v_x, v_{xx}) = \sup_{a \in \mathcal{A}} \{\mathcal{L}^a v(t, x) + f(t, x, a)\}$, in which $v_x = \partial v/\partial x$ and $v_{xx} = \partial^2 v/\partial x \partial x^\top$ are the gradient vector and the Hessian matrix of v evaluated at (t, x), is called the Hamiltonian of the associated control problem. The HJB equation "requires naturally the finiteness of the Hamiltonian H, which is typically satisfied when the set of controls $\mathcal{A}$ is bounded." When $\mathcal{A}$ is unbounded, H may take the value ∞ in some domain. In this case, the control problem is called singular. For singular

---

[3]The infinitesimal generator $\mathcal{L}^L$ of a controlled Markov process is introduced in Supplement 5 of Section 3.4. Variational and quasi-variational inequalities that extend the HJB equation to situations without the usual regularity conditions are introduced in the next paragraph, which also explains the concept of viscosity solutions in these situations.

stochastic control problems, the HJB equation is replaced by the variational inequality

$$\min\left\{-\frac{\partial v}{\partial t}(t,x) - H(t,x,v_x,v_{xx}), G(t,x,v_x,v_{xx})\right\} = 0,$$

where G is a continuous function such that $G \geq 0$ if and only if $H < \infty$. This variational inequality actually consists of two inequalities $-\partial v/\partial t - H \geq 0$ and $G \geq 0$ such that if one inequality is strict, then the other becomes an equality. A further extension, called quasi-variational inequality, replaces G above by other functionals such as $v - \mathcal{M}v$, where $\mathcal{M}$ is an operator associated with impulse controls as in (7.14) or with jumps as in a regime-switching model; see Bensoussan and Lions (1982) and Davis et al. (2010) for further details.

Existence of solutions of the HJB equation, a nonlinear PDE, requires strong regularity conditions that are violated in many stochastic control problems. These technical difficulties can be circumvented by considering viscosity solutions of the variational (or quasi-variational) inequality (instead of the HJB equations) for the value function v. Without requiring v to be continuous, the basic idea is to consider smooth test functions φ bounded below (respectively, above) by a lower (respectively, upper) semi-continuous envelope of v such that these smooth test functions satisfy the inequality $-\partial\varphi/\partial t - H(t,x,\varphi_x,\varphi_{xx}) \leq 0$ (respectively, ≥ 0), in which case v is called a viscosity subsolution (respectively, supersolution) of the variational (or quasi-variational) inequality. If v is both a viscosity subsolution and supersolution, then it is a viscosity solution; see Section 3 of Pham (2005) whose Section 5 also describes numerical methods to compute viscosity solutions.

7.4 Smart order routing and dark pools

Smart order routing

Order routing means taking an order from the end user to an exchange. Smart order routing (SOR), as illustrated by Figure 7.1, attempts to achieve "best" placement of orders by utilizing multiple choices of trading platforms based on price, cost, spread, and other factors. In 2010, SEBI (Securities and Exchange Board of India) approved the launch of SOR for every class of investors, and BSE (Bombay Stock Exchange) followed up by introducing SOR in September 2010. The availability of SOR was expected to help achieving better trade executions and better price discovery, and to increase electronic trading volume.

As we have pointed out in Sections 6.5, 7.2, and 7.3 on the optimal placement and market making, one of the main risks in algorithmic trading is the execution risk, which is the risk when traders cannot execute their intended orders during a specific time period and with the full size, and which is closely related to inventory risk in the operations research literature. In practice, in order to mitigate the execution risk in SOR, most traders utilize a combi-

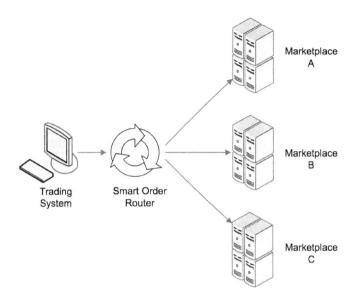

FIGURE 7.1: Illustration of smart order routing.

nation of visible (or primary) markets and dark pool markets that we will explain next.

Dark pools

One of the main difficulties in executing large-volume trades in visible markets is the difficulty to conceal such trades and dealing with consequent price impacts. Dark pools provide an alternative trading system that addresses these problems faced by large-volume traders. In a typical dark pool, neither the submitted orders nor the identities of the trades are publicly displayed. Instead buyers and sellers submit their orders by simply specifying the total volumes of shares they wish to buy or sell, with the transacted price exogenously determined according to matching rules of the specific dark pool. Upon submitting a buy (respectively, sell) order, a trader is put in a queue of buyers (respectively, sellers). Matching between buyers and sellers occurs according to the order of arrivals. However, unlike the primary markets that provide LOB information, a "dark pool" provides no information about how many parties or shares are available in the pool at any given moment. At any given period, the trader is informed only about the number of shares (up to the maximal capacity of the dark pool) that have been executed.

Dark pools have gained market shares and become increasingly popular. In the US, it is estimated that about 10–20% of overall equity volume is through dark pools, with approximately over forty dark pools for the US equity market alone. There are recognized benefits with dark pools. Besides trading with confidentiality and reduced market impact, the traders can save the cost of bid-ask spread as the trade price is usually at or close to the mid-point of the best bid and ask price. Many dark pools often offer lower execution costs as well. Every dark pool is different from the others in the rebate and fee structures and in the liquidity levels of the pool. Dark pools can be roughly categorized according to the matching rules for the price. One type is the non-displayable limited order books that could be agency/broker or exchange owned. Operated mostly by large broker-dealers, these dark pools are essentially continuously non-displayed limit order books, where the execution prices are derived from exchanges such as the midpoint of the National Best Bid and Offer (NBBO) or Volume-Weighted Average Price (VWAP). There is no price discovery in this case. Examples of agency/broker dark pools include Instinet, Liquidnet, and ITG Posit. Examples of exchange-owned dark pools include those offered by BATS Trading and NYSE Euronext. Another type of dark pools consists of those set up (and owned) by large broker-dealers for their clients and may also include proprietary traders. These dark pools derive their prices from order flows, so there can be price discovery. Examples include Credit Suisse Cross Finder, Goldman Sachs Sigma X, Morgan Stanley MS Pool, and Citigroup's Citi Cross. A third type of dark pools is offered by independent electronic market makers such as KCG (the merger of Knight Capital and Getco Holdings) that also derive their prices from order flows.

In the next three sections we consider the problem of optimal order splitting among exchanges (Section 7.5) and among dark pools (Sections 7.6 and 7.7). For the case of dark pools, an important feature is that certain important information that is needed to design the optimal order splitting scheme is not available (hence the adjective "dark"). Sections 7.6 and 7.7 show how ideas from stochastic adaptive control and reinforcement learning can be used to address this difficulty. The basic idea is to consider first the oracle case where the missing information is available, for which an optimal rule can be found, and then to modify it sequentially using suboptimal surrogates that generate data to estimate the missing information. The sequential nature of these adaptive control rules leads to recursions for updating them as new data come in. Similar recursions are used in Section 7.5 even for SOR among exchanges, basically to circumvent computational difficulties that arise when implementing the optimal order splitting procedure.

7.5 Optimal order splitting among exchanges in SOR

Since every exchange has its own reward and fee structure for handling limit and market orders — and even under identical market conditions exchanges could show different market depths and have different speeds in filling orders — how should one split a buy order of decent size among a group of exchanges to optimally reduce the execution risk together with the minimal cost within a limited horizon? Cont and Kukanov (2016) have studied the optimal order splitting problem in the simplified setting in which (a) a trader needs to buy N shares of a stock by time T and (b) a market order of any size up to N can be filled immediately at any of these exchanges. At time 0, the trader can send limit buy orders to the best bid positions of K trading venues. Let L_i be the trader's limit buy order size at time 0 at the ith venue for $i = 1, 2, \ldots, K$. The rebates (per share) for using limit orders and the fees (per share) for using the market orders at the ith venue are denoted by r_i and f_i, respectively. In addition, letting η_i denote the total outflow that consists of the order cancellations plus the trade executions from the front of the bid-queue at venue i in the period $[0, T]$, assume that $\boldsymbol{\eta} = (\eta_1, \ldots, \eta_K)$ is a random vector with a known joint distribution function F. At time 0, besides limit orders, the trader can also send a marketable buy order of size M to the best ask position of the venue whose f_i is the least among all venues. If the total amount of executed shares is less than N right before time T, the trader can then submit a final market order (with fees) to buy the remaining amount.

7.5.1 The cost function and optimization problem

The trader's order placement can be represented by $\boldsymbol{\Theta} = (M, L_1, \ldots, L_K) \in \mathbb{R}_+^{K+1}$. At time 0, limit orders with quantities $L_1, \ldots, L_K$ join queues of $Q_1, \ldots, Q_K$ orders at the K exchanges, with $L_i \geq 0$ and $Q_i \geq 0$ for all i. Cont and Kukanov assume that all these limit orders have the same price, which is the National Best Bid, i.e., the highest bid price across venues. The case $Q_i = 0$, corresponds to placing limit orders inside the bid-ask spread at venue i. Assuming also that limit orders are not modified before time T, they explicitly calculate the filled amounts (full or partial) of these limit orders at time T as a function of their initial queue position and future order outflow $\boldsymbol{\eta}$:

$$\min\left[\max(\eta_i - Q_i, 0), L_i\right] = (\eta_i - Q_i)^+ - (\eta_i - Q_i - L_i)^+$$

for $i = 1, \ldots, K$. Figure 7.2 illustrates the evolution of limit order fills in a FIFO queue from time 0 to T. Thus, the total executed quantity just before time T is given by

$$A(\boldsymbol{\eta}, \boldsymbol{\Theta}) = M + \sum_{i=1}^{K} \left[(\eta_i - Q_i)^+ - (\eta_i - Q_i - L_i)^+\right]. \qquad (7.15)$$

Using the mid-quote price as the benchmark, the execution cost associated with (7.15) is

$$(h + f)M - \sum_{i=1}^{K} (h + r_i) \left[(\eta_i - Q_i)^+ - (\eta_i - Q_i - L_i)^+\right], \qquad (7.16)$$

where h is half of the bid-ask spread at time 0, and f is the lowest market order fee (i.e., the lowest liquidity fee). There is an undershooting and overshooting penalty of λ_u and λ_o per share, respectively, which occurs when $A(\boldsymbol{\eta}, \boldsymbol{\Theta})$ in

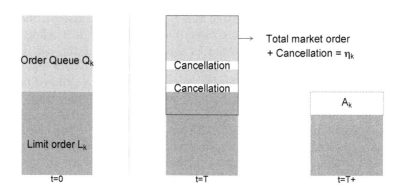

FIGURE 7.2: Evolution of of a limit order placed at venue k. The size of executed L_k is given by $A_k = (\eta_k - Q_k)^+ - (\eta_k - Q_k - L_k)^+$. Here L_k is partially filled, thus $A_k = \eta_k - Q_k$.

(7.15) differs from N. In addition, Cont and Kukanov model the market impact created at time 0 by $\theta(M + \sum_{i=1}^{K} L_i + [N - A(\boldsymbol{\eta}, \boldsymbol{\Theta})]^+)$, and call $\theta > 0$ the market impact factor. The total cost $c(\boldsymbol{\eta}, \boldsymbol{\Theta})$ is a sum of the execution cost (7.16) and these additional costs, i.e.,

$$c(\boldsymbol{\eta}, \boldsymbol{\Theta}) = (h + f)M - \sum_{i=1}^{K}(h + r_i)\left[(\eta_i - Q_i)^+ - (\eta_i - Q_i - L_i)^+\right]$$
$$+ \lambda_u[N - A(\boldsymbol{\eta}, \boldsymbol{\Theta})]^+ + \lambda_o[A(\boldsymbol{\eta}, \boldsymbol{\Theta}) - N]^+$$
$$+ \theta(M + \sum_{i=1}^{K} L_i + [N - A(\boldsymbol{\eta}, \boldsymbol{\Theta})]^+). \quad (7.17)$$

Therefore, the optimization problem is to choose $\boldsymbol{\Theta} = (M, L_1, \ldots, L_K)$ at time 0 to minimize $E[c(\boldsymbol{\eta}, \boldsymbol{\Theta})] = \int c(\boldsymbol{y}, \boldsymbol{\Theta}) \, dF(\boldsymbol{y})$. The optimal choice is denoted by $\boldsymbol{\Theta}^* = (M^*, L_1^*, \ldots, L_K^*)$.

7.5.2 Optimal order placement across K exchanges

Cont and Kukanov (2016, Sect. 3.2) assume that F is continuous, and

$$\max_i \{F_i(Q_i + N)\} < 1, \quad \lambda_u < \max_i \left\{(2h + f + r_i)[F_i(Q_i)]^{-1} - (h + r_i + \theta)\right\},$$

where $F_i(x) = P(\eta_i \leq x)$, and that

$$\min_i r_i + h > 0, \quad \lambda_u > h + f,$$
$$\lambda_o > \max\left\{-(h + f), \ h + \max_i r_i\right\}. \quad (7.18)$$

They point out that the three assumed inequalities in (7.18) have the following economic interpretations: "Even if some effective rebates r_i are negative, limit order executions let the trader earn a fraction of the bid-ask spread" and "the trader has no incentive to exceed the target quantity N." The following are their main results.

- If $\lambda_u \geq (2h + f + \max_i r_i)/F(Q_1, \ldots, Q_K) - (h + \max_i r_i)$, then $M^* > 0$.

- If $(h + r_i - \lambda_o)P(\eta_i > Q_i) > \theta$, then $L_i^* > 0$. Moreover, for $\boldsymbol{\Theta}^*$ to be the minimizer of $E[c(\boldsymbol{\eta}, \boldsymbol{\Theta})]$, it is necessary and sufficient that the following hold for $j = 1, \ldots, K$:

$$P\left\{M^* + \sum_{i=1}^{K}\left[(\eta_i - Q_i)^+ - (\eta_i - Q_i - L_i^*)^+\right] < N\right\} = \frac{h + f + \lambda_o + \theta}{\lambda_u + \lambda_o + \theta},$$

$$P\left\{M^* + \sum_{i=1}^{K}\left[(\eta_i - Q_i)^+ - (\eta_i - Q_i - L_i^*)^+\right] < N \,\middle|\, \eta_j > Q_j + L_j^*\right\}$$
$$= \frac{[\theta/P(\eta_j > Q_j + L_j^*)] + \lambda_o - (h + r_j)}{\lambda_u + \lambda_o + \theta}.$$

7.5.3 A stochastic approximation method

Since $E[c(\boldsymbol{\eta}, \boldsymbol{\Theta})]$ is a K-dimensional integral when F is absolutely continuous, Cont and Kukanov (2016, Sect. 4) note that it becomes "progressively difficult" to compute as the number of venues increases and finding the minimizer $\boldsymbol{\Theta}^*$ is even more difficult, whereas "practical applications require a fast and flexible method for optimizing order placement across multiple trading venues." To overcome this difficulty, they use a stochastic approximation method proposed by Nemirovski et al. (2009). Let $\mathbf{g}(\boldsymbol{\eta}, \boldsymbol{\Theta}) = \partial c(\boldsymbol{\eta}, \boldsymbol{\Theta})/\partial\boldsymbol{\Theta}$. The stochastic approximation method is a recursive procedure that initializes with a preliminary estimator $\boldsymbol{\Theta}^{(0)} \in \mathbb{R}_+^{K+1}$. It uses a step size $\gamma_m > 0$ that depends on m and draws i.i.d. samples $\boldsymbol{\eta}_j^*$ from the distribution F. The recursion is of the form

$$\boldsymbol{\Theta}_j = \boldsymbol{\Theta}_{j-1} - \gamma \mathbf{g}(\boldsymbol{\eta}_j^*, \boldsymbol{\Theta}_{j-1}), \qquad j = 1, \ldots, m. \tag{7.19}$$

An estimate of the optimal rule $\boldsymbol{\Theta}^*$ is given by $\hat{\boldsymbol{\Theta}}_m^* = \sum_{j=1}^{m} \boldsymbol{\Theta}_j/m$. Supplements 5 and 6 of Section 7.8 present a nonparametric variant of (7.19) and the convergence properties of $\hat{\boldsymbol{\Theta}}_m^*$ to $\boldsymbol{\Theta}^*$, and the statistical background of (7.19) and its variant.

7.6 Censored exploration-exploitation for dark pools

Ganchev et al. (2010) introduce a sequential learning and adaptive control approach to order placement in dark pools, highlighting the difference from SOR in the exchanges in the preceding section, which they call "light" (to contrast "dark") exchanges. They say:

> In a typical dark pool, buyers and sellers submit orders that simply specify the total volume of shares they wish to buy or sell, with the price of the transaction determined exogenously by "the market" (which is taken as the midpoint between the bids and asks in the light exchanges). Upon submitting an order to buy (or sell) ν shares, a trader is put in a queue of buyers (or sellers) awaiting transaction. Matching between buyers and sellers occurs in sequential arrival of orders, similar to a light exchange. However, unlike a light exchange, no information is provided to traders about how many parties or shares might be available in the pool at any given moment. Thus in a given time period, a submission of ν shares results only in a report of how many shares up to ν were executed.... How should one optimally distribute a large trade over the many independent dark pools? ... we analyze a framework and algorithm for a more general multi-venue exploration problem (based on censored observations).

7.6.1 A greedy allocation algorithm

Suppose at each time $t = 1, \ldots, T$, a trader allocates ν_t shares of a stock to K trading venues to sell these shares. The ith venue can only sell a maximum number of $U_t^i \in \mathbb{N}$ shares at time t, but the trader does not have this information since these venues are dark pools. It is assumed that $U_1^i, \ldots, U_T^i$ are i.i.d. with an unknown survival function T_i, i.e., $T_i(u) = P(U_t^i \geq u)$. Let $\boldsymbol{n}_t = (n_t^1, \ldots, n_t^K) \in \mathbb{N}^K$ be the trader's allocation at time t, with $\sum_{i=1}^K n_t^i = \nu_t$. The trader's objective, therefore, is to achieve

$$\max_{\boldsymbol{n}_t \in \mathbb{N}^K} \sum_{i=1}^K E\left[\min(n_t^i, U_t^i)\right] \qquad \text{subject to } \sum_{i=1}^K n_t^i = \nu_t, \qquad (7.20)$$

for every $t = 1, \ldots, T$. Noting that $E[\min(n, U_t^i)] = \sum_{s=1}^n T_i(s)$, Ganchev et al. (2010) express (7.20) as

$$\max_{\boldsymbol{n}_t \in \mathbb{N}^K} \sum_{i=1}^K \sum_{s=1}^{n_t^i} T_i(s) \qquad \text{subject to } \sum_{i=1}^K n_t^i = \nu_t,$$

which can be solved by using the following arguments.

- If $\nu_t = 1$, the optimal venue for placing this share is $j_1 = \arg\max_{i=1,\dots,K} T_i(1)$ and therefore the optimizing $\boldsymbol{n}_t = (n_t^1, \dots, n_t^K)$ satisfies $n_t^{j_1} = 1$ and $n_t^i = 0$ for $i \neq j_1$.

- If $\nu_t = 2$, using $j_1 = \arg\max_{i=1,\dots,K} T_i(1)$ as the venue for placing the first share, and $j_2 = \arg\max_{i=1,\dots,K} T_i(\nu_i + 1)$ as the venue for placing the second share, is optimal, where $\nu_i = 1$ if $i = j_1$ and $\nu_i = 0$ otherwise.

Proceeding inductively in this way, the optimal allocation rule $\boldsymbol{n}_{t,\text{opt}}$ that attains the maximum in (7.20) is a greedy algorithm which allocates one share at a time sequentially to maximize the tail probabilities over the K venues and which has the following pseudocode.

Greedy allocation algorithm

INPUT: Volume ν_t, tail probability distribution function $T_i, i = 1 \dots K$
OUTPUT: Optimal allocation $\boldsymbol{n}$
$\boldsymbol{n} = (n_1, \dots, n_K) \leftarrow \boldsymbol{0}$
for $\ell \in \{1, \dots, \nu_t\}$ **do**
$\quad j \leftarrow \arg\max_i T_i(n_i + 1)$
$\quad n_j \leftarrow n_j + 1$
end for
return $\boldsymbol{n}$

7.6.2 Modified Kaplan-Meier estimate $\hat{T}_i^t$ of T_i

To handle the issue of T_i being unknown, Ganchev et al. (2010) estimate T_i sequentially from the observed censored data $R_t^i = \max(n_t^i, U_t^i)$ by the Kaplan-Meier estimator, which is the nonparametric maximum likelihood estimator (NPMLE) of T_i based on the censored observations R_s^i (instead of the i.i.d. U_s^i) for $s \leq t$. Letting

$$M_{t,j}^i = \sum_{s=1}^t I_{\{R_s^i = j, n_s^i > j\}},$$

$$N_{t,j}^i = \sum_{s=1}^t I_{\{R_s^i \geq j, n_s^i > j\}},$$

the NPMLE of $h_j^i = P(U_t^i = j | U_t^i \geq j)$ is $\hat{h}_{t,j}^i = M_{t,j}^i / N_{t,j}^i$. Using the convention $0/0 = 0$, the NPMLE of $T_i(\ell)$ is

$$\hat{T}_i^t(\ell) = \prod_{j=0}^{\ell-1} \left(1 - \hat{h}_{t,j}^i\right); \tag{7.21}$$

see Section 2.4 of Lai and Xing (2016). Ganchev et al. (2010) note that $\hat{T}_i^t(\ell)$ can be updated over time via the recursions

$$
\begin{aligned}
M_{t+1,j}^i &= M_{t,j}^i + I_{\{R_{t+1}^i = j, n_{t+1}^i > j\}}, \\
N_{t+1,j}^i &= N_{t,j}^i + I_{\{R_{t+1}^i \geq j, n_{t+1}^i > j\}}.
\end{aligned} \tag{7.22}
$$

Lai and Ying (1991a) have pointed out that uniform consistency of $\hat{T}_i^t(\ell)$ only holds when the "risk set size" $N_{t,\ell}^i$ is sufficiently large, and have proposed a modification of the Kaplan-Meier estimate that has better convergence properties. Ganchev et al. (2010) propose a different modification that capitalizes on the sequential nature (over time t) of $\hat{T}_i^t(\ell)$. Their modification uses a cut-off value c_i^t so that $\hat{T}_i^t(\ell)$ estimates $T_i(\ell)$ well for $\ell \leq c_i^t$. Their choice of c_i^t involves two tuning parameters $\varepsilon > 0$ and $\delta > 0$:

$$
c_i^t = \max\left\{\ell \leq \nu : \ell = 0 \text{ or } N_{t,\ell-1}^i \geq 128[(\ell\nu)/\varepsilon]^2 \ln\left[(2\nu)/\delta\right]\right\}, \tag{7.23}
$$

recalling that $\nu = \sum_{t=1}^T \nu_t$ is the total number of shares to be sold by the trader by time T. They define the modified Kaplan-Meier estimator as

$$
\tilde{T}_i^t(\ell) = \begin{cases} \hat{T}_i^t(\ell) & \text{if } \ell < c_i^t + 1, \\ \hat{T}_i^t(c_i^t) & \text{if } \ell \geq c_i^t + 1. \end{cases}
$$

Their Lemmas 1 and 2 show that with probability that is at least $1 - \delta$,

$$
\begin{aligned}
|\tilde{T}_i^t(\ell) - T_i(\ell)| &\leq \varepsilon/(8\nu) & \text{for all } \ell \leq c_i^t; \tag{7.24}\\
|\tilde{T}_i^t(\ell) - T_i(\ell)| &\leq \varepsilon/(2\nu) & \text{for all } c_i^t < \ell \leq \nu, \text{ if } \hat{T}_i^t(c_i^t) \leq \varepsilon/(4\nu). \tag{7.25}
\end{aligned}
$$

7.6.3   Exploration, exploitation, and ε-optimal allocation

Ganchev et al. (2010, p. 102) say: "By optimistically modifying the tail probability $\hat{T}_i^t(c_i^t + 1)$ for each venue, we ensure that no venue remains unexplored simply because the algorithm unluckily observes a low demand a small number of times." Making use of this idea, they show that by replacing $T_i(\cdot)$ in the greedy allocation algorithm $n_{t,\text{opt}}$ by $\hat{T}_i^t(\cdot)$ at every time t, the adaptive allocation rule $\hat{n}_t$ is ε-optimal. This argument involves an "exploitation" lemma and an "exploration" lemma that are stated in their Lemmas 3 and 4.

Exploitation Lemma

The Exploitation Lemma says: *At any time t, if* (7.24) *holds and either* (a) $\hat{n}_i^t \leq c_i^t$ *or* (b) $\hat{T}_i^t(c_i^t) \leq \varepsilon/(4\nu)$ *for each venue i, then* $\max_{i=1,\dots,K} |\hat{n}_i^t - n_{t,\text{opt}}^i| < \varepsilon$. It implies that if (7.24) holds and the number of shares sold does not exceed the cut-off value or if (b) holds for each venue, then $\hat{n}_t$ is provably ε-optimal.

Exploration Lemma

Lemma 4 of Ganchev et al. (2010) shows that at any time t for which $\hat{\boldsymbol{n}}_t$ is not ε-optimal, there is a positive lower bound for the probability that $\hat{\boldsymbol{n}}_t$ generates "a useful observation", thereby increasing the relevant risk set size by 1. Their proof, which we restate using our notation, of this lemma explains the basic exploration idea to generate this useful observation:

> Suppose the allocation is not ε-optimal at time t. By the Exploitation Lemma, it must be the case that there is some venue i for which $\hat{n}_t^i > c_i^t$ and $\hat{T}_i^t(c_i^t) > \varepsilon/(4\nu)$, i.e., a venue in which the algorithm has allocated units past the cut-off but for which the tail probability at the cut-off is not too close to 0. Let ℓ be a venue for which this is true. Since $\hat{n}_t^i > c_i^t$, it will be the case that the algorithm obtains a useful observation for exploration of this venue (i.e., an observation causing the risk set size $N_{t,c_\ell^t}^\ell$ to be incremented).... Since $\hat{T}_\ell^t(c_\ell^t) > \varepsilon/(4\nu)$, (7.24) implies that $T_\ell^t(c_\ell^t) > \varepsilon/(8\nu)$.

Combining exploration and exploitation into an ε-optimal rule

The Exploitation and Exploration Lemmas suggest the name "censored exploration-exploitation rule" for $\hat{n}_t$. Theorem 3 of Ganchev et al. (2010) combines these lemmas to conclude that with probability at least $1 - \delta$, there exists a random burn-in time τ which is a polynomial in K, ν, $1/\varepsilon$, and $\ln(1/\delta)$ such that

$$P\left\{\max_{i=1,\dots,K} |\hat{n}_t^i - n_{t,\mathrm{opt}}^i| < \varepsilon\Big|\tau\right\} \geq 1 - \varepsilon \qquad \text{for all } t > \tau. \qquad (7.26)$$

7.7 Stochastic Lagrangian optimization in dark pools

Laruelle et al. (2011) introduce some additional features in the optimal order splitting problem of the preceding section. The most important distinguishing feature is that ν_t and U_t^i are random variables taking values in $(0, \infty]$, instead of $\mathbb{N}$ assumed by Ganchev et al. (2010). Therefore orders to the dark pools do not need to be integers and one can use r_t^i to denote the fraction of ν_t allocated to the ith dark pool such that $\sum_{i=1}^K r_t^i = 1$. Another main difference is that the rebate structure of the ith dark pool is modeled by using $\theta_i \in (0, 1]$ such that the bid price in the ith dark pool is $\theta_i b_t$, where b_t is the best bid price of the asset in the primary market at time t. Therefore, the cost $c(t)$ of splitting the purchase of ν_t shares of a stock among the K venues is

$$c(t) = b_t \sum_{i=1}^{K} \theta_i \min(r_t^i \nu_t, U_t^i) + b_t \left[\nu_t - \sum_{i=1}^{K} \min(r_t^i \nu_t, U_t^i) \right]$$

$$= b_t \left[\nu_t - \sum_{i=1}^{K} \rho_i \min(r_t^i \nu_t, U_t^i) \right],$$

where $\rho_i = 1 - \theta_i > 0$ is the coefficient of attraction of the ith dark pool. Under such a framework, the optimal order splitting problem becomes minimizing the expectation of $c(t)$ given b_t. Assuming $(\nu_t, U_t^1, \ldots, U_t^K)$ to be independent of b_t, the problem is equivalent to the optimization problem

$$\max_{r_t \in \mathbb{S}} \sum_{i=1}^{K} \rho_i E \left[\min(r_t^i \nu_t, U_t^i) \right], \tag{7.27}$$

where $r_t = (r_t^1, \ldots, r_t^K)$ and $\mathbb{S}$ is the K-dimensional simplex, i.e.,

$$\mathbb{S} = \left\{ (r_1, \ldots, r_K) \in \mathbb{R}_+^K : \sum_{i=1}^{K} r_i = 1 \text{ and } 0 \le r_j \le 1 \text{ for all } j \right\}.$$

Whereas Ganchev et al. (2010) use the representation $E[\min(n, U_t^i)] = \sum_{s=1}^{n} T_i(s)$ on which their greedy algorithm described in Section 7.6.1 is based, Laruelle et al. (2011) derive their stochastic Lagrangian algorithm from the concavity and differentiability of

$$\varphi_i(r) = E \left[\min(r \nu_t, U_t^i) \right], \tag{7.28}$$

which they call the mean execution function at the ith venue. Although $\min(r\nu, U)$ is not a smooth function of $r \in [0, 1]$, they show that φ_i is differentiable on $[0, 1]$ under the assumptions

$$P\{\nu_t > 0\} = 1, \quad P\{U_t^i > 0\} = 1, \quad \text{and } U_t^i / \nu_t \text{ is continuous.}[4] \tag{7.29}$$

Under (7.29), φ is differentiable on $[0, 1]$ with

$$\varphi_i'(r) = E[\nu_t I_{\{r\nu_t \le U_t^i\}}], \qquad \varphi_i'(0) = E[\nu_t I_{\{U_t^i > 0\}}] > 0. \tag{7.30}$$

7.7.1   Lagrangian approach via stochastic approximation

Relaxing the integer constraint on the allocations $L_1, \ldots, L_K$ in Section 7.5.2 provides the flexibility of using derivatives to solve the optimization problem

---

[3] A random variable is said to be continuous if it has a continuous distribution function.

(7.27). In fact, $\sum_{i=1}^{K} \rho_i E[\min(r_t^i \nu_t, U_t^i)] = \sum_{i=1}^{K} \rho_i \varphi_i(r_i)$. The constrained optimization problem (7.27) can be solved by using the Lagrangian approach that yields the equation $\partial G / \partial r_t = 0$, where

$$G(r_t) = \sum_{i=1}^{K} \rho_i \varphi_i(r_t^i) - \lambda \sum_{i=1}^{K} r_t^i$$

and λ is the Lagrange multiplier. The optimal allocation rule $r_{t,*} = (r_{t,*}^1, \ldots, r_{t,*}^K)$, in which we use $*$ to denote "optimal", is therefore the solution of the system of equations

$$\rho_j \varphi_j'(r_{t,*}^j) = \frac{1}{K} \sum_{i=1}^{K} \rho_i \varphi_i'(r_{t,*}^i), \qquad j = 1, \ldots, K. \tag{7.31}$$

Although (7.31) typically requires $r_{t,*} \in \text{Int}(\mathbb{S})$, Laruelle et al. (2011, Proposition 1) give the following condition for (7.31) to hold even when $r_{t,*} \in \partial \mathbb{S}$:

$$\min_{1 \leq i \leq K} \varphi_i'(0) \geq \max_{1 \leq i \leq K} \varphi_i'\left[(K-1)^{-1}\right]. \tag{7.32}$$

They discuss the interpretation of this condition and show that it holds under (7.29) when all the ρ_i's are equal, i.e., when all K venues give the same rebates. They also show that if strict inequality holds in (7.32), then $r_{t,*} \in \text{Int}(\mathbb{S})$.

Implementation via stochastic approximation (SA)

Combining (7.30) with (7.31) yields

$$E\left[\nu_t \left(\rho_j I_{\{r_{t,*}^j \nu_t \leq U_t^j\}} - \frac{1}{K} \sum_{i=1}^{K} \rho_i I_{\{r_{t,*}^i \nu_t \leq U_t^i\}}\right)\right] = 0, \qquad j = 1, \ldots, K. \tag{7.33}$$

Direct implementation of (7.33) would require specification of the joint distribution of $(\nu_t, U_t^1 \ldots, U_t^K)$ to evaluate the expectation, which can be carried out by Monte Carlo methods. Section 2.1 of Laruelle et al. (2011) discusses some possible models, but this clearly is not practical because the functional form of $r_{t,*}$ in terms of $(\nu_t, U_t^1 \ldots, U_t^K)$ is yet to be determined. Their Section 2.2 discards this direct approach and appeals to stochastic approximation[5] which is a recursive scheme over time t. Because the solution $r_{t,*}$ is constrained to belong to $\mathbb{S}$, their Section 3.3 uses an additional technique to convert it to an unconstrained problem so that SA can be applied. Specifically the SA recursion is

$$r_{t+1} = r_t + \gamma_t H(r_t, \nu_t, U_t^1, \ldots, U_t^K), \tag{7.34}$$

where $\gamma_t > 0$ is a sequence of learning rates, r_0 is an initial estimate and $H(r_t, \nu_t, U_t^1, \ldots, U_t^K)$ is a K-dimensional vector whose jth entry is given by

---

[5]Supplement 6 in Section 7.8 provides further background and details of SA.

a function of the form

$$H_j(\boldsymbol{r}, \nu, U_1, \ldots, U_K)$$
$$= \nu \left[\rho_j I_{\{r_j \nu \le U_j\}} - \frac{1}{K} \sum_{i=1}^{K} \rho_i I_{\{r_i \nu \le U_i\}} + \xi_j(\boldsymbol{r}, \nu, U_1, \ldots, U_K) \right], \quad (7.35)$$

where ξ_j has a "mean-reverting" effect to pull $\boldsymbol{r}$ into the simplex $\mathbb{S}$:

$$\xi_j(\boldsymbol{r}, \nu, U_1, \ldots, U_K) = \rho_j \left[(1 - r_j) I_{\{0 \le U_j, r_j < 0\}} + \frac{1}{r_j} I_{\{\nu \le U_j, r_j > 1\}} \right]$$
$$- \frac{1}{K} \sum_{i=1}^{K} \rho_i \left[(1 - r_i) I_{\{0 \le U_i, r_i < 0\}} + \frac{1}{r_i} I_{\{\nu \le U_i, r_i > 1\}} \right], \quad (7.36)$$

which is derived by extending φ_j on the unit interval to a concave and differentiable function on the real line. Section 7.7.2 discusses why such recursion converges to the desired limit as $t \to \infty$.

7.7.2 Convergence of Lagrangian recursion to optimizer

Laruelle et al. (2011, p. 1054) give the following heuristic interpretation of the stochastic approximation recursion (7.34)–(7.36):

> As long as r is a true allocation vector, i.e., lies in the simplex... assume first that all the factors ρ_i are equal (to 1). Then the dark pools which fully executed the sent orders ($r_i \nu \le U_i$) are rewarded proportionally to the numbers of dark pools which did not fully execute the request they received. Symmetrically, the dark pools which could not execute the whole request are penalized proportionally to the number of dark pools which satisfied the request.

Hence, if only the first L dark pools have fully executed the requests at time t, then $r_{t+1}^i = r_t^i + \gamma_t(1 - L/K)\nu_t$, for $i = 1, \ldots, L$, i.e., these venues are rewarded by increasing their proportions at time $t + 1$. On the other hand, $r_{t+1}^j = r_t^j - \gamma_t(L/K)\nu_t$ for $j = L+1, \ldots, K$, i.e., these venues are penalized by reducing their proportions at time $t+1$. This interpretation can be extended to the situation in which the ρ_i's may not be all equal, by weighting the penalties and rewards by ρ_i.

Laruelle et al. (2011, Theorem 4.1) study the convergence properties of the stochastic approximation recursion (7.34) under the assumption that $(\nu_t, U_t^1, \ldots, U_t^K)$ are i.i.d. having the same distribution as $(\nu, U_1, \ldots, U_K)$. In addition to the i.i.d. assumption, they also assume (7.32) and the following conditions:

(i) $\nu > 0$ almost surely with $E(\nu^2) < \infty$;

(ii) $F_i(x) = P\{U_i/\nu \le x\}$ is a continuous function on $\mathbb{R}_+$ and $P(U_i > 0) > 0$ for all $i = 1, \ldots, K$;

(iii) $\sum_{t=1}^{\infty} \gamma_t = \infty$ and $\sum_{t=1}^{\infty} \gamma_t^2 < \infty$.

Under these assumptions, they prove that (7.34) converges almost surely to

$$\boldsymbol{r}_* = \arg \max_{\boldsymbol{r} \in \mathbb{S}} \sum_{i=1}^{K} \rho_i E \left[\min(r_i, \nu, U_i)\right] \tag{7.37}$$

as $t \to \infty$. Moreover, if strict inequality holds in (7.32), then $\boldsymbol{r}_* \in \mathrm{Int}(\mathbb{S})$.

Section 5 of Laruelle et al. (2011) replaces the stringent i.i.d. assumption (which is made to facilitate the application of conventional SA theory) by the much weaker assumption that $(\nu_t, U_t^1, \ldots, U_t^K)$ is an ergodic sequence that converges weakly to $(\nu, U_1, \ldots, U_K)$. Assumption (i) now takes the form $\nu > 0$ and $\sup_t E\nu_t^2 < \infty$. Moreover, they replace (iii) by the assumptions

(a) $T^{-1} \sum_{t=1}^{T} \nu_t I_{\{y\nu_t \le U_t^j\}} \to E[\nu I_{\{y\nu \le U_j\}}] = O(T^{-\alpha_j})$ almost surely for all $y \in \mathbb{R}_+$ and $j = 1, \ldots, K$, and for some $\alpha_j \in (0, 1]$;

(b) $\sum_{t=1}^{\infty} \gamma_t = \infty$, $\gamma_t = o(t^{a-1})$, and $\sum_{t=1}^{\infty} t^{1-a} \max(\gamma_t^2, |\gamma_t - \gamma_{t+1}|) < \infty$,

where $a = \min_{1 \le i \le K} \alpha_j$. In this more general setting, they again prove that $\boldsymbol{r}_t \to \boldsymbol{r}_*$ almost surely; see Supplement 6 in Section 7.8 for further details.

7.8 Supplementary notes and comments

1. **Inventory control in market making** Whereas Sections 7.1–7.3 present specific models and results on optimal market making in the literature, this supplement describes variants of those models and inventory control therein, particularly those used in empirical studies or that are data-driven. In Section 7.3.2 the market maker is required to liquidate the inventory by terminal date T, using market orders at T if necessary. However, no such requirement is imposed in Sections 7.1 and 7.2 that only consider limit orders. In the empirical studies of Hasbrouck and Sofianos (1993), Madhavan and Smidt (1993), and Hendershott and Seasholes (2007), market makers (or "specialists" in NYSE) often hold inventories. In particular, Hendershott and Seasholes (2007) study the inventories of the specialists in NYSE from 1994 through 2004, using internal NYSE data from the specialist summary file (SPETS) that provides specialist closing inventories data for each stock on each day, and examine the relationship between inventories at the close of the trading day and future stock returns at daily and weekly horizons. They find asymmetry in long- and short-inventory positions, and their results show that "the marginal

additional dollar of inventory appears profitable" for the specialist's investment portfolio, that "most of the large long (short) inventory positions occur on days when prices fall (rise)", and that "prices then show small mean reversion relative to the pre-formation return, making these large inventory positions appear unprofitable overall for the specialist."

Huang et al. (2012) focus on market making in currency exchange markets. They point out that "most previous work investigates how inventories influence the market makers' behavior when quoting bid and ask prices" and cite in particular Ho and Stoll's work that we have described in Section 7.1. They say, however, that "empirical studies suggest that the impact of inventory levels on pricing is rather weak compared with the impact of other components" and that "a survey of US foreign exchange traders indicates that the market norm is an important determinant of bid-ask spreads and only a small proportion of bid-ask spreads differ from the conventional spread." They therefore focus on identifying "effective strategies for a market maker who does not control prices and can merely adjust inventory through active trading", and use similarities to the classical inventory control problem to derive "the optimal trade-off between the risk to carry excessive inventory and the potential loss of spread by trading actively with other market makers" (to make up for insufficient inventory). They also note important differences from classical inventory theory, in which "the order quantity must be non-negative, the unit inventory holding cost is deterministic, and the ordering cost is linear in the order quantity". Their formulation of the optimization problem involves the profit obtained from trading with the market maker's clients and the cost incurred from trading with other market makers to control the inventory. The controls are q_t, which represents the amount the market maker buys (in which case $q_t > 0$) or sells (resulting in $q_t < 0$) for periods $t = 1, \ldots, T$. Assuming exponential utility of the total profit (minus the cost) generated up to time T, they derive the Bellman equation for the value function and show that the optimal policy is a threshold rule that involves (a) no trading (with other market makers) if the inventory lies between two thresholds that depend on t, (b) selling shares if the inventory exceeds the upper threshold, and (c) buying shares if the inventory falls below the lower threshold.

Cartea et al. (2014), however, point out that "today computers manage inventories and make trading decisions" in equity markets, whereas quote-driven market making, "which means that market makers quoted buy and sell prices and investors would trade with them" (as in the scenario considered by Huang et al., 2012), is giving way to the limit order markets, at least for equities, where all participants can "behave as market makers in the old quote-driven market." They say: "During the trading day, all orders are accumulated in the LOB until they find a counterparty for execution or are canceled by the agent who posted them. The counterparty is a market order. ..." Therefore, market making in this framework is LOB-driven and closer in spirit to what we describe in Section 7.3 concerning the approach of Guilbaud and Pham (2013), which they also cite. They further assume that market orders (MOs) arrive in two types:

> The first type of orders are influential orders, which excite the market and induce other traders to increase the number of MOs they submit. For instance, the arrival of an influential market sell order increases the probability of observing another market sell order over the next time step and also increases (to a lesser extent) the probability of a market buy order to arrive over the next time step. On the other hand, when noninfluential orders arrive, the intensity of the arrival of MOs does not change. This reflects the existence of trades that the rest of the market perceives as not conveying any information that would alter their willingness to submit MOs. In this way, our model for the arrival of MOs is able to capture trade clustering which can be one-sided or two-sided and allow for the activity of trading to show the positive feedback that algorithmic trades seem to have brought to the market environment. Multivariate Hawkes processes have recently been used in the financial econometrics literature .... However, this paper is the first to incorporate such effects into optimal control problems related to Algorithmic Trading.... The arrival of MOs is generally regarded as an informative process because it may convey information about subsequent price moves and adverse selection risks. Here we assume that the dynamics of the midprice of the asset are affected by short-term imbalances in the number of influential market sell and buy orders; ...the arrival of influential orders has a transitory effect on the shape of both sides of the LOB. More specifically, since some market makers anticipate changes in the intensity of both the sell and buy MOs, the shape of the buy and sell sides of the book will also undergo a temporary change due to market makers repositioning their LOs in anticipation of the increased expected market activity and adverse selection risk.

2. **Market making in derivatives markets** Supplement 1 has alluded to the fact that the constraints and objectives of market makers may differ across markets and that the models and methods in Sections 7.1–7.3 have been developed in the context of equity markets. For derivatives markets, market making naturally assumes that there are market frictions that violate the classical Black-Scholes-type assumptions (Lai and Xing, 2008, p. 183), the most important violation being the presence of transaction costs. Option pricing in the presence of transaction costs involves singular stochastic control of the type discussed in Supplement 5 of Section 3.4; see

Davis et al. (1993) and Lai and Lim (2009). Stoikov and Saglam (2009) point out that for European options, "the state space may be decomposed into a trade and a no-trade region" and the market maker cannot "trade continuously and may make markets in the stock as well." The market maker's controls are the premiums δ_t^a, δ_t^b around the stock's mid-price S_t, and $\tilde{\delta}_t^a, \tilde{\delta}_t^b$ around the option's mid-price $C_t = C(S_t, t)$,

$$dC(S_t, t) = \Theta_t dt + \Delta_t dS_t + \Gamma_t (dS_t)^2/2,$$

where Δ_t, Γ_t, and Θ_t are the option's delta, gamma, and theta; see Lai and Xing (2008, p. 313). The total returns Z_t generated from the transactions are given by

$$dZ_t = \delta_t^a dN_t^a + \delta_t^b dN_t^b + \tilde{\delta}_t^a d\tilde{N}_t^a + \tilde{\delta}_t^b d\tilde{N}_t^b$$

and inventory value I_t given by $dI_t = q_t dS_t + \tilde{q}_t dC_t$, in which N_t^a, N_t^b, and q_t are the same as those defined in Section 7.1 and $\tilde{N}_t^a, \tilde{N}_t^b$, and $\tilde{q}_t$ are the corresponding values for the option. They use a mean-variance approach to find the optimal $(\delta^a, \delta^b, \tilde{\delta}^a, \tilde{\delta}^b)$ with value function

$$v(z, s, q, \tilde{q}) = \sup_{(\delta^a, \delta^b, \tilde{\delta}^a, \tilde{\delta}^b):t \le s \le T} \{E_t(Z_T) - \gamma \mathrm{Var}_t(I_T)\},$$

in which E_t stands for conditional expectation given $(Z_t, S_t, q_t, \tilde{q}_t) = (z, s, q, \tilde{q})$.

3. **Survival analysis and a Kaplan-Meier estimator** As pointed out by Ganchev et al. (2010, p. 103), (7.25) holds because "the tail probability at these large values of ℓ must be smaller than the true probability at c_i^t", for which (7.24) also holds. Their proof of (7.24) uses their Theorem 2, which says that for any $\delta > 0$ and any $i = 1, \ldots, K$,

$$P\left\{|\hat{T}_i^t(\ell) - T_i(\ell)| \le \ell \sqrt{\frac{2 \ln(2v_{\max}/\delta)}{N_{t,\ell-1}^i}} \text{ for all } 1 \le \ell \le \nu\right\} \ge 1 - \delta. \quad (7.38)$$

They prove this result by first considering the case in which $n_1^i, \ldots, n_t^i$ are i.i.d., for which they refer to the bounds of Földes and Rejtö (1981) using empirical process theory. Gu and Lai (1990) already sharpened these bounds and combined them with martingale theory to obtain better results for the modified Kaplan-Meier estimator of the type in Lai and Ying (1991a). Ganchev et al. (2010) also realized the martingale structure and used Azuma's inequality (see Theorem A.5 of Appendix A) to remove this i.i.d. assumption. Survival analysis is also closely related to the intensity models in Sections 4.6.2 and 5.3 and has many applications in credit scoring and rating; see Lai and Xing (2016) for further details.

4. **Parametric survival models and the empirical study of Ganchev**

et al. (2010, Sect. 5.1) This study uses data from a major US broker-dealer on four active dark pools. For each dark pool, the study focuses on a dozen relatively actively traded stocks, thus yielding 48 distinct stock-pool data sets. The average daily trading volume of these stocks across all exchanges (light and dark) ranges from 1 to 60 million shares. To investigate the performance of the exploration and exploitation scheme in Section 7.6.3, the study estimates the distribution of U_t^i for each i. Ganchev et al. observe a very important feature: a vast majority of submissions (84%) to the dark pools results in no shares being executed. Therefore, $P\{U_t^i = 0\}$ should be modeled separately from $P\{U_t^i > 0\}$, leading to the nonparametric Kaplan-Meier and the zero-inflated uniform, Poisson, exponential, and power laws that they consider, in which zero-inflation refers to a mixture distribution with one component of the mixture being a point mass at 0; the power law is the same as the Pareto distribution in Section 2.6.1. The zero-inflated power law seems to give the best fit to the data. They also compare the performance, in terms of the fraction of submitted shares executed and the half-life of submitted orders, of the exploration-exploitation scheme with two naïve rules.

5. **Stochastic gradient method and variants of** (7.19) The recursion (7.19) requires knowledge of F from which the η_j^* are sampled. Without assuming F to be known, Cont and Kukanov (2016, Sect. 4.1) propose to estimate certain functionals of F sequentially and use these estimates to replace those functionals with known F in a nonparametric variant of (7.19). Specifically, they write $\partial c(\boldsymbol{\eta}, \boldsymbol{\Theta})/\partial \boldsymbol{\Theta} = \mathbf{g}(\boldsymbol{\eta}, \boldsymbol{\Theta}) = (g_1, \ldots, g_{K+1})$, where

$$g_1 = h + f + \theta - (\lambda_u + \theta)I_{\{A(\boldsymbol{\eta},\boldsymbol{\Theta})<N\}} + \lambda_o I_{\{A(\boldsymbol{\eta},\boldsymbol{\Theta})>N\}},$$

$$g_{j+1} = \theta + I_{\{\eta_j > Q_j + L_j\}} \left[-(h + r_j) - (\lambda_u + \theta)I_{\{A(\boldsymbol{\eta},\boldsymbol{\Theta})<N\}} \right.$$
$$\left. + \lambda_o | I_{\{A(\boldsymbol{\eta},\boldsymbol{\Theta})>N\}} \right],$$

for $j = 1, \ldots, K$. Hence $\mathbf{g}(\boldsymbol{\eta}, \boldsymbol{\Theta})$ is related to $\boldsymbol{\eta}$ only through the types indicator variables $I_{\{A(\boldsymbol{\eta},\boldsymbol{\Theta})<N\}}$, $I_{\{A(\boldsymbol{\eta},\boldsymbol{\Theta})>N\}}$, and $I_{\{\eta_j > Q_j + L_j\}}$. In particular, $\{A(\boldsymbol{\eta}, \boldsymbol{\Theta}) < N\}$ and $\{A(\boldsymbol{\eta}, \boldsymbol{\Theta}) > N\}$ are the events of undershoots and overshoots, respectively, while $\{\eta_j > Q_j + L_j\}$ is the event of the limit order of size L_j can be executed fully on venue j whose best bid size is Q_j. Hence, instead of using a prespecified F to model the distribution of η, they propose to use the data accumulated so far on these indicator variables, and sample with replacement from the dataset, which is similar to the bootstrap procedure described in Section 2.4.3.

6. **Stochastic approximation (SA) and the recursions in Supplement 5 and Sections 7.5.3 and 7.7.1** The SA method has its origins in the recursive scheme of Robbins and Monro (1951), called stochastic approximation, to find the unique root x_* of a regression function M in

the model

$$y_t = M(x_t) + \varepsilon_t, \qquad t = 1, 2, \dots, \tag{7.39}$$

where ε_t represent unobservable random errors. Assume that M is smooth and $M(x_*) \neq 0$. Applying Newton's scheme $x_{t+1} = x_t - y_t/M'(x_t)$ as in the deterministic case (with $\varepsilon_t \equiv 0$) leads to

$$x_{t+1} = x_t - M(x_t)/M'(x_t) - \varepsilon_t/M'(x_t). \tag{7.40}$$

Hence, if x_t should converge to x_* so that $M(x_t) \to 0$ and $M'(x_t) \to M'(x_*)$, then (7.40) entails that $\varepsilon_t \to 0$, which is not possible for many kinds of random errors (e.g., i.i.d. ε_t with positive variance). To average out the errors ε_t, Robbins and Monro (1951) propose to use a recursive scheme of the form

$$x_{t+1} = x_t - \gamma_t y_t, \tag{7.41}$$

where γ_t are positive constants such that $\sum_{t=1}^{\infty} \gamma_t^2 < \infty$ and $\sum_{t=1}^{\infty} \gamma_t = \infty$, assuming that $M(x) > 0$ if $x > x_*$ and $M(x) < 0$ if $x < x_*$. The assumption of $\sum_{t=1}^{\infty} \gamma_t^2 < \infty$ ensures that $\sum_{t=1}^{\infty} \gamma_t \varepsilon_t$ converges in L_2 and almost surely for many stochastic models of random noise (such as i.i.d. ε_t). Under certain regularity conditions, this in turn implies that $x_t - x_*$ converges in L_2 and almost surely, and the assumption of $\sum_{t=1}^{\infty} \gamma_t = \infty$ then ensures the limit of $x_t - x_*$ is zero. Other than solving for the root of $M(x)$, SA also provides a recursive method to evaluate the maximum of the regression function. The method, first introduced by Kiefer and Wolfowitz (1952), chooses two design levels $x_t^{(l)}$ and $x_t^{(r)}$ such that $x_t^{(l)} = x_t - \delta_t$ and $x_t^{(r)} = x_t + \delta_t$, and uses a recursion of the form

$$x_{t+1} = x_t + \gamma_t \frac{y_t^{(r)} - y_t^{(l)}}{2\delta_t}, \tag{7.42}$$

with $\lim_{t \to \infty} \delta_t = 0$. To dampen the effect of the errors $\varepsilon_t^{(r)}$ and $\varepsilon_t^{(l)}$ in

$$\frac{y_t^{(r)} - y_t^{(l)}}{2\delta_t} = \frac{M(x_t^{(r)}) - M(x_t^{(l)})}{2\delta_t} + \frac{\varepsilon_t^{(r)} - \varepsilon_t^{(l)}}{2\delta_t},$$

Kiefer and Wolfowitz choose $\gamma_t > 0$ such that $\sum_{t=1}^{\infty} (\gamma_t/\delta_t)^2 < \infty$, $\sum_{t=1}^{\infty} \gamma_t = \infty$, and $\sum_{t=1}^{\infty} \gamma_t \delta_t < \infty$. Blum (1954) has provided multivariate extensions of the Robbins-Monro and Kiefer Wolfowitz schemes. Lai (2003) gives an overview of SA and how it has evolved to "become an important and vibrant subject in optimization, control and signal processing."

The SA recursion (7.31)

Consider the system of equations (7.30), in which $(\nu_t, U_t^1, \dots, U_t^K)$ are i.i.d. with the same distribution as $(\nu, U_1, \dots, U_K)$. This system can be written as $\boldsymbol{M}(\boldsymbol{r}) = 0$, where $\boldsymbol{r} = (r_1, \ldots, r_K) \in \mathbb{S}$, $\boldsymbol{M}(\boldsymbol{r}) = (M_1(\boldsymbol{r}), \ldots, M_K(\boldsymbol{r}))$,

$$M_j(\boldsymbol{r}) = E\left[\nu\left(\rho_j I_{\{r_j\nu \leq U_j\}} - \frac{1}{K}\sum_{i=1}^{K}\rho_i I_{\{r_i\nu \leq U_i\}}\right)\right], \qquad j = 1, \ldots, K.$$

(7.43)

Letting $\boldsymbol{r}_t = (r_t^1, \ldots, r_t^K)$ and $\boldsymbol{y}_t = (y_t^1, \ldots, y_t^K)$, where

$$y_t^j = \nu_t\left(\rho_j I_{\{r_t^j\nu_t \leq U_t^j\}} - \frac{1}{K}\sum_{i=1}^{K}\rho_i I_{\{r_t^i\nu_t \leq U_t^i\}}\right), \qquad j = 1, \ldots, K, \quad (7.44)$$

we have the regression model

$$\boldsymbol{y}_t = \boldsymbol{M}(\boldsymbol{r}_t) + \boldsymbol{\varepsilon}_t, \tag{7.45}$$

where $\boldsymbol{\varepsilon}_t = (\varepsilon_t^1, \ldots, \varepsilon_t^K)$ represents the zero-mean random error. For this case, the SA recursion (7.41) has the form $\boldsymbol{r}_{t+1} = \boldsymbol{r}_t + \gamma_t \boldsymbol{y}_t$, which may not satisfy the constraint $\boldsymbol{r}_t \in \mathbb{S}$. Laruelle et al. (2011) remove this constraint by redefining the regression function as

$$M_j(\boldsymbol{r}) = E\left[\nu\left(\rho_j I_{\{r_j\nu \leq U_j\}} - \frac{1}{K}\sum_{i=1}^{K}\rho_i I_{\{r_i\nu \leq U_i\}} + \xi_j\right)\right] \tag{7.46}$$

for $j = 1, \ldots, K$, where ξ_j is given by (7.36). The i.i.d. assumption on $(\nu_t, U_t^1, \ldots, U_t^K)$ was made above for convenience to connect (7.30) to the regression model (7.45)–(7.46). In fact, the convergence theory also applies to stochastic regression models in which ε_t is a martingale difference sequence such that $\sup_n(\varepsilon_n^2|\mathcal{F}_{n-1}) < \infty$ a.s. The martingale structure of the recursive scheme (7.40) was already noted by Blum (1954); see Section 3 of Lai (2003) that also gives an introduction to the Robbins-Siegmund lemma used by Laruelle et al. (2011, p. 1059) in extending the convergence theory of the SA recursion from i.i.d. to ergodic $(\nu_t, U_t^1, \ldots, U_t^K)$, which we have described in the last paragraph of Section 7.7.3.

The SA recursion (7.19) or its nonparametric variant

For the recursion (7.19), Cont and Kukanov (2016, Sect. 4.1) say that "the optimal step size" is $\gamma_m = C/m$, where

$$C = \max_{\Theta, \Theta' \in \mathbb{S}} \|\Theta - \Theta'\| \left/ \left(\max_{\Theta \in \mathbb{S}} E\left[\|\boldsymbol{g}(\boldsymbol{\eta}, \Theta)\|\right]\right)^{1/2}\right. ,$$

$$\mathbb{S} = \left\{\Theta \in \mathbb{R}_+^{K+1} : M \leq N, M + \sum_{i=1}^{K}L_i \geq N,\right.$$

$$\left. L_i \leq N - M \text{ for } i = 1, \ldots, K\right\}.$$

They do not, however, explain in what sense this choice is optimal. Note that $\sum_{m=1}^{\infty} \gamma_m = \infty$ and $\sum_{m=1}^{\infty} \gamma_m^2 = \infty$, which satisfies the criterion for a.s. convergence of SA recursions. They then propose to use another step size $\gamma_m = C'm^{-1/2}$, with some other explicit constant C', that "scales appropriately with problem parameters." However, for this choice, $\sum_{m=1}^{\infty} \gamma_m^2 = \infty$ and they do not state convergence results (if any) of this "scalable" choice. In fact, their paper denotes the step size by γ and gives explicit formulas in terms of K, m, and other parameters of the model for the "optimal" and "scalable" step sizes, but the important role played by m for convergence, however, is not emphasized. Section 4 of Lai (2003) reviews previous work on the asymptotically optimal choice of γ_m in SA schemes. Chapter 6 of Powell (2007) gives further discussion on the choice of step sizes in SA and more general stochastic gradient algorithms in approximate dynamic programming.

7.9   Exercises

1. Consider the market making problem when the mid-price process $S(t)$ follows

$$dS(t) = b(t, S(t))dt + \sigma(t, S(t))dW(t),$$

for some $b(\cdot, \cdot)$ and $\sigma(\cdot, \cdot)$ functions such that the dynamics allow a unique solution. Now let

$$u(t, s, q, x) = \max_{\delta^a, \delta^b} E_{t,s,q,x} \left\{ \phi\left[S(T), Q(T), X(T) \right] \right\},$$

for some appropriate choice of ϕ. Derive via the dynamic programming principle the associated HJB equation for $u(t, s, q, x)$,

$$(\partial_t + \mathbb{L})u + \max_{\delta^+} \lambda^+(\delta^+) \left\{ u\left[t, s, q - 1, x + (s + \delta^+)\right] - u(t, s, q, x) \right\}$$
$$+ \max_{\delta^-} \lambda^-(\delta^-) \left\{ u\left[t, s, q + 1, x - (s - \delta^-)\right] - u(t, s, q, x) \right\} = 0,$$

with the boundary condition $u(T, s, q, x) = \phi(s, q, x)$. Here $\mathbb{L} = b(t, x)\partial_s + (2\sigma^2(t, x))^{-1}\partial_{ss}$, where $\partial_s = \partial/\partial s$, $\partial_{ss} = \partial^2/\partial s^2$.

2. (a) Let X be a random variable on $\mathbb{Z}_+$ with $E(X) < \infty$. Show that $E(X) = \sum_{s=1}^{\infty} P(X \geq s)$.

   (b) Using (a) or otherwise, show that $E[\min(n, U_t^i)] = \sum_{s=1}^{n} T_i(s)$ in Section 7.6.1.

   (c) Prove by induction that the optimal solution of (7.20) is provided by the greedy algorithm in Section 7.6.1.

3. (a) Let $\varphi(r) = E[\min(r\nu, U)]$ as defined in Section 7.7. Show that $\varphi(r)$ is differentiable everywhere on $[0, 1]$ if $E[\nu I_{\{U>0\}}] > 0$ and the distribution function of U/ν is continuous on $\mathbb{R}_+$.

   (b) Show that $\varphi'(r) = E[\nu I_{\{r\nu \leq D\}}]$ for all $r \in [0, 1]$ under the conditions of (a).

   (c) Show that $\varphi(r)$ can be extended to the whole real line and maintain its concavity by using

$$\varphi(r) = \begin{cases} (r - r^2/2)\,\varphi'(0) & \text{if } r < 0, \\ \varphi(1) + \varphi'(1)\log r & \text{if } r > 1. \end{cases}$$

   What is the relationship of this extension to the ξ_j in (7.36)?

4. Let X be a random variable with mean μ. Let $x_1, x_2, \ldots$ be a sequence of i.i.d. realizations of X. Consider the sequence $\mu_1 = x_1$ and for $i \geq 2$,

$$\mu_i = (1 - \alpha_i)\mu_{i-1} + \alpha_i x_i = \mu_{i-1} + \alpha_i(x_i - \mu_{i-1}).$$

   The α_i are parameters for the step size. Show that if $\sum_{i=2}^{\infty} \alpha_i = \infty$ and $\sum_{i=2}^{\infty} \alpha_i^2 < \infty$, then $\mu_i \to \mu$ almost surely.

8

Informatics, Regulation and Risk Management

Informatics and infrastructure refer to the computational components, especially the software and hardware aspects, of a typical automated trading setup. Broadly speaking, they can be categorized into two main pillars: strategy and exchange. The strategy pillar refers to components that are under the trading entity's control and is developed and maintained by the entity. A trading strategy reacts to market events, such as a price change or news arrival, and makes trading decisions based on the information content of these events. The exchange pillar refers to components under the domain of the exchange, and consists of components that process incoming order requests, validate order attributes, match buy and sell side orders and finally, disseminate matching results to market data subscribers. From an infrastructure standpoint, the two pillars are connected via some standard transmission protocol that facilitates the communication between them. Figure 8.1 illustrates the relationship between these two pillars.

Before the advent of electronic exchanges and automated trading, securities transactions were conducted mainly via three mechanisms: dedicated market makers such as the NYSE specialists, trading pits such as the CME futures pit, and over the counter (OTC) such as the brokered market for interest rate swaps trading. Over time the trading pits have evolved into electronic exchange platforms, and the functions of the specialists, pit traders and, to some extent, of the OTC brokers and traders as well, are carried out by automated trading algorithms. The traditional role of the broker is replaced by the limit order book and the matching engine, while many dealers and specialists

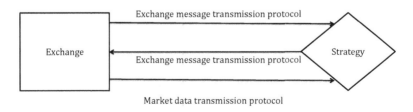

FIGURE 8.1: Illustration of the exchange pillar and the strategy pillar.

are superseded by market making algorithms that are capable of posting various types of orders in split seconds. Nevertheless, the broker's role in helping the originating dealer seek a counterparty dealer anonymously still exists for OTC transactions[1], particularly for specialized or structured products such as non-standard interest rate swaps.

On a typical electronic exchange, all approved participants can submit orders in the form of market, limit, or cancellation orders via the gateways. As long as these orders are admissible, they will be processed by the exchange matching engine. Then, the exchange platform would send a message via the gateway to inform the order sender according to the outcome of the matching process. For example, the sender of a cancellation order will receive a rejection message if the limit order that was planned to be canceled is executed before the order reaches the matching machine. Moreover, if any order is successfully implemented, the updated limit order book is disseminated to all market data subscribers through the broadcasting system of the exchange platform. The clearing is done by a third-party clearing house, to which the two counterparties in the transaction post collateral to cover initial and maintenance margin requirements. The margin amount and any special cross-margining concessions are predefined per product by the exchanges. Nowadays, on most exchanges, a large proportion of the daily trading is attributable to some automated trading algorithm, acting either as an execution engine for a human-originated order, or as a fully automated trading algorithm that generates buy or sell orders following a triggering event.

To motivate the discussion on infrastructure, Section 8.1 reviews some quantitative strategies in Chapters 1 to 3 and describes several high-frequency trading strategies. In particular, we explain why a good understanding of the exchange and the strategy pillars is critical to implementing these strategies as automated algorithms. Section 8.2 provides an overview of an exchange's infrastructure by using NASDAQ and CME as examples and discusses the roles of the matching engine, the matching mechanics of some commonly used order types, and some market data protocols. The infrastructure and informatics on the trading strategy side are introduced in 8.3. Current regulations and rules for trading on the US exchanges are described in Section 8.4, and Section 8.5 discusses issues in managing the operational, modeling, and trading risks. Supplementary material on related topics and contemporary developments is presented in Section 8.6, and then exercises are given in Section 8.7.

---

[1] Participants in an OTC market usually negotiate with each other in a pairwise manner. Without LOB information, the participants other than the major dealers do not have much information about prices that are currently available elsewhere in the market. Hence, the OTC market is sometimes referred to as the dark market because of the lack of price information; see Duffie (2012) for quantitative analysis of the matching process, asset pricing, and information dynamics in such a market.

8.1   Some quantitative strategies

Broadly speaking, algorithmic and systematic quantitative trading can be categorized into relative value trading, statistical arbitrage, and latency arbitrage. Although terms such as statistical arbitrage are sometimes used to describe a generic modeling paradigm based on statistical methods, the following terminologies have been reasonably standardized among practitioners to mean a specific style of trading.

Relative value trading

Relative value trading, as discussed in Section 1.2, often refers to a style of trading in fixed income where trading strategies are based on analyzing historical relationships among different parts of the yield curve; see Chapter 10 of Lai and Xing (2008) for an introduction to yield curve analytics and interest rate markets. These strategies seek out dislocations on the curve and take positions based on the speed and probability of mean reversion. Often, a trade consists of multiple parts, known as the *legs* of the trade, with different maturities. Typically, low-frequency relative value strategies rely on models that process historical bond yields or interest rate swaps data to identify long-term stable relationships, and take positions when these relationships deviate from threshold levels. High-frequency relative value strategies have a similar flavor, driven by comparable factors that affect temporary supply and demand. The essence of these strategies is to find suitable data for model fitting and to determine the timing of getting in and out of a trade. Typical instruments used in fixed-income relative value strategies are sovereign bonds, interest rate swaps, and inflation-linked bonds. Options, such as swaptions, caps and floors, are also often used to express a view of market conditions for or against a relative value trade. Outside of fixed income, relative value trading describes a generic style of trading driven by *rich-cheap analysis* of one stock or a basket of stocks versus other stocks. This style of trading could be both low-frequency and high-frequency.

Statistical arbitrage

Statistical arbitrage sometimes refers to the statistical underpinnings of Markowitz's portfolio optimization and its variants. Sections 2.3 to 2.5 as well as Section 2.7 have already studied this style of trading from the classical approach to some recent developments. It also refers to other styles of trading described in Section 2.8. In addition, Chapter 3 has analyzed the multi-period portfolio optimization problem that takes transaction costs into consideration. It also describes the modern approach that uses stochastic control theory and approximate dynamic programming, filtering, big data analytics and machine learning. There are also other nonparametric methodologies such as log-optimum strategies and the expected utility theory in lieu of mean-variance portfolio optimization as in the classical formulation. The trading style can be both low-frequency and high-frequency; see Fan et al. (2012) for how to use high-frequency data for trading a basket of stocks. Statistical arbitrage is also used as a generic term to describe quantitative trading that uses some statistical analysis.

Latency arbitrage

This refers to trading assets that are highly correlated and sometimes even equivalent. For example, the ETF SPY and the E-mini S&P 500 futures contract (denoted by ES under CME) are both based on the S&P index, ignoring the dividend and interest rate parts. However, ES is traded on the CME's platform in Chicago and SPY is traded on the platforms of several exchanges based in the East Coast. When ES moves up, SPY should move up too, albeit with a delay. This delay is a function of the speed at which the arbitragers can operate. This latency can be in the range of tens of milliseconds, mostly due to the geographic separation of the two exchanges: the distance between Chicago and New York is roughly 714 miles, the speed of light is 670,616,629 mile per hour; this equates to about 3.8 milliseconds[2] if one can connect the two by a straight beam of light. Thus, there is a considerable delay between the movements of SPY and ES. Latency arbitrage strategies focus on this type of opportunities, and they serve as a facilitator for information transfer between the two trading hubs. Another type of latency arbitrage is related to tradable assets that can be decomposed into other tradable components. On some futures exchanges, the *outright* contracts A, B, C, and *spreads*[3] {A-B}, {B-C} and the *bundle*[4] {A, B, C} can all be tradable. Eurodollar futures provide an example in which the spreads, bundles, and their packs are all actively traded. High-frequency arbitragers monitor pricing consistency between these tradable assets with common components and arbitrage away any short-term inconsistency.

All of these approaches are popular in the industry. To implement them as automatic trading algorithms, understanding of the exchange infrastructure is crucial. For example, although relative value trading can be applied to both fixed income securities and equities, their matching rules are very different; see Section 8.2.2 for details of the pro-rata rule which is used in fixed income matching, and of the price-time priority rule which is commonly used on the platform of stock exchanges. High-frequency implementation of the ap-

---

[2]The unit of millisecond is abbreviated as ms in this chapter.

[3]Buying one contract of the spread {A-B} means the simultaneous transaction of buying one contract of A and selling one contact of B at a price that is based on the difference between A and B. As such, spread prices can be negative; see Section 8.2.1 for more discussion on outright and spread markets.

[4]Buying a bundle {A, B, C} often means the simultaneous transaction of buying one contract each of A, B, C, at a price that is calculated based on the average change in price from the previous closing prices of these assets.

TABLE 8.1: Key US exchanges and their most traded products, with the open interest or daily volume of each as of November 29, 2015. For example, the traded volume of AAPL on that day is 72,056,345 shares while the open interest for ES is 2,920,713 contracts.

Exchange	Most traded product	Symbol	Open Interest / Daily Volume
CME	Eurodollar Futures	ED	11,089,332
	E-mini S&P Futures	ES	2,920,713
	10-Year T-Note Futures	ZN	2,764,946
NASDAQ	Apple	AAPL	72,056,345
	PowerShares QQQ Trust	QQQ	46,277,773
	Facebook	FB	41,294,632
NYSE	Bank of America	BAC	79,806,275
	Sprint Corp	S	47,012,913
	Ford Motor Company	F	44,414,913

Source: NASDAQ OMX, CME.

proaches for currency trading requires monitoring LOB of many independent platforms. Moreover, all quantitative trading implementations require substantial research on market data. It is important to store the market data so that the strategy can be backtested in a simulated environment which mimics the rhythm of the market.

8.2 Exchange infrastructure

Some of the key US exchanges are listed in Table 8.1, together with the daily volume and open interest[5] of their most popular equities and derivative products, respectively. Table 8.2 shows the top ten algorithmically traded futures on CME. To handle such huge volumes of transactions, exchanges have to be equipped with powerful electronic platforms that process inbound orders, match the buy-side and sell-side orders, and disseminate the most updated LOB via the outbound channels to all subscribers of market data. Loosely speaking, there are three main components in the exchange infrastructure: order gateway, matching engine, and market data broadcasting engine, as illustrated in Figure 8.2.

---

[5]Open interest is the total number of options and/or futures contracts that are not closed or delivered at the end of a particular trading day.

TABLE 8.2: Breakdown, in percentage of total, of top ten algorithmically traded futures classes on CME Globex. Numbers are based on Trade Indicator (FIX Tag 1028). "Automated" orders are submitted by a computer algorithm, without human intervention; "Manual" orders are trades submitted with human intervention; "Other" orders include trades transacted in the pit or block trades. Refer to Haynes and Roberts (2015) for complete breakdown and description of the CME transaction dataset, November 12, 2012 to October 31, 2014.

Product Group/Subgroup Name	Automated	Manual	Other
FX/G10	80.7	16.4	2.8
FX/Emerging Markets	70.3	17.1	12.5
Equities/International Index	69.3	30.3	0.4
Equities/Equity Index	66.5	32.6	0.9
FX/E Micros	66.3	33.7	0.0
Interest Rate/US Treasury	64.0	30.2	5.9
Interest Rate/STIRS	60.3	37.5	2.2
Equities/Select Sector Index	59.2	22.0	18.8
Energy/Crude Oil	54.3	40.2	5.5
Metal/Base	49.2	47.7	3.1

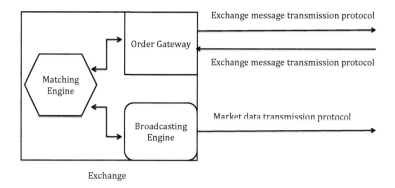

FIGURE 8.2: Components of a prototypical exchange platform.

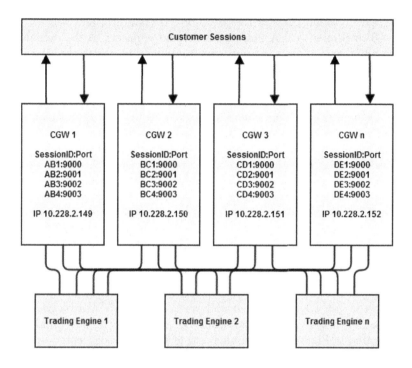

FIGURE 8.3: Schematic diagram of the iLink order routing interface and the matching engines of CME Globex. CGW1, CGW2, CGW3, CGW4 are 4 gateway nodes; Trading Engine 1, Trading Engine 2, Trading Engine 3 are 3 matching engines. CGW1 is handling 4 sessions: AB1:9000, AB2:90001, ..., AB4:9003, where 9000 to 9003 are the ports to which the sessions are connected. Session AB1 can exist on only one gateway; AB1 has a single sequence stream comprised of messages to/from all market segments. (Used with permission of CME.)

8.2.1 Order gateway

A real exchange platform is more complex than the simplified model shown in Figure 8.2 and the order gateways are usually arranged in a system called the order routing interface. For Globex, the exchange platform of CME, this interface is called *iLink*; the iLink schematic diagram is shown in Figure 8.3. The box at the top of the diagram represents the trading firm and the strategy processes, such as the book-building engine, alpha engine, and order management engine, which are all parts of the setup of a trading firm and will be further discussed in Section 8.3. A market participant connects to the exchange via a node point known as a *session*, which uniquely identifies the participant. Orders are sent through specific sessions to the order gateway which is represented as the second row in the diagram. The customer can have multiple sessions and the exchange often has multiple order gateways to process inbound order messages. The matching engines, where the actual matching of buy and sell orders takes place, are represented by the boxes at the bottom of the diagram. Each gateway processes and verifies inbound order messages from the market participants then routes the orders according to specific rules. For example, under the CME iLink order messaging framework, there are multiple order gateways, each working independently to decode and route incoming customer orders according to the product group of the order. These product groups are defined based on the sector of the underlying asset of the futures or options contracts. For example, West Texas Intermediate (abbreviated as WTI) Crude Oil futures and Brent Crude Oil futures would be routed to the same matching engine, whereas Wheat and Corn futures would be routed to another.

Once the gateway has parsed and verified the basic attributes of an incoming order, it will send out an acknowledgment via the outbound channel as illustrated in Figure 8.2, announcing that the order is accepted or rejected by the exchange. If the order is accepted, it will be submitted to the matching engine. Such acknowledgment is received by the order management engine on the strategy side, which then passes the order reply event to the alpha engine for strategy-level processing, as will be described in Section 8.3.

8.2.2 Matching engine

The matching engine takes orders routed by the order gateway and performs the actual matching of buy and sell orders based on some predetermined algorithms. Upon arrival of a market order or marketable limit order (which is collectively known as *aggressive* or *aggressor* orders), the matching engine determines the quantity based on the matching rule. Some of the commonly used matching algorithms supported on US exchanges are the following.

- *First-In-First-Out* (FIFO) The most common algorithm which uses price and time priority for filling an order, so that orders at the same price are filled according to time priority; also known as the price-time priority rule which has been referred to in Section 5.5.2. Most US equity exchanges match orders based on this algorithm.

- *Pro-rata* This algorithm fills orders according to price, order lot size, and time priority, where each order in the queue is filled as a function of the percentage of the order size relative to the total queue size, with appropriate rounding. Excess lots as a result of the rounding down of the original allocated trade quantity is then allocated via FIFO.

- *Allocation* This is a modification of the pro-rata algorithm, such that an order having a price that betters the market at the time the order is received (i.e., the order that establishes a more aggressive price level of the order book) is given time priority in the matching process. A slight variation of this algorithm is currently used for Eurodollar futures, one of the most liquid futures traded on CME Globex.

For the CME, and for similar exchanges where spreads between one or more assets are traded as an asset by itself, the concepts of outright market and spread market can be explained by using ES as an illustration. In particular, ESM6 that matures in June 2016 is a typical example of an outright market. The spread market refers to instruments that consist of multiple legs of outright instruments. For example, ESM6-ESU6 consists of two *legs*, ESM6 and ESU6, where ESU6 refers to ES that matures in September 2016. The quoting convention for the spread market is the price of ESM6 minus the price of ESU6, and therefore prices for spread markets can be negative. A trade in the spread represents the simultaneous purchase and sale of two or more outright futures instruments. For example, buying the ESM6-ESU6 spread means buying one contract of ESM6 and selling one contract of ESU6. In this particular example, upon execution the exchange will send three execution reports, one for the spread, and one each for the two legs. Since the trade is on the spread, the prices for the separate legs will need to be determined via some exchange-defined rule, so that the legs are assigned prices close to the price of the underlying asset or index.

8.2.3 Market data dissemination

Exchanges make public order book events, such as the cancellation and execution of an existing order or the addition of new orders, via the exchange market data dissemination system. As the LOB events happen with very high frequency, optimal utilization of network resources to deliver data in the shortest time to every subscriber, while avoiding network congestion, is a critical task. To achieve high-speed broadcast and scalability in terms of the spectrum of products and the number of subscribers, many exchanges broadcast their market data as network packets using multicast technology rather than in the unicast style[6] see Supplement 2 in Section 8.6.

In addition to considering network utilization, exchanges also need to choose the market data transmission protocol — the scheme for encoding and decoding order information — that is most suited to their market, the matching engine infrastructure, and the target users of the data. Two popular protocols are FIX/FAST and ITCH. The FIX/FAST (FIX Adapted for STreaming) protocol is a technology standard developed by FIX Protocol Ltd. specifically aimed at high-volume and low-latency market data feeds. It is currently adopted by exchanges such as NYSE and Eurex. ITCH is the outbound direct-access market data protocol initially developed and adopted by NASDAQ; ITCH is also used by other exchanges and dark pools such as LSE and Chi-X.

The broadcasting system of CME is called the CME Group Market Data Platform (MDP) which uses its own protocol MDP 3.0.[7] This protocol includes (i) Simple Binary Encoding (SBE) and (ii) Event Driven Messaging. Both (i) and (ii) have the ability to deliver multiple events in single packets to achieve good efficiency in bandwidth utilization. Figure 8.4 illustrates a MDP 3.0 packet containing two FIX messages which carry data for two events. MDP 3.0 also allows a single event to be delivered by multiple packets. When that happens, MDP 3.0 ensures the strategy-side parsing procedure[8] can start as soon as the first packet is received. Supplements 3 and 4 in Section 8.6 give more details about MDP 3.0 packets.

8.2.4   Order fee structure

An exchange's fee structure varies based on venue, asset class, order type, and Total Consolidated Volume (TCV).[9] Table 8.3 summarizes the fees levied by some representative exchanges in the US. Most exchanges charge a fee for orders that deplete liquidity and offer rebates to those that add liquidity. The rationale of this fee-rebate structure is to promote liquidity of the order book and to compensate the market-maker for warehousing risk.

However, on some exchanges, such as Boston Stock Exchange (abbreviated as BX), this structure is reversed: liquidity takers are given a rebate and

---

[6]If there is any data packet loss during the transmission process, the unicast protocol usually can detect such loss and performs retransmissions automatically while the multicast protocol may demand more effort from the subscribers to manage the situation.

[7]CME MDP 3.0 is the latest protocol. The official document for protocol and message specifications can be accessed at http://www.cmegroup.com/confluence/display/ EPICSANDBOX/CME+MDP+3.0+Market+Data. Note that other than MDP 3.0, MDP also supports "Streamlined FIX/FAST" for non-actionable price data; see the official document for details.

[8]A parsing procedure is the computer subroutine that translates the network packets into data.

[9]TCV is calculated as the volume reported by all exchanges and trade reporting facilities to a consolidated transaction reporting plan for the month for which the fees apply.

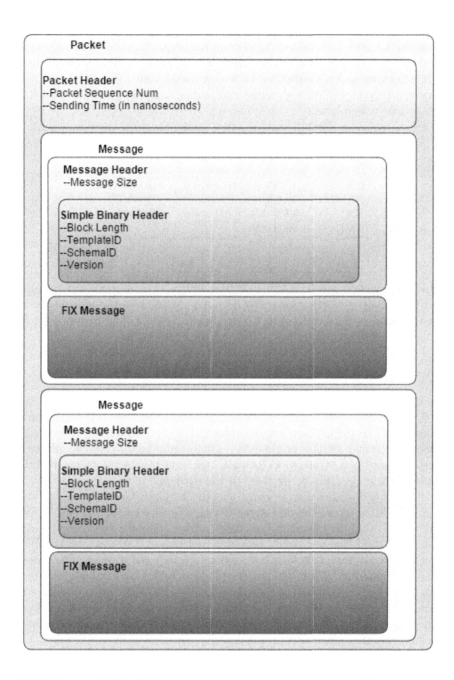

FIGURE 8.4: CME MDP3.0 packet with two FIX messages. (Used with permission of CME.)

TABLE 8.3: Fee structure as of October 5, 2015. Fees in brackets signify a rebate to the liquidity provider: NYSE fees for taking and providing displayed liquidity; NASDAQ for taking liquidity through any single MPID and providing displayed designated retail liquidity; BX rebate for taking liquidity from firms removing at least 0.20% of TCV and for providing displayed and non-displayed liquidity; CME for taking and adding liquidity for E-mini S&P as non-exchange members.

Exchange	Taking	Providing Displayed	Providing Non-displayed
NYSE	$0.00270	$(0.00140)	$(0.00000)
NASDAQ	$0.00295	$(0.00340)	$(0.00250)
BX	$(0.0017)	$0.00140	$0.00240
CME	$0.40000	–	–

liquidity providers are levied fees to post orders. The economic intuition is that those market participants who do not want to cross the bid-ask spread but still want to gain order queue advantage may be willing to pay a fee for being be able to stay close to the front of the order queue.

8.2.5   Colocation service

For high-frequency trading, a few microseconds can make a significant difference. Therefore the proximity of the server boxes (where the strategy components reside) to the exchange data center (where the matching process takes place and the market data originates) is a significant factor that can affect the overall latency of a strategy. Colocation is a term that refers to the practice of placing trade servers at or close to the exchange data center. For example, CME offers colocation services in both the CME Group center located in Aurora, Illinois and the downtown Chicago facility located at 350 E. Cermak. NASDAQ offers similar colocation service at its Carteret, New Jersey center and also offers microwave connectivity between Carteret and Aurora.[10] Exchanges that provide colocation services are required to offer equal access to all participants.[11] In particular, the physical cables that link all servers to their gateways are required to be of the same length, and hence all colocated clients are subject to the same inbound and outbound latencies. For colocation services, exchanges charge a fee that can depend on the space and power consumption that client servers require.

---

[10]According to NASDAQ, the one-way latency for the microwave connection is around 4.13ms versus 6.65ms on the fastest fiber route.

[11]The professional standard of exchange data centers is described in "FISC Security Guidelines on Computer Systems for Banking and Related Financial Institutions"; FISC refers to The Center for Financial Industry Information Systems.

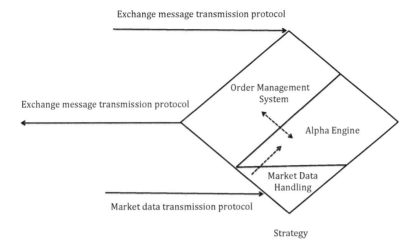

FIGURE 8.5: Main processing units of a typical strategy infrastructure. The dashed arrows show the direction of information flow between the three processing units.

8.2.6 Clearing and settlement

Settlement and clearing refer to the post-trade processes initiated once the buyer and seller orders are matched. The *clearing house* performs centralized clearing by acting as a centralized counterparty, stepping in between the original buyer and seller, and assuming the liability for covering the future settlement of the contract between the two original counterparties. This *novation* process effectively replaces one contract with two separate contracts, with the centralized counterparty facing the original buy and sell counterparties. For the CME, the definition of a clearing house is an institution that provides settlement and clearing services, and is comprised of members who keep funds on deposit with the clearing house to cover trading and margining requirements. Similar definitions are used by other exchanges. The clearing house, acting as centralized counterparty, helps reduce default risk and also the economic impact in the case of default by requiring members to maintain sufficient collateral deposit to cover potential losses, and by providing funds to absorb any excess amount over the posted collateral in the event of a default. The clearing house also helps improve capital efficiency via the netting of offsetting transactions between multiple counterparties. Clearing members represent the trades of their own accounts and those of their customers, and perform trade-related services such as margin calculation and inventory tracking for a fee, often based on the volume traded.

8.3   Strategy informatics and infrastructure

Strategy informatics and infrastructure deal with processes and procedures of the trading firm. There are three main processing units: market data handling, alpha engine processing, and order management. Each unit is responsible for a specific task in the overall process. Figure 8.5 illustrates a typical strategy infrastructure.

8.3.1   Market data handling

The market data handling unit listens to the market data broadcast channels of the exchange, decodes raw market data, and updates order book states and processes trade information. There are two main streams of market data:[12] quotes and ticks. Quotes data contain information pertaining to the states of the order book such as the prices, volumes and number of orders at each level of the order book. Ticks data contain information related to executions. The market data handling unit decodes market data from wire format to specific information relevant to the states of the order book and the execution. The decoded information set consists of general order event attributes such as time-stamp, type of event, symbol, price, size. Depending on the message protocol, it could also contain information such as the type of price,[13] the matching sequence number of an execution, and so on. Using the decoded messages, the unit then updates the order book.

There are two main types of quotes data, namely order-based and level-based. Some exchanges, such as NASDAQ, offer both types of quote streams, while others, like CME, offer only level-based quotes and are called level-driven exchanges; see Section 4.1. Order-based streams allow the reconstruction of the order book, which consists of order details for specific orders at each price level, so the strategy can track the life cycle of the orders and deduce accurate queue transitions when new orders are added and existing orders executed or canceled. Level-based market data, on the other hand, only provide snapshot updates of affected book levels. For example, upon order cancellation, the exchange will send information on the new level of total volume and number of orders, but no information on which order is canceled. This adds considerable complexity to deduce the queue position of a strategy's own orders, and requires some non-trivial and subjective modeling to obtain a reasonable estimate.

Since market data handling and the book-building processes are at the top

---

[12]There are other streams of data related to the other aspects of the matching process, such as auction status and opening and closing imbalance, but the bulk of the market data are related to quotes and ticks.

[13]For some futures contract traded on CME's platform, the tradable price could be an outright price or an implied price.

node of the critical path of an event trigger,[14] the efficiency of its implementation is important for the overall performance of the strategy. Software engineering expertise is required to ensure the book building process is carried out in the most efficient manner. The type of data structure, the memory management method, and the bookkeeping algorithm all need to be carefully designed using the most efficient computer language. C++ is a popular language for latency-sensitive tasks. Java is also widely used because it is considered more user-friendly and less error-prone. However, the memory management aspect of Java, e.g., the garbage collection algorithm, has to be fine-tuned to achieve high stability of the overall latency.

Other than feeding high-frequency algorithms with real-time market data, the unit of market data handling is also responsible for collecting data for off-line alpha research.[15] To develop high-frequency algorithms, storing market data at the raw packet level, rather than at the reconstructed LOB level, is common practice. These raw data are played back for back-testing with either a standard read-file procedure or a suitable application that rebroadcasts via multicasting. The advantage of using read-file is that the speed of playback is easily controlled by the researcher; the downside of this method is that the playback does not mimic the environment in which actual trading took place. The raw packet rebroadcast is able to mimic the environment at the cost of flexibility in playback speed control.

Another issue with market data collection is time-stamp synchronization. This is particularly important for some strategies that rely on cross-exchange information aggregation. For example, a latency arbitrage strategy that relies on aggregated LOB data over ten exchanges can be highly sensitive to the synchronization of different market data feeds. For accurate collection and playback of such market data, collection has to be performed by colocation servers that are synchronized by Precision Time Protocol, usually abbreviated as PTP. However, even with PTP the probability of suffering from misplacement errors is still non-negligible, especially during peak trading hours. Section 4.4.2 and Supplement 9 in Section 4.7 have described generalized synchronization methods to handle misplacement errors.

8.3.2 Alpha engine

Once the market data are decoded and the order book is updated with trade information, the market data handling unit then passes the result, often via a callback mechanism, to the strategy alpha processing unit. The alpha engine implements callback functions of the market data handling unit, and is responsible for making trading decisions, based on the most recently updated market data and other strategy attributes such as position inventory, outstanding or-

---

[14]Event trigger is the subroutine in a high-frequency algorithm that detects trading signals from the data generated from the event packets; see Supplement 4 in Section 8.6 for further discussion.

[15]See Supplement 5 of Section 8.6 for a systematic introduction to alpha research.

ders and risk parameters. The trading decision is driven by a model, or a series of models, that takes the current strategy states as input and that outputs actionable directives such as to submit a new order, to cancel or replace an existing order, or to do nothing and wait for the next event.

To illustrate, consider a canonical market-making strategy for which the alpha unit performs short-term price prediction and optimal quote spread estimation. After a new market event is decoded and the order book updated, the alpha engine first updates the parameter estimates of the price model and, in the process, also calculates a new price prediction together with error bounds. Once it has calculated the predicted fair-value price, it checks for the latest inventory of the strategy, updates the spread model, and then calculates the optimal quote prices and sizes. The actionable directive is to adjust, via cancel or replace, the existing quotes by the new optimal quote. Note that the trigger of the alpha engine does not need to be any market event from the market data feed. Order events, such as an existing order filled, can also trigger some parts of the alpha engine to recalculate and resubmit new orders. In that case, the action of the alpha engine is generated by the acknowledgment information sent from the order gateway of the exchange.

Given the tick-to-trade requirement of a high-frequency strategy, all necessary calculations of the strategy unit need to be completed within a few microseconds. It takes the combined effort of quantitative researchers and software engineers to find the optimal tradeoff between model complexity, accuracy and latency impact. Once the strategy unit has made a trading decision, it passes the necessary order attributes such as side, size and price, to the order management unit. Alpha research will be discussed further in Supplements 5-7 of Section 8.6, where Supplements 8-10 will provide additional details and background on latency reduction techniques and hardware setup for algorithmic trading.

8.3.3   Order management

The order management unit receives real-time information from the order messaging channels of the exchange, and updates strategy inventory and order status. It also interfaces with the alpha engine unit and encodes strategy orders into wire format before sending these orders to the exchange. With interfaces to both the external exchange order messaging channel and the internal alpha engine, the order management unit is well suited as a place to keep track of order status and strategy inventory, and to perform risk checks before an order is submitted to the exchange.

8.3.4   Order type and order qualifier

Each outgoing order from the strategy side is characterized by its order type and its order qualifier. As described in the introductory paragraph of Chapter 5, order types can be generally classified as limit order, market order and cancellation order. Order qualifiers are additional flexibility provided by the exchanges to cater for various needs of the market participants. For example, a commonly used order qualifier is GTC which means "good-till-cancel" allows the accompanied limit order to stay on the LOB until the order is completely executed, canceled or when the instrument expires (for derivatives). Without using GTC, all orders are considered as day orders which are removed from the LOB at the end of the trading session, i.e., when the market close.

Each exchange offers its own order types and order qualifiers. For example, the NASDAQ platform offers more than ten variants of orders while that of CME provides more than five. The following paragraphs describe three kinds of order qualifiers.

Midpoint-peg order

Midpoint-peg (abbreviated as MP) orders allow market participants to submit orders that are pegged to the midpoint of the National Best Bid and Offer (NBBO) market; see Section 8.4.1 for the background of the NBBO market. Once the matching machine of the exchange has accepted a pegged order, it will remain hidden in the LOB. Only the execution of these orders will reveal their presence. For odd spreads, the pegged orders can execute in half-penny increments, effectively reducing the cost to less than the minimum tick size. The key benefits of this order qualifier are price improvement and anonymity. This qualifier is available on a number of US equity exchanges, such as NASDAQ, BATS global markets (abbreviated as BATS), Archipelago Exchange (abbreviated as Arca), Boston Stock Exchange (BX as in Section 8.2.4), and Philadelphia Stock Exchange (abbreviated as PSX).

To illustrate, consider the two cases in Table 8.4. In the first case, the strategy submits an order: "Buy 200 shares of AAPL, on NASDAQ, Midpoint-peg." This pegged order is accepted when the NBBO is ($114.49, $114.50) for (*bid, ask*). Therefore, to price relative to the NBBO, the order is placed in the LOB as a hidden order with price $114.495 which is equal to ($114.49 + $114.50)/2. After the national best ask order at $114.50 is completely canceled, the new national best ask price becomes $114.51. Such cancellation causes the pegged order to change its price from $114.495 to $114.50 because $114.50 is the average of the second-best bid price and the new best ask price. That is, the cancellation causes the hidden bid-ask spread to increase from a half-tick to 1 tick and the displayed bid-ask spread from 1 tick to 2 ticks. The second case in Table 8.4 corresponds to the situation that a hidden order is on the ask side at $114.495 while the national best bid is $114.49. Hence, a newly accepted pegged buy order would result in immediate execution which gives a traded volume of 200 shares and a transacted price of $114.495. Note that $114.495 is a better price than the one obtained by using a market order in an exchange that does not offer the MP qualifier.

TABLE 8.4: Midpoint-peg (MP) order to buy 200 shares. NASDAQ is at the NBBO. Top panel: MP buy order resulting in the order being added to the book as hidden order priced at the mid market at $114.495; it then reprices up to the new mid price of $114.50 immediately after the L1 (level 1) ask orders are cancelled and a new L1 ask level established. Bottom-panel: MP buy order that is filled immediately at $114.495.

		114.52	900
		114.51	500
		114.50	300
200	114.49		
600	114.48		

ADD BUY ⇒ *200@MP*

		114.52	900
		114.51	500
		114.50	300
200	114.495		
400	114.49		
600	114.48		

REPRICE ⇓ *200@114.50*

		114.53	970
		114.52	900
		114.51	500
200	114.50		
400	114.49		
600	114.48		

		114.52	900
		114.51	500
		114.50	300
		114.495	250
200	114.49		
600	114.48		

FILL ⇒ *200@MP*

		114.52	900
		114.51	500
		114.50	100
		114.495	50
200	114.49		
600	114.48		

Iceberg order

Like MP orders, an iceberg order also has an invisible component. On the CME Globex, the qualifier takes the form of the *FIX Tag 210 MaxShow* that specifies the display quantify of the order. Upon execution of the displayed quantity, the order is replenished automatically, until the full amount is filled. For example, consider the order: "Buy 50 contracts of CLH6[16] with *MaxShow* set to 10, on CME, with a limit price of $48.65, GTC." Only 10 contracts will be visible to the market. Once the 10 contracts are executed, another 10 are then shown to the market. This process is repeated until the entire order size of 50 is filled. Note that the replenishment order is treated as a new order in queue priority.

Post-only

NASDAQ post-only orders allow market makers to control how quotes interact with existing liquidity. A post-only order only executes if the price crosses the market by an amount that is economically advantageous for the order to execute, otherwise the order is repriced to one tick away from the best bid and ask prices. An important advantage of this order qualifier for market makers who use the rebate to control trading costs is that even when the snapshot of the market maker's order book is delayed, the risk of taking liquidity and paying a fee is reduced. Post-only orders must be marked as display and limit, or else are rejected by the exchange. This qualifier is available on a number of US equity exchanges, such as NASDAQ, BATS, Arca, BX, and PSX. To illustrate, consider the two cases in Table 8.5. In the first case, the strategy submits an order: "Buy 200 shares of AAPL, on NASDAQ, with a limit price of $114.50, Post-only." This order will be accepted and added to the book at $114.49. In the second case, for which the order price is $114.51, the order will be filled at $114.50, because the forgone rebate (net of fee) is $0.0041 which is less than $0.01, so it is economically advantageous for the order to be executed.

8.4 Exchange rules and regulations

8.4.1 SIP and Reg NMS

In the US, the Securities Act Amendments of 1975 introduced the security information processor (SIP) and the equity national market structure (NMS). The SIP links all exchanges by processing all protected quotes and trades from the participating centers and consolidating these data into one datastream.

---

[16]CLH6 is the CME symbol for crude oil (WTI) futures that expire in March 2016.

TABLE 8.5: Post-only order to buy 200 shares. Top panel: limit price of $114.50, resulting in the order being added to the book and priced one tick away from the ask side at $114.49. Bottom panel: limit price of $114.51, resulting in the order being filled at $114.50.

		114.52	900				114.52	900
		114.51	500	ADD BUY			114.51	500
		114.50	300	⇒			114.50	300
200	114.49			200@$114.50	400	114.49		
600	114.48				600	114.48		

		114.52	900				114.52	900
		114.51	500	ADD BUY			114.51	500
		114.50	300	⇒			114.50	100
200	114.49			200@$114.51	200	114.49		
600	114.48				600	114.48		

On behalf of the Consolidated Tape Association (CTA),[17] Securities Industry Automation Corporation (SIAC) operates and maintains two separate computer environments to process trade and quote information: trade information is processed by the Consolidated Tape System (CTS) and quote information by the Consolidated Quotation System (CQS). The CTA SIP collects quotes and trades from all market centers trading *Tape A* and *Tape B* securities, and disseminates real-time trades and quotes information. Tape A securities are those listed on NYSE LLC, and Tape B securities are those listed on BATS, NYSE Arca, NYSE MKT[18] and other regional exchanges. For NASDAQ-listed securities (Tape C), the UTP[19] SIP makes available the UTP Level 1 information on data feeds – trade information processed by the Consolidated Tape System, and quote information by the Consolidated Quotation System. There is a considerable amount of traffic going in and out of the SIPs every day, and according to the Q1 2015 CTA SIP Metrics Report, the median latency to process all incoming data was close to 400 ms for that quarter.

In 2007, SEC adopted the Regulation National Market System (Reg NMS) for equity exchanges.[20] To ensure that market data dissemination is fair for SIP and direct market feed consumers, Reg NMS affirms the standard of simultaneous data distribution so that the quotes and trades information leaves

---

[17]CTA is officially the operating authority for both the Consolidated Quotation System (CQS) and the Consolidated Tape System (CTS).

[18]NYSE MKT is formerly known as American Stock Exchange.

[19]UTP is the abbrevitaion of Unlisted Trading Privileges, which essentially means Tape C securities.

[20]Note that this rule is strictly applied to equity exchanges only. Futures, such as those traded on the CME platform, are not subject to Reg NMS rules.

TABLE 8.6: Order book snapshot of XYZ that trades on Exchange A, Exchange B, and Exchange C; the NBBO is \$114.90 bid and \$114.50 offer.

Exchange A				
			114.53	980
			114.52	900
			114.51	400
	400	114.49		
	600	114.48		
	750	114.47		

Exchange B				
			114.52	350
			114.51	200
			114.50	100
	150	114.48		
	170	114.47		
	270	114.46		

Exchange C				
			114.52	300
			114.51	100
			114.50	50
	110	114.48		
	210	114.47		
	300	114.46		

the exchange at the same time for different feeds. Once the market data leaves the exchanges, it can go through multiple network hops and different paths, and therefore the arrival of market data can vary widely depending on the particular setup and proximity to the source. For this reason, Reg NMS does not require data to be synchronized to arrive at a destination at the same time. It intends to connect market places and contains a provision preventing standing orders on an automated market from being bypassed in favor of inferior orders submitted elsewhere. The key initiative of Reg NMS[21] is the *Order Protection* or *Trade Through* Rule (Rule 611) together with a series of exceptions, such as Rule 611(b) which establishes the inter-market sweep orders (ISO) that are designed to provide efficient and workable intermediate price priority.

Rule 611 requires trading centers to enforce written policy and procedures to prevent execution of trades at prices inferior to *protected* quotations, i.e., to prevent what is known as "trade-throughs". For an NMS stock price quotation to be considered as protected, it must be immediately and automatically accessible[22] and be the best bid or best offer. The exchange can either reject

---

[21]The other three notable initiatives are the Access Rule (Rule 610), Sub-Penny Rule (Rule 612) and the Market Access Rule.

[22]Note that reverse or hidden orders are not immediately accessible and therefore are not protected, even if the hidden prices are better than the displayed price.

marketable orders or route them to exchanges displaying the best price. Registered exchanges in the US are required to provide the best bids and offers to be included in the SIPs of the primary exchanges that establish the NBBO. Alternative trading systems such as *dark pools* are not required to provide best quotes to any SIP, but are required to check and match trades within the NBBO, and to report trades immediately to the Trade Reporting Facility of FINRA (Financial Industry Regulatory Authority, Inc.)

ISO orders provide a mechanism to bypass the NBBO and trade-through protection checks that are otherwise required under Rule 611. The broker-dealer who originates an ISO order is required to trade across all protected venues and assume the exchange's liability of Rule 611 compliance. One of the advantages of using ISO orders is reduced latency and reduced likelihood of missing liquidity as a result of stale NBBO prices. To illustrate how this works, consider XYZ, an NMS security that trades on exchanges A, B, and C. Table 8.6 shows the state of the order book of the three exchanges. The NBBO is \$114.90 bid and \$114.50 offer. Dealer D wants to pay up to \$114.51 for 500 shares of XYZ. If Dealer D enters a buy order of \$114.51 for 500 shares to Exchange A, the order will be rejected or routed to another exchange. If the order is submitted to Exchange B or Exchange C, the order will be partially executed with the remaining added as an improving bid level. Alternatively, Dealer D can first send two orders simultaneously, one paying \$114.50 for 100 shares to Exchange B, and another one paying \$114.50 for 50 to Exchange C. Upon execution and subsequent SIP update, the new NBBO will then be \$114.49 bid and \$114.51 offer. Dealer D can then submit a third buy order paying \$114.51 for the remaining 350 shares. On the other hand, using ISO, Dealer D can send three orders simultaneously, marked as ISO, to all three exchanges at once.

8.4.2   Regulation SHO

Short selling, which refers to selling a security without owning, is possible via loaning the security from another party, delivering the security to the buyer, and then keeping the loan with the same or different counterparties until the short position is eventually bought back. In doing so, the short seller aims to profit from a decrease in the value of the security so that the profit is large enough to offset any losses in what is effectively a collateralized loan to the security lender. Short selling provides a valuable mechanism to facilitate price discovery. In response to the market developments since short sale regulation was first introduced in 1938, Reg SHO provides the updates necessary to safeguard the market from potential side effects of this mechanism. It was first approved in 2004 and, after successive iterations, the SEC adopted the current amendments in 2011. Rule 200 of Reg SHO requires the proper marking of orders as "long" if the selling owns the security, "short" if the seller does not own the security at time of sale, and "short exempt" if it is an over-allotment or lay-off sale by an underwriter. Rule 203(b) specifies a threshold that triggers forced close-out in case of delivery failure.

The alternative uptick rule, Rule 201, was widely discussed and eventually amended after a number of iterations. The rule pertains to the conditions under which short selling is permitted, and prohibits the display or execution of short selling at or below the prevailing bid price of the listed exchange if a covered security decreases by 10% or more from the closing price. This serves as a circuit breaker that helps to mitigate a downward price spiral due to an increase in short selling pressure. Note that long selling (i.e., owning and selling the security) is still allowed to execute at the best bid. As an example, consider XYZ, an NMS security listed on NASDAQ with a closing price of $100 the previous day. Suppose that because of some news event, the NBBO of XYZ has moved rapidly down to $89.99 bid and $90.00 offer, causing NASDAQ to trigger the circuit breaker per Rule 201. Dealer A, who does not own XYZ stock and is looking to short sell, submits a limit order to sell at $89.99 on NASDAQ. The order will be repriced and displayed at $90.00 as XYZ is now a covered security under Rule 201. Dealer B, who owns 100 shares of XYZ and senses a continued downward pressure on the stock, decides to sell 50 shares and submits a limit sell order of 50 shares at $89.99, marking the order as "long". This order goes through and results in immediate execution.

8.4.3 Other exchange-specific rules

NYSE Rule 48

Designated Market Maker (DMM) of a particular NYSE symbol has the obligation to disseminate opening price indication – these prices will need to be approved by stock market floor managers before the trading can actually begin. However, under extreme circumstances, the exchange has the power to override this obligation in order to speed up the opening of the market by appealing to Rule 48. A recent example was in the last week of August 2015, when the Dow Jones Industrial Index was at one point down by 1,000 points because of significant market decline in China. NYSE invoked Rule 48 in order to speed up the market opening in anticipation of significant market volatility.

NASDAQ Rule 5550(a)(2)

Under this rule, for a company to satisfy the continued listing requirement, it must ensure that the best bid price of its stock stays above $1. If the best bid price stays below $1 for over 30 consecutive days, NASDAQ will send the company a "deficiency notice". The company will be given 180 days to regain compliance, i.e., the best bid price needs to be above $1 for at least 10 consecutive days in the 180-day period. The company will receive a delisting letter if it fails to comply within those 180 days.

8.4.4 Circuit breaker

The alternative uptick rule of Reg SHO is a circuit breaker that limits the downside price movement after a significant drop in price by specifying a set of conditions under which short selling is permissible. There are other more explicit circuit breakers currently in effect at many registered exchanges. These circuit breakers, also known as trading curbs, can actually halt trading completely for a specific period of time in order to allow the market to reach equilibrium prices in a more orderly fashion. They can be market-wide or security-specific. For each NMS stock, the circuit breaker monitors a price band, continuously over a specified time window, based on the reference price calculated by the SIP. The width of the price band depends on the liquidity tier of the security. Supplement 11 in Section 8.6 provides more details on circuit breakers.

8.4.5 Market manipulation

The following are examples of abusive behavior monitored by the exchanges and regulatory agencies that consider such behavior as violation of genuine trading activity or as market manipulation ploys.

Layering

Layering creates an artificial imbalance of the order book so as to skew the perceived supply and demand, with the goal of inducing a favorable fill or influencing the direction of the market.

Spoofing

Spoofing develops a false sense of liquidity by submitting an excessive number of orders; it also refers to intentionally adding heavy workload to the exchange infrastructure so as to increase the processing latency of competitors.

Marking the close

The intent of this violation is to influence the closing price of an asset so as to gain a more favorable margin, or to influence other algorithms that rely on closing prices for future trading decisions.

8.5 Risk management

Given the fact that a significant percentage of trade volume on global exchanges is now attributable to some form of automated trading, this trading method can have far-reaching impact on global financial markets. Proper risk management and implementation of compliance procedures within the trading firm and some degree of regulatory oversight are beneficial to all participants of the market in the long run. At the trading firm level, the role of the risk officer is to ensure that all trading activities done by the firm adhere to a set of predefined good practices and limits, and to update these practices and limits based on the performance of the strategy, the volume traded and assessment of the robustness of the strategy. Some of the risk limits are also monitored and enforced by the exchange. For example, the CME has a strict limit on the message rate so that any order that exceeds this threshold is dropped. Many exchanges also monitor some definition of the ratio of the number of fills to the total number of messages. The role of the compliance officer at the firm level is to monitor all trading activities to ensure that proper exchange and regulatory guidelines are followed and to investigate any potential problems which, if left unresolved, could lead to compliance and legal issues. For quantitative trading, two main sources of risk at the firm level are operational risk and strategy risk.

8.5.1 Operational risk

As trading decisions are made at split-second intervals, the task of ensuring that the algorithm and the infrastructure are error free pertains to both software engineering and quantitative analytics. Trading algorithms are required to undergo multiple stages of testing and certification, with duplicate checks in place. Seemingly inconspicuous bugs such as integer overflow or underflow could have a significant impact on the strategy. Operational risk in the context of an algorithmic trading process mainly refers to the risk stemming from infrastructure disruptions and software bugs. As with any large-scale software development of sophisticated and mission-critical applications, software testing methods are indispensable.

Software error risk

Software testing techniques such as unit testing, regression testing and non-regression testing are widely used as an integral part of the software development process. *Unit tests* aim at isolating specific classes or class member functions so that the test is independent of the other part of the code. Sometimes, the developer would also need a way to check if a new bug fix or a new release does not break the existing code. Therefore, in addition to unit tests on the specific modifications, the developer also performs *regression testing*, which compares the output of the actual application instead of some abstraction parts of the code. The comparison is carried out by (a) saving the output of the application to a file, (b) implementing the modification, and then (c) carrying out a simple comparison of the two output files to ensure that the modifications do not affect the existing correct performance. Whereas regression testing of software aims at assuring that no existing functionality has been broken in the process of fixing a defect, *non-regression testing* is performed for a new release of the software to ensure that any newly added functionality can produce the intended effect.

Order transmission protocol risk

Under CFTC (Commodity Futures Trading Commission) directives, in order for trading firms to have direct access to the exchange matching and market data infrastructure, they need to have passed a set of *conformance tests*. For conformance testing, trading firms connect to exchange-managed testing facilities that mimic the real production environment. The goal is for the trading firm to test their complete end-to-end trading processes in a production environment. For example, CME provides dedicated testing environments for current production (Certification Environment) and for future release (New Release Environment). Clients who wish to test their trading infrastructure can connect to the environment from a dedicated data center (Cert Data Center) colocated in the same production facility, with physical cross-connect to the test gateways and matching engines.

Other infrastructure-related risk

The operating system kernel can drop network packets if the incoming message rate is higher than the rate at which messages are processed; the kernel drops all incoming packets once the capture buffer is full. When this happens, some information is lost and the book state can become corrupted. This is often manifested in the form of stale prices when the best bid and ask prices appear to be locked or crossed. Proper mechanisms in place are needed to deal with scenarios in which the book state is either outright incorrect or is suspect as a result of stale information. The former is reasonably straightforward to handle. For example, the algorithm can request an order book snapshot from the exchange or playback pre-recorded data from another independent data capturing device.[23] The latter, however, is problematic for cases where the book state appears to be normal but there is uncertainty as to whether the price is stale or real.

Unexpected shutdown or network failure is another source of infrastructure risk. This is especially important for market-making strategies which can have many outstanding orders in the order book queue and all these orders require constant monitoring and adjusting. An expected shutdown of strategy or order management engines, or network failure, can lead to unintended execution of outstanding orders which will require manual intervention to close out or to hedge. One way to deal with this at the exchange is via automated cancel if the gateway detects a potential shutdown of the client engines or a network

---

[23]Exchanges, such as NASDAQ and CME, also provide Drop Copy services, which send client order reply messages to a designated location that serves as backup copy of the order reply messages to the client via the normal production connection.

failure. Similar mechanisms can be put in place between the strategy and order management engines.

Regulatory compliance risk

The role of a compliance officer is to help detect and prevent potential regulatory violations. Detecting violations requires large-scale data analysis at microsecond resolution. It is a challenge to monitor trading activities that can span multiple time zones at multiple exchanges and over multiple asset classes. Given the sheer amount of orders transmitted by a typical high-frequency strategy, automation is needed to process the incoming and outgoing messages to uncover potential violations. This can be done by playing back strategy orders overlaid on top of historical market data, looping through a list of compliance policy checks, and looking for potential violations. While such policy as excess message rate is well defined by most exchanges and is simple to detect, others such as momentum ignition or layering require sophisticated models that keep track of a number of internal states and provide the necessary capability to differentiate market-making activity from liquidity-taking activity.

8.5.2 Strategy risk

For strategy implementation, besides having sound procedures in place, it is also important to take into account the risks that can adversely affect the overall integrity of the trading algorithm.

Modeling risk

For models that require real-time parameter updates, there should be mechanisms in place to monitor and correct the errors in model fitting (for example, due to deteriorating condition numbers of matrices that need to be inverted), as these errors can corrupt future predictions and estimations.

Trading risk

From the viewpoint of risk management, high-frequency trading has a number of common risk parameters that have been considered for more traditional trading. Checks of the following aspects should be performed to manage trading risk:

- Position limit: This caps the net position of a particular portfolio. It could be in notional terms for simple equity, delta terms for options, or (dollar) duration terms for bond portfolios. For a plain equity portfolio, there might be limits on specific sectors of the industry so as to limit systematic exposure to a particular sector. For bond portfolios, there might be specific limits for different parts of the yield curve to limit the exposure of the trading firm to changes in the shape of the yield curve.

- P&L (Profit and Loss) limit: This caps the amount of drawdown of a particular strategy.

- Options' Greeks exposure: This caps the exposure of an option's delta, vega, gama and other Greeks.

- Scenario limit: This caps the P&L impact given some economic scenario, either based on historical realizations or based on some theoretical worst-case economic environment. It is closely connected to the commonly used VaR (value at risk) limit, where a specific threshold is put on the tail events or correlated tail events.

- Inventory reconciliation: At the beginning of the day (BOD), it is the traders' duty to make certain that the BOD position is as expected. This involves checking the inventories from the previous end of day (EOD) close against the BOD positions, and to check the opening position against the position according to the clearing firm. Such checking ensures that the automated process run by the system indeed picks up the correct positions and that there are no outstanding trade breaks.

For automated trading, additional relevant parameters that need to be considered include the following.

- Order size: This caps the size of each order submitted.

- Order rate: This caps the number of orders that the strategy can submit within a second.

- Message rate: This caps the ratio of the total number of messages and the number of orders filled.

- Potential position: This caps the potential position that the portfolio can take if all outstanding orders are filled. It is relevant to market making strategies where there might exist multiple orders at different levels of the limit order book, with decreasing probability of ever getting executed.

Before sending each order, there needs to be pre-trade risk checking, in which the strategy loops through necessary checks to ensure that the outgoing instruction does not violate any of the preceding risk caps.

In addition, the strategy has to monitor all incoming order reply messages, to check rejects or other messages that can affect trading. Rejects can be originated from the exchange or from the order management engine. The former is commonly referred to as *exchange reject*, and the latter as *internal reject*. An exchange can reject an order for a number of reasons such as incorrect order attributes, excessive message rate, or insufficient margin. A strategy needs to have a mechanism to deal with both exchange and internal rejects so that it does not try to resend continuously, which will have negative impact on the overall latency and, more importantly, will risk clogging up the exchange bandwidth.

8.6 Supplementary notes and comments

1. **OTC transactions** The OTC market mentioned in the second introductory paragraph mainly deals with more complex and customized products. An OTC transaction often involves at last three counterparties: an
originating dealer, a counterparty dealer, and a broker who acts as the me-
diator. The role of the broker is to assist the originating dealer in seeking
out necessary liquidity, anonymously. For this service, the broker earns a
commission. For interest rate swaps, this commission amounts to a fraction of a *running basis point*.[24] To illustrate OTC transactions, consider a
simple OTC traded interest rate swap involving two fictitious counterparties: Bank of Zeus (Dealer A) and Bank of Janus (Dealer B). The following
dialogue outlines how the OTC trading mechanism works in real life:

```
Dealer A to Broker: where is 2yr semi-annual-bond in 100MM?
          Broker: let me check.
Broker to Dealer B: where is your market in 100MM
                    2yr semi-annual-bond?
        Dealer B: I deal one and a quarter,
                  one and three quarters in 100MM
          Broker: thanks.
Broker to Dealer A: one and a quarter,
                    one and three quarters in 100MM
        Dealer A: yours in 100MM.
Broker to Dealer B: yours at one and a quarter in 100MM.
                    It's Bank of Zeus.
Dealer B to Broker: done.
Broker to Dealer A: done. It's Bank of Janus.
Dealer A to Broker: done.
Broker to Dealer A: you sold 100MM of 2yr semi-annual-bond
                    at one and a quarter to Bank of Janus.
Broker to Dealer B: you bought 100MM of 2yr semi-annual-bond
                    at one and a quarter from Bank of Zeus.
```

2. **Network layer transport protocol** There are two main types of the
network transport protocol: Transmission Control Protocol (TCP) and
User Datagram Protocol (UDP). TCP is connection-oriented as it establishes a one-to-one connection, or *uni-cast*, between two hosts so that once
a connection is established, data can be sent bidirectionally. A TCP connection has multiple levels of checks and handshakes to ensure transmission
reliability. UDP, on the other hand, is a connectionless protocol. Instead
of requiring a one-to-one connection as in the case for TCP, UDP can be
used in a multicast setting, there one host broadcast the data to multiple

---

[24]A basis point is $1/100$ of a percent and a running basis point refers to a periodic
stream of basis points, e.g., a running annualized 25bps over two years represents two equal
cashflows of 0.25% times notional each, one at the end of the first year, and another at the
end the second year.

hosting listening to the specific multicast channel. The advantage is UDP multicast is bandwidth optimization and throughput maximization. However, once a UDP packet leaves the host, it is not guaranteed to be received by the designated listeners in full or in part. The onus is on the listeners to detect any missing packets and to request the host to rebroadcast. Unless there are multiple hops between the broadcaster and the listener, the occurrence of dropped packets is often rare if the listener is able to keep up with processing the arriving packets.

3. **CME MDP 3.0 messages**   Here we give more details about the CME MDP network packets described in Section 8.2.3. The packet follows a standard layering format:

   - The Binary Packet Header gives the unique sequence number for the time-stamp of the packet, for example,

     – 10897345 20151015123452345632488

     which means the packet whose sequence number is 10897345 was sent by the Gateway at time 2015-10-15 12:34:52.345632488.

   - The Message Header gives methods of decoding the MDP messages; see Figure 8.4 in which Message Size refers to the length of the message including the binary header, Block Length refers to the length of the rest of the message, TemplateID is the ID used to encode the message, SchemaID is the ID of the system that published the message, and Version refers to the Schema version.

   - The FIX Message Header gives the type of FIX message, e.g., 35=X in Figure 8.6.

   - The FIX Message Body gives the type of match event conditional on the actual FIX message type.

   - The FIX Message Block gives the details of the match event.

   Two main types of MDP message important to trading are Trade Summary and Book Management. Trade Summary messages are for trade events, i.e., whenever a trade or a series of trades have occurred as a result of the matching process, one or more of these messages will be disseminated by the exchange. Book Management messages have two subtypes: *Multiple Depth Book*, and *Implied Book*. Book update events often follow trade events. For example, consider the case that an aggressor buy order takes out three levels of the book on the ask side. The resulting MDP3.0 message is represented by Figure 8.6. For the full MDP3.0 message protocol format, consult CME online resources.

4. **Market event, strategy trigger, and CME MDP 3.0**   Event trigger deals with the sequence of events that triggers strategy calculation and order submission. Market data come in network packets, which can contain information for an atomic book transition, multiple atomic book

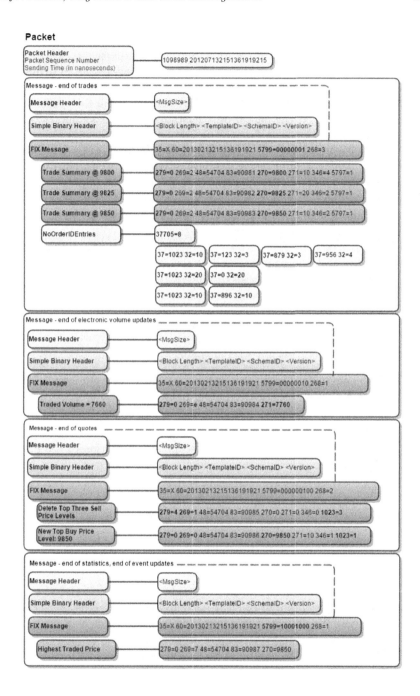

FIGURE 8.6: MDP3.0 message for a buy-side aggressor order that takes on three levels on the order book on the ask side. (Used with permission of CME.)

transitions, or a partial or incomplete transition, as discussed in Section 8.2.3. An atomic book transition consists of all book and order events as a result of a triggered event, so that the snapshot of the order book used as input information for the strategy matches that of the matching engine at the end of the trigger event. For example, Figure 8.6 defines an atomic book transition triggered by the arrival of the buy order for 40 futures contracts. If the packet only contains the trade summary but not the subsequent book management updates, then the packet only contains a partial book transition because at the end of the packet, the state of the book does not fully reflect the events triggered by the incoming buy order. This distinction is important from a strategy standpoint. Although the strategy needs to react to events soon after the raw network packet is parsed, incomplete book information can lead to incorrect or inconsistent update of the LOB state. To illustrate, consider a market-making strategy that recalculates internal model states upon detecting a trade event. If the event spans multiple packets, model calculation and quote orders should wait until all packets are parsed. One of the advantages of CME MDP 3.0 is that the parsing procedure can start right after the first packet has arrived and does not need to wait for all packets.

5. **Alpha research** Strategy alpha model research typically consists of seven key steps.

- Problem scope formation: This defines the nature and scope of the research project; for example, the problem scope could be to "refine the current model to increase level-based book queue position inference accuracy".

- Literature review: Based on the scope and definition of the research topic, and on the existing work done in the area and related fields;

- Theoretical foundation: Based on the literature review, the research team can choose the most suitable modeling framework and lay down the theoretical foundation of the modifications needed.

- Prototype: The research team codes up the theoretical model using a scripting language (e.g., R, Python) and performs simulation studies as proof of concept.

- Production code implementation and testing: Once satisfied with the result of the proof-of-concept simulation, the research team can implement the prototype in a compiled language, such as C++ or Java, followed by testing with a more accurate matching engine level simulator. This is often followed by production trading with conservative risk limits.

- Production launch: The model is finally released into production with normal risk limits and, gradually, the risk parameters are relaxed to normal settings.

- Refinement: Using the trade debug logs, the research team studies the model behavior, identifies technical issues that need to be addressed and describes additional features for future research.

6. **Programming platforms for alpha research** In quantitative research, there is often the need to quickly prototype an idea, i.e., code up the model, fit it to some historical data and assess the result. Scripting languages, such as R, Python, or Matlab, are well suited for this type of rapid prototyping. The advantage of a scripting language over a compiled language such as C++ and Java is the ease with which a researcher can modify the model, see the result and query variables on the fly. R has been widely adopted by the statistics community, which also includes machine learning and general data analysis. It is open source and the code base is updated frequently. While the main code base provides a large collection of basic data analysis functionality, the power of R comes from the many packages that cover a wide range of research needs, from cutting-edge machine learning libraries such as **kernlab**, **nnet** and **glmnet**, to sophisticated optimization routines such as **nloptr** and **optim**. For handling large datasets, R has packages, for example **biglm**, **ff**, that can perform iterative model updates so that the whole dataset does not need to be loaded all at once. Parallel computing on a single multicore computer or over a grid of computers can also be implemented via R packages such as **snow** and **parallel**. Additional R packages, such as **gputools** and **gmatrix**, are also available for parallel computing on graphics processing unit (GPU[25]) which is now a standard component in modern computers.[26] Python is a popular choice among researchers who prefer the object-oriented programming language syntax of Python. Together with the data analysis library **pandas**, Python rivals R in terms of model sophistication and capability.

Once the prototyping is complete, the model will need to be implemented in a compiler language in order to take full advantage of the low-latency capabilities of the kernel and computer hardware. Ideally, the research model and the production model share the same code or call the same core set of libraries, so as to minimize error in the translation phase. One way to do this is to code the prototype in the final language of choice and to conduct research in that language. This may be inefficient because the process of compiling and debugging may hinder the research process. Another method is to architect a suite of C++ libraries via well-designed syntactic sugar.

The approach that seems to offer the best tradeoff is to extend already well-supported and widely-adopted languages such as R and Python, in the

---

[25]GPU stands for Graphics Processing Unit. Because of their highly parallel structure, modern GPUs are used not only in manipulating computer graphics and image processing but also in parallel computing.

[26]See Matloff (2015) for more information on paralell computation used in modern data analysis by R and C++.

form of packages and libraries. Research groups in trading firms can implement libraries using highly optimized code, written in C++ or Java, that supports specific models or parts of a model, and can add proprietary functions to support firmware implementation of a computation grid, GPU, or co-processor clusters, as well as easy access to historical market data and trade logs. R can load these libraries for prototyping, and the same library will then be used in production code.

7. **Numerical accuracy** While trading algorithms are derived by using operations on the set of real numbers, they are implemented on computational systems that perform numerical operation on floating point numbers that can be represented in the form

$$\pm \left[d_0 + d_1\beta^{-1} + d_2\beta^{-2} + \cdots + d_{p-1}\beta^{-(p-1)} \right] \beta^e, \tag{8.1}$$

where both $p, \beta \in \mathbb{N}$, $\beta > 1$, $e \in \mathbb{Z}$, $d_i \in \mathbb{Z}$ and $0 \leq d_i < \beta$ for $i = 0, \ldots, p - 1$. This is called a floating point number representation with *precision* p, *base* (or *radix*) β and *exponent* e. The number $d_0 + d_1\beta^{-1} + \cdots + d_{p-1}\beta^{-(p-1)}$ (denoted by m) in (8.1) is known as the *significand* or the *mantissa* of the representation. The usual convention of (8.1) also assumes $d_0 \neq 0$ which makes $1 \leq m < \beta$. d_0 and d_{p-1} are called the *most significant digit* and the *least significant digit*, respectively. IEEE 754 binary format is the standardized rule for the bit representation (i.e., $\beta = 2$) of floating point number. Such format also deals with zero, negative values and NaNs.[27] While decimal numbers such as $0.25 (= 1.0 \times 2^{-2})$ and $0.5 (= 1.0 \times 2^{-1})$ have exact representation in binary format, many others such as $0.1 (= 1.1001100110011 \cdots \times 2^{-4})$ cannot be exactly represented and may lead to an approximation error that has significant impact in ways often least expected; typical examples are loss of precision in unintended cancellation and decision errors created by comparing floating point numbers.

8. **Hardware setup for algorithmic trading** The appropriate computer hardware setup largely depends on the type of strategy. For high-frequency trading, the demand on computational hardware can be degrees of magnitude higher than that of medium- to low-frequency strategies. Generally speaking, faster clock speed means more lines of code executed per second, which in turn means faster reaction to market events. However, clock speed is only one aspect of an optimal setup. Factors such as CPU architecture, motherboard and chipset design, memory profile, and network card characteristic can all have a significant impact on the overall performance of

---

[27] NaN stands for "not a number" and is an undefined value caused by numerical operations such as $0/0$ or $\sqrt{-1}$. Also, note that the IEEE 754 format not only uses $\beta = 2$ but also chooses p and integers e_{min} and e_{max} (with $e_{min} < 0 < e_{max}$) so that all representable numbers would have $e \in [e_{min}, e_{max}]$, and numerical overflow and underflow refer to the case of $e > e_{max}$ and $e < e_{min}$, respectively, i.e., the tolerance on the approximation error determines the choice of p, e_{min} and e_{max}.

the strategy. Modern CPU architecture allows multiple processing cores per CPU die,[28] and each core can have multiple hardware threads. The difference between cores and hardware threads is that each core has separate hardware resources, whereas physical threads share almost all the spawning core's resources, such as actual registers.

9. **Latency reduction techniques: Cache optimization and vectorization** As the execution efficiency and clock speed of CPUs increase, the bottleneck of a trading application is often in memory access. Therefore, some understanding of the memory architecture, especially in memory caching, is essential to implement low-latency applications. Cache memory, also known as the CPU memory, is the random access memory (RAM) that a computer microprocessor can access more quickly than it can access regular RAM. The speed difference is due to the fact that (a) the cache memory content has a higher priority to be handled by the microprocessor than the data in the regular memory and, (b) the cache memory is implemented by the semiconductor SRAM (static RAM) that can be accessed faster than DRAM (dynamic RAM) which makes the regular memory. In addition, since SRAM is considerably more costly than DRAM, cache memory is substantially smaller than regular memory in most servers and is usually in the form of levels with level 1 (abbreviated as L1) having the highest priority to be accessed and level 2 having the second highest priority, and so on; see Drepper (2007) for more details about computer memory architecture. Cache optimization uses software engineering techniques that improve spatial and temporal locality which, at a high level, means that data accessed frequently and in succession should be close in memory. It also involves other methods of utilizing cache memory in order to attain low latency. In particular, noting that cache is a scarce resource, Loveless et al. (2013) suggest to keep only a relatively small amount of data and parameters in the L1 cache during runtime by using on-line one-pass algorithms (such as the scalar form of the Kalman filter in lieu of the Kalman gain matrix in Supplement 9 of Section 3.4).

Vectorization is a form of code optimization, either done by the developer manually using kernel built-in functions known as "intrinsics", or by the complier automatically. It takes advantage of Single Instruction, Multiple Data (SIMD) operations, by using vector registers over static registers and can boost performance significantly. Intel's latest Streaming SIMD Extensions (SSE), SSE4, is able to handle 128bit in one operation. Vectorization is enabled by default with the compile flag `-O3`, and can be turned on or off explicitly by the flag `-ftree-vectorize` and `-fno-tree-vectorize`. For compiler vectorization options and examples of functions that can

---

[28]Fabrication produces integrated circuits in the form of wafers. Each wafer is further diced into small blocks such that each block contains one copy of the circuit and is called a processor die. The unit "die" is originated from the wafer dicing process.

or cannot be optimized, see `https://gcc.gnu.org/projects/tree-ssa/vectorization.html`.

10. **Static versus dynamic dispatch** In Object Oriented Programming, polymorphism is often achieved by using virtual functions. This enables the compiled code to execute the appropriate function in the derived class via dynamic, or runtime, binding. In low-latency application, this method of dispatch has a number of drawbacks, such as the inability to use inlining, extra indirection for each virtual method call, and the extra overhead of eight bytes per object for the virtual table pointer which can lead to cache misses. The remedy is usage of *static dispatching* to mimic its dynamic counterpart by template design patterns such as CRTP (Curiously Recurring Template Pattern).

11. **History of circuit breaker mechanisms** These mechanisms were put in place after Black Monday on October 19, 1987, when stock markets around the world crashed, shedding a huge value in a very short time. The Dow Jones Industrial Average dropped by 508 points to 1738.74. The objective of circuit breaker mechanisms is to reduce market volatility and massive panic sell-offs, giving traders time to reconsider their transactions.

    The NYSE circuit breaker set up in 1987 has the following features:

    - At the beginning of each quarter, NYSE sets three circuit breaker levels, usually at 7% (Level 1), 13% (Level 2), and 20% (Level 3) of the average closing price of the S&P 500 for the month preceding the start of the quarter, rounded to the nearest 50-point interval.

    - Level 1 and Level 2 declines can result in 15-minute trading halts unless they occur after 3:25pm, when no trading halts apply.

    - A Level 3 decline results in trading being suspended for the remainder of the day.

    The SEC asked all US exchanges to implement an additional security-level circuit breaker for all S&P 500 Index stocks in June 11, 2010, after the Flash Crash on May 6, 2010. In particular, instead of setting the quarterly average closing prices as benchmarks, the new stock-by-stock circuit breaker uses the transaction price in the preceding five minutes as a rolling benchmark. If the price of an equity experiences a more than 10% move in a five-minute period, all transactions related to this equity, including its trades at other US exchanges as well as its derivatives, will pause for at least five minutes. After that, the exchange that issued the pause could consider extending that halt if significant imbalances still exist in the affected equity. If the pause lasts longer than 10 minutes, the other exchanges are free to resume trading.

    The security-level circuit breaker was first extended to cover Russell 1000 stock indexes and 344 of the most heavily traded ETFs on September 13,

2010. Since June 2011, it has covered all NMS securities. On May 31, 2012, the SEC refined the security-level circuit breaker by using a *Limit Up-Limit Down* mechanism to address market volatility by preventing trades in listed equity securities when triggered by large, sudden price moves. On the same day, the SEC also revised the Market Wide Circuit Breakers rule which set the trigger to be computed daily based on the prior day's closing price of the S&P 500 index. The new trigger is much more sensitive to the market conditions than the former one which is determined only on a quarterly basis. Such revised rule was implemented in February 2013.

12. **Flash crashes and operational risk in HFT**   A flash crash is a deep drop in security prices occurring rapidly within a very short time period. The Flash Crash of 2010 occurred on May 6 when the US stock markets were on a downward trend due to worries about the debt crisis in Greece. At 2:42pm, with the Dow Jones Industrial Average (DJIA) down by over 300 points for the day, a rapid drop of over 600 points occurred within five minutes, resulting in a loss of nearly 1000 points for the day by 2:47pm. Twenty minutes later, DJIA had regained most of the 600-point drop. On September 30, 2010, SEC and CFTC issued a joint report *Findings Regarding the Market Events of May 6, 2010* to identify what led to the Flash Crash. The report describes that a large mutual fund firm selling an unusually large number of E-Mini S&P index futures first exhausted available buyers and then led high-frequency traders to sell aggressively, driving down the price of E-Mini S&P 500 approximately 3% from 2:41pm through 2:44pm. During this time, cross-market arbitrageurs who bought the E-Mini S&P 500 sold simultaneously equivalent amounts in the equities markets, driving the price of SPY down approximately 3% also. The report also points out that shortly after trading on the E-Mini was paused for five seconds by the CME Stop Logic Functionality, prices stabilized and the E-Mini began to recover, followed by the SPY. Within a day after the release of the 104-page report, a number of critics noted that blaming a single order (from Waddel & Reed) for triggering the Flash Crash was highly problematic, and in particular, CME argued against the report's explanation of what caused the Flash Crash. A working paper (Kirilenko et al., 2015) by an economist at the Federal Reserve Board and academic collaborators provides an empirical analysis of the trade data from the Flash Crash. Similar to the SEC-CFTC report, the working paper calls the cascade of selling shown by these data "hot potato trading" as high-frequency firms rapidly acquired and then liquidated positions among themselves in response to rapidly declining prices. On April 21, 2015, the US Department of Justice laid "22 criminal charges, including fraud and market manipulation" against a high-frequency trader who was alleged to have used spoofing algorithms to place thousands of orders for E-Mini S&P 500 index futures, amounting to about "$200 million worth of bets that the market would fall" and to modify them 19,000 times before can-

celing them; this conduct "was at least significantly responsible for order imbalance that in turn was one of the conditions that led to the flash crash."

In October 2013, a flash crash occurred on the Singapore Exchange (SGX), Southeast Asia's biggest bourse, and wiped out $6.9 billion in capitalization, with three stocks BLUM, ACAP and LionGold Corp losing at least 87% of their value.

On October 15, 2014, the US bond market had a flash crash when the yields on the benchmark 10-year notes plunged about 30 basis points from its previous-day close of 2.2% to 1.9% from about 9am to 9:36am ET, but then bounced back to above 2% within 15 minutes. A joint report in July 2015 by the New York Fed, the US Treasury Department, then SEC and the CFTC said that the flash crash could not be attributed to a specific cause, but was due to a number of factors, including a decline in order book depth, an unprecedented number of short positions, and a heightened level of self-trading during portions of that event window. The report defines self-trading as a transaction in which the same entity takes both sides of the trade.

On August 24, 2015, the DJIA dropped 1100 points in the first five minutes of trading. On Sunday night, August 23, a large drop in equities in Asia triggered a drop of index futures on stocks in the US and Europe, with the US index futures down 7% prior to opening of the US market.

Risk management to guard against potential big losses in the event of flash crashes is therefore particularly important in high-frequency trading. Failure to protect against software error risk mentioned in Section 8.5.1 on operational risk can result in catastrophic losses. A case in point was the 2012 software error of the Knight Capital Group, whose largest business is market making in US equities and options and which also engages in institutional sales and electronic execution of trades. In 2011, the company was worth US$1.5 billion and had about 1450 employees. On August 1, 2012, it deployed untested software that contained an obsolete function PowerPeg because a technician forgot to copy the new RLP (Retail Liquidity Program) code to one of the eight servers for automated routing of equity orders. RLP repurposed a flag that was formerly used to activate PowerPeg, which was designed to move prices. Therefore, orders sent to that server triggered the defective PowerPeg, causing a major disruption in stock prices and resulting in 4 million executions in 154 stocks within 45 minutes. After Knight Capital took a pre-tax loss of $440 million, its stock price dropped by over 70%. On August 5 it raised around $400 million from several major investors to stay in business, and was subsequently acquired by the Global Electronic Trading Company (Getco LLC) which then formed the new company KCG Holdings, Inc.

The Flash Crash of 2010 has spurred much interest from the public in high-frequency trading (HFT) and many myths and debates on HFT. Principal

Traders Group of The Futures Industry Association (2015) has provided a summary of these debates and further discussion of the issues. The regulatory response to the debates and to improve the equity market structure is still work in progress; the speeches by the chair of SEC, Mary Jo White, and commentaries by industry representatives and regulators, are posted at the Harvard Law School Forum on Corporate Governance and Financial Regulation for July 9, 2014 (`http://corpgov.law.harvard.edu`).

8.7 Exercises

NASDAQ uses the TotalView-ITCH, a direct outbound market data protocol developed by the NASDAQ OMX Group, for both their software and the FPGA hardware versions of market data feed services.[29] ITCH message fields, unlike FIX, are fixed-length without delimiter or tag. The message start and end positions are indicated by an offset relative to the start of the message, and the length of the field. This feature allows fast direct casting of the message into structure by the parser, which in turn allows rapid look up of specific fields in the message. TotalView-ITCH features order level data with full order depth using the standard ITCH format, and uses a series of order messages to track the life of a customer order, from inception to execution. The feed is made up of a series of sequenced messages, each variable in length based on the message type. NASDAQ broadcasts the market data from the US primary data center facility in Carteret, New Jersey.

Table 8.7 and Table 8.8 show the specification for ADD and EXECUTE order messages, respectively. All integers fields are big-endian binary encoded numbers, and all alpha fields are left justified and right padded ASCII fields. Prices are integer fields, supplied with associated precision. Table 8.9 shows the ITCH representation of an inbound *Add Order Message* for a buy side order for 500 shares of TEST with a price of $114.500.

1. Estimate the transmission time between Aurora, Chicago, and Secaucus, New Jersey, based on the following transmission transport: (i) copper wire, (ii) optic fiber, (iii) microwave, (iv) milliwave.

2. In C++, the compiler translates a two-dimensional array using row-major indexing; that is, the following matrix,

$$\begin{bmatrix} 1 & 2 & 3 \\ 4 & 5 & 6 \\ 7 & 8 & 9 \end{bmatrix}$$

---

[29]`http://www.nasdaqtrader.com/content/technicalsupport/specifications/dataproducts/NQTVITCHspecification.pdf`

TABLE 8.7: ITCH5.0 Add Order Message specification. This message will be generated for unattributed orders accepted by the NASDAQ system. The Add Order message has a fixed length of 36 Bytes.

Name	Offset	Length	Value	Message
Message Type	0	1	"A"	Add Order – No MPID attribution Message
Stock Locate	1	2	Integer	Locate code identifying the security
Tracking Number	3	2	Integer	NASDAQ OMX internal tracking number
Time-stamp	5	6	Integer	Nanoseconds since midnight
Order Reference Number	11	8	Integer	The unique reference number assigned to the new order at the time of receipt
Buy/Sell Indicator	19	1	Alpha	The type of order being added. "B" = buy order. "S" = sell order
Shares	20	4	Integer	The total number of shares associated with the order being added to the book
Stock	24	8	Alpha	Stock symbol, right padded with spaces
Price	32	4	Price (4)	The display price of the new order

TABLE 8.8: ITCH5.0 Order Executed Message specification. This message is sent whenever an order on the book is executed in whole or in part.

Name	Offset	Length	Value	Message
Message Type	0	1	"A"	Add Order – No MPID attribution Message
Stock Locate	1	2	Integer	Locate code identifying the security
Tracking Number	3	2	Integer	NASDAQ OMX internal tracking number
Time-stamp	5	6	Integer	Nanoseconds since midnight
Order Reference Number	11	8	Integer	The order reference number associated with the executed order
Executed Shares	19	4	Alpha	The number of shares executed
Matching Number	23	8	Integer	The NASDAQ generated day-unique Match Number of this execution

TABLE 8.9: An ITCH5.0 Add Order Message. "A" denotes an Add message, "B" a buy order, "S" a sell order; "Stock Locate" takes integer values and is the locate code identifying the security; "Tracking Number" refers to the NASDAQ OMX internal tracking number; "Time-stamp" is the number of nanoseconds since midnight.

Add Order Message	
Message Type	A
Stock Locate	12345
Tracking Number	12233
Time-stamp	1222444555
Order Reference Number	3321123456
Buy/Sell Indicator	B
Shares	500
Stock	TEST
Price	114.500

```
0000 41 30 39 2F C9 00 00 48 DD 06 0B 00 00 C5 F4 52 A|12345|12233|1222444555|3321123456
0001 80 42 00 00 01 F4 54 45 53 54 20 20 20 20 00 11 B|500|TEST
0002 78 A8                                              114.500
```

is stored in contiguous memory in the row ordering 1, 2, 3, ..., 8, 9. Write a C++ class, MatrixMultiplier, which has the following interface:

```
template<class T>
class MatrixMultiplier {
public:
    // initialize m_matrix
    MatrixMultiplier(int a_row, int a_col);

    // return the sum of elements by looping
    // over first the columns then the rows
    T rowMajorSum();

    // return the sum of elements by looping
    // over first the rows then the columns
    T colMajorSum();
private:
    // two dimensional array of
    // a_row - by - a_col
    T** m_matrix;
};
```

Implement the member functions and then instantiate this class and create a matrix of size 10 by 50. Call the two member functions and time their execution times. Repeat with larger size in increments of 10% increase, up to 1000 by 5000. Plot your result. Does your result vary significantly depending on data type: int32_t, int64_t, and double? Note: The implementation needs to ensure that between two successive tests, all data caches are properly invalidated.

3. Let $x \in \mathbb{R}$ and $\hat{x}$ be its floating point representation with precision p and base β as in (8.1) with $d_0 \neq 0$. Suppose $\hat{x}$ is obtained by rounding. Let $\delta = |x - \hat{x}|$ be the absolute approximation error. For $x \neq 0$, the corresponding relative error is defined as $\varepsilon = \delta/|x|$.

   (a) Find $b > 0$ such that all real numbers in the interval $[\hat{x}-b, \hat{x}+b)$ share the same floating point representation. Then, show that $\delta \leq \beta^{e-p+1}/2$ and $\varepsilon \leq \beta^{1-p}/2$. When will these upper bounds be tight? [30]

   (b) Let x and y be two positive real numbers with $x > y$. Let the computing difference of $\hat{x}$ and $\hat{y}$ with p digits be $x \ominus y$.

      i. Let the absolute approximation error of such difference operator be $\delta_\ominus = |x - y - (x \ominus y)|$ and $\varepsilon_\ominus = \delta_\ominus/|x-y|$. Show that there exists two real numbers ε_x and ε_y such that

$$\delta_\ominus = x\varepsilon_x - y\varepsilon_y \text{ with } |\varepsilon_x| \leq \beta^{1-p}/2 \text{ and } |\varepsilon_y| \leq \beta^{1-p}/2.$$

---

[30] β^{e-p+1} is called the *unit in the last place* or the *unit of least precision* and is commonly denoted by ulp. $\beta^{1-p}/2$ is called the *unit roundoff* or the *relative precision* and is denoted as rp. Note that while both rp and ulp are functions of the representation parameters, p and β, ulp also depends on the exponenet e in (8.1) which makes it also a function of the real number being approximated.

Hence, show that $\varepsilon_\ominus$ can be as large as $\beta - 1$.

ii. Show that $|\varepsilon_\ominus| \le \beta^{1-p}/2$ if the difference operator $\ominus$ uses one guard digit, i.e., $\hat{x} - \hat{y}$ is computed by using $p + 1$ digits.

4. Write the 32-bit IEEE 754 floating point representation of the following numbers: Newtonian constant of gravitation $G = 6.6784 \times 10^{-11}$, Planck constant $h = 6.626069 \times 10^{-34}$, and elementary charge $e = 1.602176565 \times 10^{-19}$. Check whether there are numbers that could have the underflow issue stated in Supplement 7.

5. CME disseminates incremental market data down two feeds[31] for each Market Data Group. This duality is in place to minimize the impact of UDP packet loss. Both channels broadcast concurrently with time difference in the region of $\pm 2\mu s$. Develop an algorithm that takes both feeds, and output the fastest raw data to a client that listens to the feed. This is commonly known as feed arbitration. Note: Refer to CME message specifications for available market data feed attributes.

---

[31] CME also provides two Market Recovery Feeds that disseminate snapshot market data.

A

Martingale Theory

This appendix provides a summary of the basic theory of discrete-time and continuous-time martingales that we have used in different parts of the book. It can be used as a reference source for readers who want some basic background on martingale strong law, quadratic variation, semimartingales and stochastic integrals.

A.1 Discrete-time martingales

Let X be a random variable defined on the probability space $(\Omega, \mathcal{F}, P)$ such that $E|X| < \infty$. Let $\mathcal{G} \subset \mathcal{F}$ be a σ-field. A random variable Y is called a version of the *conditional expectation* of X given $\mathcal{G}$, denoted by $E(X|\mathcal{G})$, if it satisfies the following two properties:

1. Y is $\mathcal{G}$-measurable,

2. $\int_A X \, dP = \int_A Y \, dP$ for all $A \in \mathcal{G}$.

The conditional expectation is therefore the Radon-Nikodym derivative $d\nu/dP$, where $\nu(A) = \int_A X \, dP$ for $A \in \mathcal{G}$. Hence it is unique except for P-null sets. The special case $Y = I(B)$ gives the conditional probability $P(B|\mathcal{G})$ of B given $\mathcal{G}$. A *filtration* is a nondecreasing sequence (i.e., $\mathcal{F}_n \subset \mathcal{F}_{n+1}$) of σ-fields $\mathcal{F}_n \subset \mathcal{F}$. A non-negative integer-valued random variable N is called a *stopping time* with respect to the filtration $\{\mathcal{F}_n\}$ if $\{N = n\} \in \mathcal{F}_n$ for every n.

Associated with a stopping time N is the σ-field

$$\mathcal{F}_N = \{A \in \mathcal{F} : A \cap \{N = n\}\} \in \mathcal{F}_n \qquad \text{for all } n \geq 1\}. \tag{A.1}$$

A sequence of random variables X_n is said to be *uniformly integrable* if

$$\sup_{n \geq 1} E\{|X_n| I(|X_n| \geq a)\} \to 0 \qquad \text{as } a \to \infty. \tag{A.2}$$

Uniform integrability is an important tool in martingale theory. Some fundamental results in martingale theory are the *optional stopping theorem* and the *martingale convergence theorem*, together with classical martingale inequalities. The sequence $\{X_n\}$ is called a *martingale* with respect to a filtration

$\{\mathcal{F}_n\}$ if X_n is $\mathcal{F}_n$ measurable and $E(X_n|\mathcal{F}_{n-1}) = X_{n-1}$ almost surely for every n, and a *submartingale* (respectively, *supermartingale*) if $=$ is replace by $\geq$ (respectively, $\leq$). The sequence $\{\varepsilon_n\}$ is called a *martingale difference sequence* with respect to $\{\mathcal{F}_n\}$ if $E(\varepsilon_n|\mathcal{F}_{n-1}) = 0$ almost surely for every n.

Theorem 1. *Let $\{X_n, \mathcal{F}_n, n \geq 1\}$ be a submartingale and $M \leq N$ be stopping times (with respect to $\{\mathcal{F}_n\}$). If $\{X_{N \wedge n}, n \geq 1\}$ is uniformly integrable, then $E(X_N|\mathcal{F}_M) \geq X_M$ a.s., and consequently, $EX_N \geq EX_M$.*

Theorem 2. *Let $\{X_n, \mathcal{F}_n, n \geq 1\}$ be a submartingale. If $\sup_{n \geq 1} E(X_n^+) < \infty$, then X_n converges a.s. to a limit X_∞ with $E|X_\infty| < \infty$.*

Theorem 3. *Let $\{X_n, \mathcal{F}_n, n \geq 1\}$ be a submartingale. Then for every $\lambda > 0$,*

$$\lambda P \left\{ \max_{1 \leq i \leq n} X_i > \lambda \right\} \leq E \left\{ X_n I \left(\max_{1 \leq i \leq n} X_i > \lambda \right) \right\} \leq EX_n^+,$$

$$\lambda P \left\{ \min_{1 \leq i \leq n} X_i < -\lambda \right\} \leq EX_n^+ - EX_1.$$

Moreover, for any $p > 1$,

$$E \left(\max_{1 \leq i \leq n} X_i^+ \right)^p \leq \left(\frac{p}{p-1} \right)^p E(X_n^+)^p.$$

Theorem 4. *Let $\{M_n = \sum_{i=1}^n d_i, \mathcal{F}_n, n \geq 1\}$ be a martingale. Then there exist finite positive constants a_p, b_p depending only on p such that*

$$a_p E \left(\sum_{i=1}^n d_i^2 \right)^{p/2} \leq E \max_{j \leq n} |M_j|^p \leq b_p E \left(\sum_{i=1}^n d_i^2 \right)^{p/2} \qquad \text{for all } p \geq 1.$$

The martingale $(M_n, \mathcal{F}_n, n \geq 1)$ is said to be *square-integrable* if $EM_n^2 < \infty$ for all n. A stochastic sequence $\{M_n\}$ adapted to a filtration $\{\mathcal{F}_n\}$ is said to be a *locally square-integrable martingale* if there are stopping times τ_m with respect to $\{\mathcal{F}_n\}$ such that $\lim_{m \to \infty} \tau_m = \infty$ a.s. and $\{M_{\tau_m \wedge n}, \mathcal{F}_n, n \geq 1\}$ is a square-integrable martingale for every $m \geq 1$. The following are Azuma's inequality (Theorem 5) and the convergence theorem and strong law (Theorem 6) for locally square-integrable or square-integrable martingales. Theorem 6 is an important tool in the theory of stochastic regression developed by Lai, Wei and Ying referred to in the second paragraph of Section 2.6.3. Theorem 5 plays an important role in the development of the ε-optimal allocation rule based on censored data from dark pools in Section 7.6, as shown in Supplement 3 of Section 7.8.

Theorem 5 (Azuma, 1967). *Let $\{M_n = \sum_{i=1}^n d_i, \mathcal{F}_n, n \geq 1\}$ be a locally* square-integrable martingale such that there exist nonrandom constants $a_i < b_i$ for which $a_i \leq d_i \leq b_i$ for all $i \geq 1$. Then for all $x \geq 0$,

$$P(|M_n| \geq x) \leq 2\exp\left(-\frac{2x^2}{\sum_{i=1}^n (b_i - a_i)^2}\right).$$

Theorem 6. *Let $\{\varepsilon_n\}$ be a martingale difference sequence with respect to an increasing sequence of σ-fields $\{\mathcal{F}_n\}$ such that $\sup_n E\{\varepsilon_n^2|\mathcal{F}_{n-1}\} < \infty$ a.s. Let u_n be an $\mathcal{F}_{n-1}$ measurable random variable for every n. Then*

$$\sum_1^n u_i\varepsilon_i \quad \text{converges a.s. on} \quad \left\{\sum_1^\infty u_i^2 < \infty\right\}, \tag{A.3}$$

$$\left(\sum_1^n u_i\varepsilon_i\right)\Big/\left\{\left(\sum_1^n u_i^2\right)^{1/2}\left[\log\left(\sum_1^n u_i^2\right)\right]^\eta\right\} \to 0$$

$$\text{a.s. on} \quad \left\{\sum_1^\infty u_i^2 = \infty\right\} \tag{A.4}$$

for every $\eta > 1/2$ and consequently with probability 1,

$$\sum_1^n u_i\varepsilon_i = o\left(\sum_1^n u_i^2\right) + O(1). \tag{A.5}$$

Moreover,

$$\sum_1^\infty |u_i|\varepsilon_i^2 < \infty \qquad \text{a.s. on} \quad \left\{\sum_1^\infty |u_i| < \infty\right\}, \tag{A.6}$$

$$\left(\sum_1^n |u_i|\varepsilon_i^2\right)\Big/\left(\sum_1^n |u_i|\right)^\rho \to 0$$

$$\text{a.s. on} \quad \left\{\sum_1^\infty |u_i| = \infty\right\} \quad \text{for every } \rho > 1. \tag{A.7}$$

If $\sup_n E(|\varepsilon_n|^\alpha|\mathcal{F}_{n-1}\} < \infty$ a.s. for some $\alpha > 2$, then (A.7) can be strengthened into

$$\limsup_{n\to\infty} \left(\sum_1^n |u_i|\varepsilon_i^2\right)\Big/\left(\sum_1^n |u_i|\right) < \infty$$

$$\text{a.s. on} \quad \left\{\sup_i |u_i| < \infty, \sum_1^\infty |u_i| = \infty\right\}. \tag{A.8}$$

A.2   Continuous-time martingales

The basic martingale theory in Section A.1 can be readily extended to continuous-time martingales/submartingales/supermartingales if the sample paths are a.s. right-continuous. In particular, such processes have left-hand limits and are therefore *càdlàg (continu à droit, limité à gauche)*. Here we summarize some of the results. Comprehensive treatments of continuous-time martingales can be found in the monographs by Elliott (1982), Karatzas and Shreve (1991) and Revuz and Yor (1999). A filtration $(\mathcal{F}_t)$ is said to be *right-continuous* if $\mathcal{F}_t = \mathcal{F}_{t+} := \bigcap_{\epsilon>0} \mathcal{F}_{t+\epsilon}$. It is said to be *complete* if $\mathcal{F}_0$ contains all the P-null sets (that have zero probability) in $\mathcal{F}$. In what follows we shall assume that the process $\{X_t, t \geq 0\}$ is right-continuous and adapted to a filtration $\{\mathcal{F}_t\}$ that is right-continuous and complete. $\{X_t\}$ is said to be *adapted* to the filtration $\{\mathcal{F}_t\}$ if X_t is $\mathcal{F}_t$-measurable for all $t \geq 0$. The σ-field generated on $\Omega \times [0, \infty)$ by the space of adapted processes which are left-continuous on $(0, \infty)$ is called the *predictable σ-field*. A process $\{X_t\}$ is *predictable* if the map $(\omega, t) \mapsto X_t(\omega)$ from $\Omega \times [0, \infty)$ to $\mathbb{R}$ is measurable with respect to the predictable σ-field. An extended random variable T taking values in $[0, \infty]$ is called (i) a *stopping time* (with respect to the filtration $\{\mathcal{F}_t\}$) if $\{T \leq t\} \in \mathcal{F}_t$ for all $t \geq 0$, (ii) an *optional time* if $\{T < t\} \in \mathcal{F}_t$ for all $t > 0$, and (iii) a *predictable time* if there exists an increasing sequence of stopping times T_n such that $T_n < T$ on $\{T > 0\}$ and $\lim_{n\to\infty} T_n = T$ a.s.

A stochastic process $X = \{X_t, t \geq 0\}$ is called *measurable* if the map $(\omega, t) \mapsto X_t(\omega)$ from $\Omega \times [0, \infty)$ to $\mathbb{R}$ is measurable. For a measurable process, there exists a predictable process $Y = \{Y_t, t \geq 0\}$ such that

$$E\{X_T I(T < \infty)|\mathcal{F}_{T-}\} = Y_T I(T < \infty) \quad \text{a.s.}, \tag{A.9}$$

for every predictable time T. The process Y is essentially unique and is called the *predictable projection* of X, denoted by X^Π. Suppose $A = \{A_t, t \geq 0\}$ is a nondecreasing, measurable process such that $E(A_\infty) < \infty$. Then there exists an essentially unique nondecreasing, predictable process $A^{(p)}$ such that for all bounded, measurable processes $\{X_t, t \geq 0\}$,

$$E(X_t^\Pi A_t) = E(X_t A_t^{(p)}). \tag{A.10}$$

The process $A^{(p)}$ is called the *dual predictable projection* of A; see Elliott (1982, p. 72).

Doob-Meyer decomposition and applications

Let $\mathcal{T}_a$ be the class of stopping times such that $P(T \leq a) = 1$ for all $T \in \mathcal{T}_a$. A right-continuous process $\{X_t, t \geq 0\}$ adapted to a filtration $\{\mathcal{F}_t\}$ is said to be of class DL if $\{X_T, T \in \mathcal{T}_a\}$ is uniformly integrable for every $a > 0$. If $\{X_t, \mathcal{F}_t, t \geq 0\}$ is a non-negative right-continuous submartingale, then it is of class DL. The *Doob-Meyer decomposition* says that if a right-continuous submartingale $\{X_t, \mathcal{F}_t, t \geq 0\}$ is of class DL, then it admits the decomposition

$$X_t = M_t + A_t, \tag{A.11}$$

in which $\{M_t, \mathcal{F}_t, t \geq 0\}$ is a right-continuous martingale with $M_0 = 0$ and A_t is predictable, nondecreasing and right-continuous. Moreover, the decomposition is essentially unique in the sense that if $X_t = M'_t + A'_t$ is another decomposition, then $P\{M_t = M'_t, A_t = A'_t \text{ for all } t\} = 1$. The process A_t in the Doob-Meyer decomposition is called the *compensator* of the submartingale $\{X_t, \mathcal{F}_t, t \geq 0\}$.

Suppose $\{M_t, \mathcal{F}_t, t \geq 0\}$ is a right-continuous martingale that is square integrable, i.e., $EM_t^2 < \infty$ for all t. Since M_t^2 is a right-continuous, non-negative submartingale (by Jensen's inequality), it has the Doob-Meyer decomposition whose compensator is called the *predictable variation* process and denoted by $\langle M \rangle_t$, i.e., $M_t^2 - \langle M \rangle_t$ is a martingale. If $\{N_t, \mathcal{F}_t, t \geq 0\}$ is another right-continuous square integrable martingale, then $(M_t + N_t)^2 - \langle M + N \rangle_t$ and $(M_t - N_t)^2 - \langle M - N \rangle_t$ are martingales, and the *predictable covariation* process $\langle M, N \rangle_t$ is defined by

$$\langle M, N \rangle_t = \frac{1}{4} \left\{ \langle M + N \rangle_t - \langle M - N \rangle_t \right\}, \qquad t \geq 0. \tag{A.12}$$

Let $\mathcal{M}$ denote the linear space of all right-continuous, square-integrable martingales M with $M_0 = 0$. Two processes X and Y on $(\Omega, \mathcal{F}, P)$ are *indistinguishable* if $P(X_t = Y_t \text{ for all } t \geq 0) = 1$. Define a norm on $\mathcal{M}$ by

$$\|M\| = \sum_{n=1}^{\infty} \frac{\min\left(\sqrt{EM_n^2}, 1\right)}{2^n}. \tag{A.13}$$

This induces a metric $\rho(M, M') = \|M - M'\|$ on $\mathcal{M}$ if we identify indistinguishable processes. A subspace $\mathcal{H}$ of $\mathcal{M}$ is said to be *stable* if it is a closed set in this metric space and has two additional "closure" properties:

1. $M \in \mathcal{H} \Rightarrow M^T \in \mathcal{H}$ for all stopping times T, where $M_t^T = M_{T \wedge t}$;

2. $M \in \mathcal{H}$ and $A \in \mathcal{F}_0 \Rightarrow MI(A) \in \mathcal{H}$.

Two martingales M, N belonging to $\mathcal{M}$ are said to be *orthogonal* if $\langle M, N \rangle_t = 0$ a.s. for all $t \geq 0$, or equivalently, if $\{M_t N_t, \mathcal{F}_t, t \geq 0\}$ is a martingale. If $\mathcal{H}$ is a stable subspace of $\mathcal{M}$, then so is

$$\mathcal{H}^{\perp} := \{N \in \mathcal{M} : N \text{ is orthogonal to } M \text{ for all } M \in \mathcal{H}\}. \tag{A.14}$$

Moreover, every $X \in \mathcal{M}$ has a unique (up to indistinguishability) decomposition

$$X = M + N, \qquad \text{with } M \in \mathcal{H} \text{ and } N \in \mathcal{H}^{\perp}. \tag{A.15}$$

We next further decompose M as a sum of its purely discontinuous part consisting of jumps and a "continuous part" M^c of $M \in \mathcal{M}$ (or a somewhat more general M which is the a.s. limit of elements of $\mathcal{M}$), which is related to the decomposition (A.15) with $\mathcal{H} = \mathcal{M}^c$, where

$$\mathcal{M}^c = \{M \in \mathcal{M} : M \text{ has continuous sample paths}\}. \tag{A.16}$$

It can be shown that $\mathcal{M}^c$ is a stable subspace of $\mathcal{M}$, and therefore (A.15) means that every $M \in \mathcal{M}$ has an essentially unique decomposition

$$M = M^c + M^d, \qquad \text{with } M^c \in \mathcal{M}^c \text{ and } M^d \in (\mathcal{M}^c)^\perp. \tag{A.17}$$

While M^c is called the continuous part of M, M^d is called its "purely discontinuous" part. Note that M^c and M^d are orthogonal martingales.

For $M \in \mathcal{M}$ and $t > 0$, let Π be a partition $0 = t_0 < t_1 < \cdots < t_k = t$ of $[0, t]$. Then as $\|\Pi\| := \max_{1 \le i \le k} |t_i - t_{i-1}| \to 0$, $\sum_{i=1}^{k}(M_{t_i} - M_{t_{i-1}})^2$ converges in probability to a limit, which we denote by $[M]_t$. The random variable $[M]_t$ is called the *quadratic variation* process of M. For $M \in \mathcal{M}^c$, $[M] = \langle M \rangle$. More generally, for $M \in \mathcal{M}$,

$$[M]_t = \langle M^c \rangle_t + \sum_{0 < s \le t} (\triangle M_s)^2, \tag{A.18}$$

where $\triangle M_s = M_s - M_{s-}$, noting that $M_{s-} = \lim_{u \uparrow s} M_u$ exists since M is càdlàg. The *quadratic covariation* of M and N, which both belong to $\mathcal{M}$, is defined by $[M, N]_t = \{[M + N] - [M - N]\}/4$, and (A.18) can be generalized to

$$[M, N]_t = \langle M^c, N^c \rangle_t + \sum_{0 < s \le t} (\triangle M_s)(\triangle N_s). \tag{A.19}$$

We can relax the integrability assumptions above by using *localization*. If there exists a sequence of stopping times T_n such that $\{M_{T_n \wedge t}, \mathcal{F}_t, t \ge 0\}$ is a martingale (or a square-integrable martingale, or bounded), then $\{M_t, \mathcal{F}_t, t \ge 0\}$ is called a *local martingale* (or *locally square-integrable martingale*, or *locally bounded*). By a limiting argument, we can again define $\langle M \rangle_t$, $\langle M, N \rangle_t$, $[M]_t$, $[M, N]_t$, M^c and M^d for locally square integrable martingales. Moreover, a continuous local martingale M_t can be expressed as time-changed Brownian motion:

$$M_t = W_{\langle M \rangle_t}. \tag{A.20}$$

If V is an adapted process with finite variation on bounded intervals, then its dual predictable projection process $V^{(p)}$ (see (A.10)) is the essentially unique predictable process having finite variation on bounded intervals and such that $V - V^{(p)}$ is a local martingale; see Elliott (1982, p. 121).

Stochastic integrals and Itô's formula for semimartingales

Let $\mathcal{M}^{\mathrm{loc}}$ denote the class of right-continuous, locally square-integrable martingales M with $M_0 = 0$. Let M, N belong to $\mathcal{M}^{\mathrm{loc}}$ and H, K be two measurable processes. For $0 \le s \le t$, let $\langle M, N \rangle_{s,t} = \langle M, N \rangle_t - \langle M, N \rangle_s$ and note

that
$$\langle M + rN \rangle_{s,t} = \langle M, M \rangle_{s,t} + 2r\langle M, N \rangle_{s,t} + r^2 \langle N, N \rangle_{s,t}$$
is a non-negative quadratic function of r. Therefore
$$|\langle M, N \rangle_{s,t}| \leq \{\langle M, M \rangle_{s,t}\}^{1/2} \{\langle N, N \rangle_{s,t}\}^{1/2}. \tag{A.21}$$
Hence, approximating the Lebesgue-Stieltjes integral below by a sum, we obtain from (A.21) the *Kunita-Watanabe inequality*

$$\int_0^t |H_s||K_s| \, d\overline{\langle M, N \rangle}_s \leq \left(\int_0^t H_s^2 \, d\langle M \rangle_s \right)^{1/2} \left(\int_0^t K_s^2 \, d\langle N \rangle_s \right)^{1/2}, \tag{A.22}$$

where we use the notation $\bar{\zeta}_t$ to denote the total variation of a process ζ on $[0, t]$.

We now define the stochastic integral $\int_0^t X_s \, dY_s$ with *integrand* $X = \{X_s, 0 \leq s \leq t\}$ and *integrator* $Y = \{Y_s, 0 \leq s \leq t\}$. If Y has bounded variation on $[0, t]$, then the integral can be taken as an ordinary pathwise Lebesgue-Stieltjes integral over $[0, t]$. If Y is a right-continuous, square-integrable martingale and X is a predictable process such that $\int_0^t X_s^2 \, d\langle Y \rangle_s < \infty$ a.s., then $\int_0^t X_s \, dY_s$ can be defined by the limit (in probability) of integrals (which reduce to sums) whose integrands are step functions and converge to X in an L_2-sense. A process Y which is adapted to the filtration $\{\mathcal{F}_t, t \geq 0\}$ is called a *semimartingale* if it has a decomposition of the form $Y_t = Y_0 + M_t + V_t$, where M is a locally square-integrable, right-continuous martingale, V_t is an adapted càdlàg process with finite variation on bounded intervals, and $M_0 = V_0 = 0$. We can therefore define $\int_0^t X_s \, dY_s$ when X is a predictable, locally bounded process and Y is a semimartingale. Moreover, the process $\{\int_0^t X_s dY_s, t \geq 0\}$ is also a semimartingale.

Theorem 7 (Itô's formula for semimartingales). *Let $X(t) = (X_1(t), \ldots, X_m(t))$, $t \geq 0$, be a vector-valued process whose components X_i are semimartingales, which can be decomposed as $X_i(0) + M_i(t) + V_i(t)$. Let $f : [0, \infty) \times \mathbb{R}^m \to \mathbb{R}$ be of class $C^{1,2}$ (i.e., $f(t, x)$ is twice continuously differentiable in x and continuously differentiable in t). Then $\{f(t, X(t)), t \geq 0\}$ is a semimartingale and*

$$f(t, X(t)) = f(0, X(0)) + \sum_{i=1}^m \int_{0+}^t \frac{\partial}{\partial x_i} f(s, X(s-)) \, dX_i(s)$$

$$+ \frac{1}{2} \sum_{i=1}^m \sum_{j=1}^m \int_{0+}^t \frac{\partial^2}{\partial x_i \partial x_j} f(s, X(s-)) \, d\langle M_i^c, M_j^c \rangle_s$$

$$+ \sum_{0 < s \leq t} \left\{ f(s, X(s)) - f(s, X(s-)) \right.$$

$$\left. - \sum_{i=1}^m (\partial/\partial x_i) f(s, X(s-)) \, \triangle X_i(s) \right\}.$$

An important corollary of Theorem 7 is the *product rule* for semimartingales X and Y: XY is a semimartingale and

$$d(X_t Y_t) = X_{t-} dY_t + Y_{t-} dX_t + d[X, Y]_t. \tag{A.23}$$

In particular, for a locally square-integrable, right-continuous martingale M, setting $X = Y = M$ in (A.23) yields

$$[M]_t = M_t^2 - 2 \int_0^t M_{s-} \, dM_s. \tag{A.24}$$

By (A.24), $M^2 - [M]$ is a martingale but $[M]$ is not predictable. Since $M^2 - \langle M \rangle$ is a martingale and $\langle M \rangle$ is predictable, it follows that $\langle M \rangle_t$ is the compensator of the nondecreasing process $[M]_t$.

B

Markov Chain and Related Topics

A stochastic process $(X_t)_{t \geq 0}$ with a discrete state space $\mathbb{S}$ is called a continuous-time Markov chain (CTMC) if for all $s, t \geq 0$, $i, j \in \mathbb{S}$,

$$P\left(X_{s+t} = j | X_s = i, \mathcal{F}_s^X\right) = P\left(X_{s+t} = j | X_s = i\right) = P_{ij}(t),$$

where $\mathcal{F}^X$ is the σ-field generated by $(X_t)_{t \geq 0}$. In this appendix we introduce the generator and discrete-time embedded chain of a CTMC, and potential theory for discrete-time Markov chains and the Bellman equation in infinite-horizon Markov decision theory. In particular, this provides the basic background for Sections 5.5.3 and 6.5.2 on LOB modeling and order placement.

B.1 Generator Q of CTMC

A CTMC is completely characterized by its generator $\boldsymbol{Q} = (q_{ij})_{i,j \in \mathbb{S}}$, where

$$P(X_{t+h} = j | X_t = i) = \delta_{ij} + q_{ij}h + o(h),$$
$$\text{with} \quad q_{ii} = -\sum_{j \neq i} q_{ij} < 0, \quad q_{ij} \geq 0.$$

Here $\delta_{ij} = 1$ (or 0) if $i = j$ (or $i \neq j$), and q_{ij} is called the transition intensity from state i to j.

Embedded chain

The discrete-time Markov chain $(\tilde{X})_{n=0,1,\dots}$ with transition matrix $\boldsymbol{P} = (p_{ij})_{i,j \in \mathbb{S}}$, where $p_{ij} = q_{ij}/\sum_{k \neq i} q_{ik}$ for $i \neq j$ and $p_{ii} = 0$, is called an embedded chain associated with $(X_t)_{t \geq 0}$. In fact, $\tilde{X}_1 = X_{\tau_1}$, $\tilde{X}_2 = X_{\tau_2}, \dots$, where $\tau_i = \min\{t > \tau_{i-1} : X_t \neq X_{\tau_{i-1}}\}$, $\tau_0 = 0$. If X_t is ergodic, it has a stationary distribution, which is the same as that of its embedded chain.

B.2   Potential theory for Markov chains

Suppose $X = (X_n)_{n \geq 1}$ is a Markov chain on a countable state space $\mathbb{S}$ with a transition matrix $P = (P_{ij})_{i,j \in \mathbb{S}}$. Let $D \subset \mathbb{S}$ and ∂D be the boundary of D. Let $\tau = \inf\{n : X_n \notin D\} = \inf\{n : X_n \in \partial D\}$. Given a non-negative running cost function $V_c = (V_c(s))_{s \in D} \geq 0$ and a non-negative terminal cost $V_T = (V_T(s))_{s \in \partial D} \geq 0$, define the expected cost starting from state s and ending up at ∂D by

$$\Phi(s) = \mathbb{E}\left[\sum_{n=1}^{\tau-1} V_c(X_n) + V_T(X_\tau) I_{\{\tau < \infty\}} | X_0 = s \right]. \tag{B.1}$$

Then Φ satisfies a linear system of equations

$$\begin{cases} \Phi = P\Phi + V_c & \text{in } D, \\ \Phi = V_T & \text{in } \partial D. \end{cases}$$

B.3   Markov decision theory

Suppose $X = (X_n)_{n \geq 1}$ is a Markov chain on a finite state space $\mathbb{S}$ with transition matrix P. Now suppose $\mathbb{A} = \mathbb{C} \cup \mathbb{T}$ is a finite set of actions, where $\mathbb{C}$ is the set of continuing actions (i.e., prior to the terminal action) and $\mathbb{T}$ is the set of terminal actions. Define $A : \mathbb{S} \mapsto 2^{\mathbb{A}}$ as the function associating a non-empty set of actions $A(s)$ to each state $s \in \mathbb{S}$. Here $2^{\mathbb{A}}$ is the power set consisting of all subsets of $\mathbb{A}$. Clearly $C(s) = A(s) \cap \mathbb{C}$ and $T(s) = A(s) \cap \mathbb{T}$. For every $s \in \mathbb{S}$ and $a \in C(s)$, the transition probability from s to s' when selecting action a is denoted as $P_{ss'}(a)$. Assume in addition the following regularity conditions.

- $V_c(s, a)$ is bounded and non-negative; it is the cost of continuation at state s with action $a \in C(s)$.

- $V_T(s, a)$ bounded and non-negative; it is the cost of termination at state s with action $a \in T(s)$.

- $\alpha = (\alpha_0, \alpha_1, \dots)$ is a sequence of actions, where $\alpha_n : \mathbb{S}^{n+1} \to \mathbb{A}$ such that $\alpha_n(s_0, \dots, s_n) \in A(s_n)$ for each $n \geq 0$ and $(s_0, \dots, s_n) \in \mathbb{S}^{n+1}$.

- $V(s, \alpha)$ is the expected cost starting at $X_0 = s$, following a policy α until termination.

A policy α^* is called optimal if for any policy α,

$$V(s, \alpha^*) \leq V(s, \alpha).$$

The optimal expected cost $V^*(s)$ is defined by

$$V^*(s) = \inf_{\alpha} V(s, \alpha).$$

Bellman equation

$V^*(s)$ is associated with solution to the *Bellman equation*

$$V(s) = \min \left[\min_{\alpha \in C(s)} V_c(s, \alpha) + \sum_{s' \in \mathbb{S}} P_{ss'}(\alpha) V(s'), \ \min_{\alpha \in T(s)} V_T(s, \alpha) \right].$$

C

Doubly Stochastic Self-Exciting Point Processes

In this Appendix, we provide the essential background material and implementation details for Hawkes processes that we have introduced in Sections 4.6.2 and 5.3 to model inter-transaction durations and arrivals of marketable orders on opposite sides of the LOB. To begin with, we relate the arrival times $t_1, t_2, \ldots$ to the process

$$N(t) = \sum_{i=1}^{\infty} I_{\{t_i < t\}} = \max\{i : t_i \leq t\}, \tag{C.1}$$

introduced in Section 4.6.2. The first equality in (C.1) explains why $N(t)$ is called a *counting process* (sum of indicator variables) and the second equality explains why it is also called a *renewal process*, which is associated with renewal theory for the inter-arrival times $t_i - t_{i-1}$. Section C.1 follows up on Appendix A and gives the martingale theory of counting processes. Section C.2 uses this theory to build doubly stochastic point process models of which the self-exciting point processes (SEPPs) in Section 4.6.2 are special cases. In Section C.3, we consider likelihood inference on SEPPs with particular focus on the linear case with $\phi(x) = x$ in (4.51) and on the multivariate setting of Section 5.3. Section C.4 describes Monte Carlo methods to simulate point processes given the process specification.

C.1 Martingale theory and compensators of multivariate counting processes

Using the theory of continuous-time martingales in Section A.2, a multivariate counting process $\mathbf{N} = (N_1, \ldots, N_k)$ is a vector of k *càdlàg* processes adapted to a filtration $\{\mathcal{F}_t\}$ such that (a) $\mathbf{N}(0) = \mathbf{0}$, (b) each N_i is nondecreasing and piecewise constant, with jumps of size 1 only, and (c) no two components of $\mathbf{N}$ can jump simultaneously. If $\mathcal{F}_t$ is generated by $\{\mathbf{N}_s, s \leq t\}$, then the filtration is called "self-exciting". Because of (c), $N_\bullet = \sum_{i=1}^{k} N_i$ is also a counting process and there exist stopping times $0 < t_1 < t_2 < \cdots$, together with random variables $J_1, J_2, \ldots$, taking values in $\{1, \ldots, k\}$, so that $N_\bullet$ jumps at time t_i and J_i is the index of the process that jumps.

To begin with, consider the case $k = 1$. Since $N(t)$ is right-continuous and nondecreasing and is adapted to the filtration $\{\mathcal{F}_t\}$, it is a submartingale. By the Doob-Meyer decomposition (A.11), it has the decomposition $N(t) = M_t + A_t$, in which M_t is a martingale with $M_0 = 0$ and A_t is the compensator that is predictable, nondecreasing and right-continuous. Suppose A_t is absolutely continuous and its derivative is continuous almost everywhere (a.e.) with respect to Lebesgue measure. Then from the definition $N(t)$ as a sum of indicator variable $I_{\{t_i \leq t\}}$, it follows that the derivative of A_t is equal a.e. to λ_t defined by (4.50). Therefore $A_t = \int_0^t \lambda(s)ds$, which is denoted by $\Lambda(t)$ and called the integrated intensity in Section 4.6.2, where it is asserted that $N(t) - \Lambda(t)$ is a martingale.

We next consider the case of general k and assume that the limit in (4.50) with N replaced by $\mathbf{N}$ exists a.e. Denote the limit by $\boldsymbol{\lambda} = (\lambda_1, \ldots, \lambda_k)$. Then for $i = 1, \ldots, k$, N_i is a counting process whose compensator is given by $\Lambda_i(t) = \int_0^t \lambda_i(s)ds$. Let $M_i(t) = N_i(t) - \Lambda_i(t)$ be the martingale in the Doob-Meyer decomposition. It follows from (b) and (c) in the first paragraph of this section that $\langle M_i \rangle_t = \Lambda_i(t)$ and $\langle M_i, M_j \rangle_t = 0$ for $i \neq j$. More generally, without assuming the compensator Λ_i of N_i to be absolutely continuous, we have

$$\langle M_i \rangle_t = \Lambda_i(t) = \int_0^t \Delta \Lambda_i(s)d\Lambda_i(s)$$

$$\langle M_i, M_j \rangle_t = -\int_0^t \Delta \Lambda_i(s)d\Lambda_j(s) \text{ for } i \neq j,$$

$$\text{(C.2)}$$

where $\Delta A(s) = A(s) - A(s-)$ and $A(s-) = \lim_{t \uparrow s} A(t)$ for a nondecreasing function A. Note that although $\mathbf{N}$ has jumps, $\boldsymbol{\Lambda}$ can be absolutely continuous. A well-known example is the Poisson process, which has independent increments that are Poisson distributed, i.e., for $k = 1$, $N(t) - N(s)$ is independent of $\mathcal{F}_s$ for $0 \leq s \leq t$, and $P(N(t) - N(s) = n | \mathcal{F}_s) = e^{-\lambda(t-s)}[\lambda(t-s)]^n/n!$. In this case (4.50) holds with $\lambda(t) \equiv \lambda$ and therefore $\Lambda(t) = \lambda t$. Watanabe (1964) gives the following martingale characterization of the Poisson process: For $\lambda > 0$, $N(t)$ is a Poisson process with intensity λ if and only if $N(t) - \lambda t$ is a martingale with respect to a filtration $\{\mathcal{F}_t\}$.

C.2 Doubly stochastic point process models

The Poisson process $N(t)$, $t \geq 0$, is sometimes called a *Poisson point process* as the arrival times $t_1 < t_2 < \cdots$ are random points on the half line $(0, \infty)$. The inter-arrival times are i.i.d. exponential in this case. For a general counting process, the inter-arrival times may not be identically distributed, and therefore the process is sometimes called a *non-homogeneous Poisson process*. When its intensity function $\lambda(t)$ is also a stochastic process, the point process is called *doubly stochastic* because not only is the counting process $N(t)$ is a stochastic process but its intensity function is also random. To model such a point process, one only needs a model for $\lambda(t)$ because whenever a jump occurs, it has to have size 1. One such model is (4.51), a special case of which is the Hawkes process defined by (4.52), which satisfies the stochastic differential equation $d\lambda(t) = \beta[\mu - \lambda(t)]dt + \alpha dN(t)$. If $\alpha/\beta < 1$, the Hawkes process (4.52) is stationary and its steady-state expectation is given by $\mu/(1 - \alpha/\beta)$. Moreover, its integrated intensity is

$$\Lambda(t) = \int_0^t \lambda(u) \, du = \mu t - \frac{\alpha}{\beta} \sum_{i:t_i < t} \left(e^{-\beta(t-t_i)} - 1 \right),$$

which can be derived by considering piecewise integration of the intensity.

Multivariate extensions

A doubly stochastic multivariate counting process $\mathbf{N} = (N_1, \ldots, N_k)$ can be modeled through its intensity processes

$$\lambda_i(t) = \phi_i \left(\mu_i(t) + \sum_{j=1}^k \int_{u < t} h_{ij}(t - u) \, dN_j(u) \right), \quad i = 1, \ldots, k, \qquad (C.3)$$

in which λ_i depends not only on the past events of the ith component of $\mathbf{N}$ through h_{ii} but also on those of the other components through h_{ij} for $j \neq i$. A special case is the k-variate Hawkes process with intensity processes

$$\lambda_i(t) = \mu_i + \sum_{j=1}^k \alpha_{ij} \int_{u < t} e^{-\beta_{ij}(t-u)} \, dN_j(u),, \quad i = 1, \ldots, k. \qquad (C.4)$$

The stationarity condition is $\|\mathbf{A}\| < 1$, where $\mathbf{A}$ is the $k \times k$ matrix whose (i,j)th entry is given by α_{ij}/β_{ij} for $1 \leq i, j \leq k$.

C.3   Likelihood inference in point process models

Let $t_1 < \cdots < t_n$ be the observed event times (or arrivals) of a univariate point process $N(t)$ with intensity function $\lambda_\theta(t)$. Since $N(t_n) = n$, it can be shown that the likelihood function based on these data is

$$L(\theta) = \exp \left(- \int_0^{t_n} \lambda_\theta(u) \, du \right) \prod_{j=1}^n \lambda_\theta(t_j),$$

hence the log-likelihood function $\mathcal{L}(\theta) = \log L$ can be expressed as

$$\mathcal{L}(\theta) = -\int_0^{t_n} \lambda_\theta(u) \, du + \int_0^{t_n} \log \lambda_\theta(u) \, dN(u); \qquad (C.5)$$

see Daley and Vere-Jones (2003, Prop. 7.2 III). For the case of a Hawkes process (4.52) with exponential kernel that has parameter $\theta = (\mu, \alpha, \beta)$, the log-likelihood function (C.5) with $\lambda_\theta(t)$ given by (4.52) reduces to

$$\mathcal{L}(\theta) = -\int_0^{t_n} \underbrace{\left(\mu + \int_{u<t} \alpha e^{-\beta(t-u)} \, dN(u) \right)}_{\text{concave wrt } \beta} dt$$

$$+ \int_0^{t_n} \log \underbrace{\left(\mu + \int_{u<t} \alpha e^{-\beta(t-u)} \, dN(u) \right)}_{\text{convex wrt } \beta} dN(u). \quad (C.6)$$

To evaluate the MLE, Ogata (1978) rewrites (C.6) as

$$\mathcal{L}(\theta) = -\mu t_n - \frac{\alpha}{\beta} \sum_{i=1}^n \left(1 - e^{-\beta(t_n - t_i)} \right) + \sum_{i=1}^n \log\left[\mu + \alpha A_i(\beta) \right], \qquad (C.7)$$

in which $A_i(\beta) = \sum_{j:t_j < t_i} e^{-\beta(t_i - t_j)}$ for $i \geq 2$ and $A_1(\beta) = 0$. The gradient vector consists of

$$\frac{\partial \mathcal{L}}{\partial \mu} = -t_n + \sum_{i=1}^n \left(\frac{1}{\mu + \alpha A_i(\beta)} \right),$$

$$\frac{\partial \mathcal{L}}{\partial \alpha} = \frac{1}{\beta} \sum_{i=1}^n (e^{-\beta(t_n - t_i)} - 1) + \sum_{i=1}^n \frac{A_i(\beta)}{\mu + A_i(\beta)},$$

$$\frac{\partial \mathcal{L}}{\partial \beta} = -\alpha \sum_{i=1}^n \left[\frac{1}{\beta}(t_n - t_i)e^{-\beta(t_n - t_i)} + \frac{1}{\beta^2}e^{-\beta(t_n - t_i)} \right] + \sum_{i=1}^n \left(\frac{B_i(\beta)}{\mu + \alpha A_i(\beta)} \right),$$

where $B_i(\beta) = \sum_{j:t_j < t_i}(t_i - t_j)e^{-\beta(t_i - t_j)}$ for $i \geq 2$ and $B_1(\beta) = 0$. The Hessian matrix has entries

$$\frac{\partial^2 \mathcal{L}}{\partial \mu^2} = \sum_{i=1}^n \left(\frac{-1}{(\mu + \alpha A_i(\beta))^2} \right), \quad \frac{\partial^2 \mathcal{L}}{\partial \alpha^2} = -\sum_{i=1}^n \left(\frac{A_i(\beta)}{\mu + \alpha A_i(\beta)} \right)^2,$$

$$\frac{\partial^2 \mathcal{L}}{\partial \mu \partial \alpha} = \sum_{i=1}^n \left(\frac{-A_i(\beta)}{(\mu + \alpha A_i(\beta))^2} \right), \quad \frac{\partial^2 \mathcal{L}}{\partial \mu \partial \beta} = \sum_{i=1}^n \left(\frac{\alpha B_i(\beta)}{(\mu + \alpha A_i(\beta))^2} \right),$$

$$\frac{\partial^2 \mathcal{L}}{\partial \alpha \partial \beta} = -\sum_{i=1}^n \left[\frac{1}{\beta}(t_n - t_i)e^{-\beta(t_n - t_i)} + \frac{1}{\beta^2}\left(e^{-\beta(t_n - t_i)} - 1 \right) \right]$$

$$+ \sum_{i=1}^n \left[\frac{-B_i(\beta)}{\mu + \alpha A_i(\beta)} + \frac{\alpha A_i(\beta) B_i(\beta)}{(\mu + \alpha A_i(\beta))^2} \right],$$

$$\frac{\partial^2 \mathcal{L}}{\partial \beta^2} = \alpha \sum_{i=1}^{n} \left[\frac{1}{\beta}(t_n - t_i)^2 e^{-\beta(t_n - t_i)} + \frac{2}{\beta^2}(t_n - t_i) e^{-\beta(t_n - t_i)} \right.$$

$$\left. + \frac{2}{\beta^2}\left(e^{-\beta(t_n - t_i)} - 1\right)\right] + \sum_{i=1}^{n} \left[\frac{C_i(\beta)}{\mu + \alpha A_i(\beta)} - \left(\frac{\alpha B_i(\beta)}{\mu + \alpha A_i(\beta)}\right)^2 \right],$$

where $C_i(\beta) = \sum_{j:t_j < t_i}(t_i - t_j)^2 e^{-\beta(t_i - t_j)}$ for $i \geq 2$ and $C_1(\beta) = 0$. Note that $A_i(\beta)$ can be computed recursively by $A_i(\beta) = e^{-\beta(t_i - t_{i-1})}[1 + A_{i-1}(\beta)]$, $B_i(\beta)$ and $C_i(\beta)$ can also be computed by similar recursions.

A simple way to deal with the non-concavity issue is to fix the parameter β *a priori*, based on some estimate of intensity decay. With β fixed, the resulting log-likelihood function is then concave in the reduced parameter vector. Then optimization over the univariate parameter β is relatively straightforward. Other optimization procedures to maximize $\mathcal{L}$ have been suggested by Ozaki (1979), including the Davidon-Fletcher-Powell method.

Multivariate extensions

For the multivariate extension of (C.5), see Daley and Vere-Jones (2003, Prop. 7.3 III). In the special case of the bivariate Hawkes process with exponential kernel in Section 5.3.1, the log-likelihood function for the bivariate process can be written as $\mathcal{L}(\theta_1, \theta_2) = \mathcal{L}(\theta_1) + \mathcal{L}(\theta_2)$, which is the background of (5.4) and a similar formula for $\mathcal{L}(\theta_2)$.

Asymptotic normality of MLE and χ^2-approximation to GLR

Under certain regularity conditions, Ogata (1978) has shown that the classical asymptotic theory of likelihood inference (see Section 2.4 of Lai and Xing, 2008) also applies to stationary Hawkes processes:

$$\sqrt{n}\left(\hat{\theta} - \theta_0\right) \xrightarrow{\mathcal{D}} N\left(0, I^{-1}(\theta_0)\right), \qquad 2\left[\mathcal{L}\left(\hat{\theta}\right) - \mathcal{L}(\theta_0)\right] \xrightarrow{\mathcal{D}} \chi_d^2, \qquad \text{(C.8)}$$

where $\xrightarrow{\mathcal{D}}$ denotes convergence in distribution, θ_0 is the true parameter, $\hat{\theta}$ is the MLE, and $I(\theta_0)$ is the Fisher information matrix. The martingale central limit theorem (Appendix D) and martingale strong law (Section A.1) can be used to prove (C.8) since the underlying data generating mechanism is that of a counting process, instead of the traditional setting of an i.i.d. sample under which the asymptotic theory of MLE and GLR statistics was originally developed. In this time series setting, although the Fisher information matrix can still be defined in terms of the Hessian matrix of $\mathcal{L}(\theta)$ (Lai and Xing, 2008, p.61), $I(\theta)$ is not easy to compute and θ_0 is also an unknown parameter in (C.8). Hence one typically appeals to the law of large numbers to replace $I(\theta_0)$ by the observed Fisher information matrix $\nabla^2 \mathcal{L}(\hat{\theta})$ in (C.8) for likelihood inference; see Lai and Xing (2008, pp 58, 328 and 329).

Software implementation

For the two R packages to fit univariate point processes, which have been referred to in Section 4.6.2, ppstat fits an intensity process of the form $\lambda_\theta(t) = \phi(\boldsymbol{\alpha}^\mathsf{T}\mathbf{u}_t + \sum_{i:t_i < t} h_\beta(t - t_i))$, in which $\mathbf{u}_t$ is a vector of exogenous covariates at time t and $h_\beta(s) = \boldsymbol{\beta}^\mathsf{T}\mathbf{B}(s)$ where $\mathbf{B}(s)$ is a vector of prescribed basis functions. It uses Riemann sum approximation to the integral $- \int_0^{t_n} \lambda_\theta(s)ds$ in the log-likelihood function (C.5). NHPoisson assumes a log-linear intensity process

$$\log(\lambda_\beta(t)) = \boldsymbol{\beta}^\mathsf{T}\mathbf{x}_t, \tag{C.9}$$

in which $\mathbf{x}_t$ includes $t_i I_{\{t_i < t\}}$ and exogenous covariates $\mathbf{u}_t$. The package's outputs include the MLE of $\boldsymbol{\beta}$, the inverse of the observed information matrix, and the confidence intervals for the intensity at different times. It also performs variable selection and residual analysis.

C.4 Simulation of doubly stochastic SEPP

Lewis' method of thinning

A popular simulation procedure of point processes is the *thinning* algorithm proposed by Lewis and Shedler (1979) who originally designed their algorithm for simulating inhomogeneous Poisson processes. From the fact that Lewis' thinning algorithm is derived from the technique of rejection sampling[1], Ogata (1981) modifies the algorithm so that it can be used to simulate SEPP whose intensity is random. The following is the outline of the thinning algorithm in the form of recursion.

1. Suppose the set of event times $\{t_i : i = 0, \ldots, j\}$ has been simulated with $t_0 = 0$. To simulate t_{j+1}, let $s = t_j$ and $\delta > 0$ be a small positive number.

2. Let $\lambda_* = \max_{t \in (s, s+\delta)} \lambda(t)$.

3. Sample ξ from the exponential distribution with rate λ_*.

4. If $\lambda(s + \xi)/\lambda_* < 1$, go to Step 5. Otherwise, no event occurs in $(s, s + \delta)$; set $s := s + \delta$ and go to Step 2.

5. Sample U from the uniform distribution over $[0, 1]$.

6. If $\lambda(s + \xi)/\lambda_* \geq U$, then $t_{j+1} = s + \xi$, set $j := j + 1$ and go to Step 1. Otherwise, go to Step 3.

---

[1] The theory of rejection sampling is commonly discussed in the literature of simulation; see Section 5.2 of Ross (2015) for an introduction.

Note that δ can be chosen arbitrarily if $\lambda(t)$ is monotonically decreasing (except at event times) because $\lambda_* = \lambda(s)$. However, the choice of δ is crucial when the conditional intensity $\lambda(t)$ is monotonically increasing. If δ is too small, then λ_* will be relatively small and ξ could be so large that it becomes greater than $s + \delta$, resulting in many small intervals, each with a very low likelihood of including an event. If δ is too large, then λ_* will be relatively large and ξ can be quite small, causing many potential "events" to be thinned out. According to Harte (2010), the simulation algorithm in `PtProcess` addresses this by setting δ to be the 70th percentile of an exponential distribution with rate equal to $\lambda(s)$. Moreover, since the thinning algorithm is a rejection sampling procedure, its efficiency is also measured by the average number of iterations between Steps 3 and 6 for a successful simulation of t_{j+1}.

The above thinning algorithm is implemented in the `R` library of `PtProcess`; see Harte (2010) for other applications of `PtProcess` in statistical inference. The modified simulation procedure for SEPP is implemented in the `R` package `SAPP` which is the abbreviation of Statistical Analysis of Point Processes. `SAPP` is built from the Fortran 77 programs of Ogata et al. (2006) and is designed to study seismicity. Another `R` library `hawkes` can be used to simulate the multivariate Hawkes processes. For simulation of the Hawkes processes with time-varying baseline intensity discussed in Section 4.6.2, one can use the `R` package `IHSEP`[2].

Recent advances in the simulation of Hawkes processes

Møller and Rasmussen (2005) point out that the thinning algorithm suffers from "edge" effects which are essentially created by the dependency of the event in $(s, s + \delta)$ and the events in $(-\infty, s]$. They propose a perfect sampling scheme (i.e., sampling the whole process in an i.i.d. manner) by using a cluster-based approach with a branching structure. Also, to eliminate the edge effects, their algorithm starts from $t = -\infty$ and requires certain stationarity conditions; see also Møller and Rasmussen (2006). Another advancement is developed by Giesecke and Kim (2007) and Giesecke et al. (2011) who sample durations $t_i - t_{i-1}$ directly via their underlying analytic distribution functions. They successfully avoid generating intensity paths as well as introducing discretization bias. However, their algorithm could be computationally intensive due to evaluating the inverse of the distribution functions via Brent's method. Dassios and Zhao (2013) also derive another exact simulation algorithm by employing the cluster-based definition of Hawkes processes rather than the intensity-based definition that we stated in (4.50). Their definition allows them to use the implicit branching structure of Hawkes processes so that its

---

[2] `IHSEP` is the abbreviation of Inhomogeneous Self-Exciting Process and is contributed by the first author of Chen and Hall (2013). Other than model simulation, `IHSEP` also performs likelihood inference, calculates mean intensity process and analyzes point process residuals for the class of SEPPVB.

simulation can be achieved via a recursion of Poisson process simulations; see Dassios and Zhao (2013) for more details.

D

Weak Convergence and Limit Theorems for Queueing Systems

The classical Strong Law of Large Numbers (SLLN) and the Central Limit Theorem (CLT) describe the convergence of the scaled nth partial sum S_n/n and $S_n - n\mu/\sqrt{\sigma^2 n}$ respectively for an i.i.d. sequence $\{X_i\}_{i \geq 1}$ with $S_n = \sum_{i=1}^{n} X_i$, $E[X_i] = \mu$, $\text{Var}(X_i) = \sigma^2$. A stochastic process limit concerns the convergence of a sequence of stochastic processes, and in particular, the limit for the entire sequence of partial sums instead of the nth partial sum in CLT. If the stochastic process is viewed as a random function, then the stochastic process limit can be translated into the convergence of probability measures on a function space, which we describe below.

First, let X be a mapping from a probability space $(\Omega, \mathcal{F}, \mathbb{P})$ to a measurable space $(A, \mathcal{A})$. We call X a *random element* if for every $B \in \mathcal{A}$, $X^{-1}B \in \mathcal{F}$. Moreover, $P = \mathbb{P}X^{-1}$ is the probability measure on $(A, \mathcal{A})$ introduced by X. Let A be a metric space and $\mathcal{A}$ be its Borel σ-field (generated by the open sets), and let $\{X_n\}_{n \geq 1}$ be a sequence of random elements with values in A. $\{X_n\}_{n \geq 1}$ is said to converge weakly to X (or converge in distribution), denoted as $X_n \Rightarrow X$, if $Ef(X_n) \to Ef(X)$ as $n \to \infty$, for any bounded, continuous real-valued function f on A.

We consider the function space of sample paths of stochastic processes, with focus on the càdlàg type. A function on $[0, 1]$ is càdlàg if it is right-continuous and has left-hand limits. We denote by $D[0, 1]$ the set of all càdlàg functions from $[0, 1]$ to $\mathbb{R}$. Let Λ be the class of strictly increasing, continuous mappings of $[0, 1]$ onto itself. For $\lambda \in \Lambda$, define

$$||\lambda|| = \sup_{s \neq t} \left| \log \frac{\lambda(t) - \lambda(s)}{t - s} \right|. \tag{D.1}$$

Then the Skorokhod metric on $D[0, 1]$ is

$$d(x_1, x_2) \equiv \inf_{\lambda \in \Lambda} \{||x_1 \circ \lambda - x_2|| \vee ||\lambda - e||\},$$

where $a \vee b = \max\{a, b\}$ and $\circ$ denotes the composition of functions. Let $\mathcal{B}$ be the Borel σ-field on $D[0, 1]$ with the Skorokhod metric. Then a stochastic process X_t, $0 \leq t \leq 1$, with càdlàg sample paths is a $D[0, 1]$-valued random element. Hence one can define the weak convergence for a sequence of càdlàg stochastic processes on $[0, 1]$ in the space $D[0, 1]$ with the Skorokhod metric.

The space $D([0,1], \mathbb{R})$ can be generalized in two ways. First, the range of the functions could be generalized from $\mathbb{R}$ to $\mathbb{R}^k$; secondly, the domain of the functions can be generalized from $[0,1]$ to $[0, \infty)$; see Whitt (2002).

In this Appendix we describe some basic results on stochastic process limits and thereby providing some basic background material for Sections 5.5.1, 5.5.2 and C.3.

D.1   Donsker's theorem and its extensions

This is a fundamental theorem in stochastic process limits, also known as the Functional Central Limit Theorem (FCLT). It is on the convergence of a sequence of scaled partial sum of i.i.d. random variables to the Brownian motion in the function space D. It is also referred to as *Donsker's Invariance Principle* because the limit only depends on the first two moments of X_i.

Theorem 8. *Let $\{X_n\}_{n \geq 1}$ be a sequence of i.i.d. random variables with $E(X_1) = \mu$ and $\mathrm{Var}(X_1) = \sigma^2$, and define the nth normalized partial-sum process S_n by*

$$S_n(t) \equiv n^{-1/2} \left(\sum_{i=1}^{\lfloor nt \rfloor} X_i - \mu nt \right). \tag{D.2}$$

Then as $n \to \infty$,

$$S_n \Longrightarrow \sigma B, \qquad \text{in } D[0, \infty) \tag{D.3}$$

with the Skorokhod metric, where B is a standard Brownian motion, and $\lfloor nt \rfloor$ means the largest integer less than or equal to nt.

There are many extensions of Donsker's Theorem, including extension to the multidimensional case via the Cramér-Wold device, or extensions via relaxing the i.i.d. conditions by the so-called weak dependence conditions. One such weak dependence condition is the uniform mixing condition for a two-sided stationary sequence. Other alternatives to uniform mixing conditions for FCLT include ergodicity and martingale structure; see Bradley (2005), Billingsley (1968), Jacod and Shiryaev (1987), and Whitt (2002). A two-sided stationary sequence $\{X_n\}_{-\infty < n < \infty}$ is said to satisfy the uniform mixing condition if $\sum_{n=1}^{\infty} \rho_n < \infty$, where

$$\rho_n \equiv \sup \left\{ |E[XY]| : X \in \mathcal{F}_k, \ E(X) = 0, \ E(X^2) \leq 1, \right.$$
$$\left. Y \in \mathcal{G}_{n+k}, \ E(Y) = 0, \ E(Y^2) \leq 1 \right\},$$

with $\mathcal{F}_n = \sigma(X_k, k \leq n)$ (σ-field generated by $\{X_k, k \leq n\}$) and $\mathcal{G}_n = \sigma(X_k, k \geq n)$ (σ-field generated by $\{X_k, k \geq n\}$).

Theorem 9. *If $\{X_n\}_{-\infty < n < \infty}$ is a two-sided stationary sequence satisfying the uniform mixing condition and*

$$E(X_n) = m < \infty, \qquad E(X_n^2) < \infty, \tag{D.4}$$

then $\sigma^2 = \mathrm{Var}X_n + 2\sum_{k=1}^{\infty} \mathrm{Cov}(X_1, X_{1+k}) < \infty$ and (D.3) holds. Here σ^2 is called the asymptotic variance.

Another extension that is very useful for inference on stochastic processes is the martingale FCLT; see Durrett (2010).

Theorem 10. *Suppose $\{X_{n,m}, \mathcal{F}_{n,m}, 1 \le m \le n\}$ is a martingale difference array. Let $S_{n,k} = \sum_{m=1}^{k} X_{n,m}$, $V_{n,k} = \sum_{m=1}^{k} E(X_{n,m}^2 | \mathcal{F}_{n,m-1})$. Assume that as $n \to \infty$,*

(i) $V_{n,\lfloor nt \rfloor} \to t$ in probability for every $0 \le t \le 1$, and

(ii) $\sum_{m=1}^{k} E(X_{n,m}^2 I_{\{|X_{n,m}| > \delta\}} | \mathcal{F}_{n,m-1}) \to 0$ in probability for every $\delta > 0$.

Let $S_n(t) = S_{n,\lfloor nt \rfloor}$. Then (D.3) holds with $\sigma = 1$.

Condition (ii) in Theorem 10 is often referred as *conditional Lindeberg*. When $X_{n,1}, \ldots, X_{n,m}$ are independent zero-mean random variables such that $\sum_{m=1}^{k} E(X_{n,m}^2) = 1$, it reduces to the classical Lindeberg condition

$$\lim_{n \to \infty} \sum_{m=1}^{n} E(X_{n,m}^2 I_{\{|X_{n,m}| > \delta\}}) = 0 \text{ for every } \delta > 0, \tag{D.5}$$

which is necessary and sufficient for $S_{n,m}$ to have a limiting standard normal distribution.

D.2   Queuing system and limit theorems

Queuing theory, a stochastic approach to analyzing the congestions and delays of waiting in line, studies various quantities including the time spent waiting in line to be served, the arrival process, the service process, the number of servers, the number of system places, and the number of "customers" (which might be people, data packets, cars, etc.).

M/M/1 queue as a CTMC

The $M/M/1$ queue is the simplest queuing system where (a) there is one server and one type of job request, (b) the arrival of requests occurs according to a Poisson process with intensity λ, and (c) the service time is exponentially distributed with rate μ. Here a single server serves customers one at a time

from the front of the queue, based on the first-come-first-serve policy. The
customer leaves the queue once the service is complete, and the buffer is of
infinite size so there is no limit on the number of customers in the system. Let
$C(t)$ denote the number of customers waiting in the system at time t. It is
not difficult to see that it is a continuous-time Markov chain with a generator
matrix Q given by

$$
Q = \begin{pmatrix}
-\lambda & \lambda & & & \\
\mu & -(\lambda + \mu) & \lambda & & \\
& \mu & -(\lambda + \mu) & \lambda & \\
& & \mu & -(\lambda + \mu) & \lambda \\
& & & \ddots & \ddots & \ddots
\end{pmatrix}.
\tag{D.6}
$$

Heavy traffic limits for queues

If one is interested in the evolution of a queuing system over a time scale that is
much larger than that of inter-arrivals, the dynamics of appropriately rescaled
queue lengths can be described in terms of simple stochastic processes such as
Brownian motion. Mathematically, heavy traffic limits are various forms of the
Functional Strong Law of Large Numbers (FSLLN) and the Functional Central
Limit Theorem (FCLT); see Iglehart (1973a,b), Borovkov (1976), Harrison
(1985), and Whitt (2002).

In particular, consider the M/M/1 queuing system with the arrival rate
λ, the service rate μ, and with $C(t)$ the number of customers waiting in the
system at time t modeled as a continuous-time Markov chain. Now, let $\rho =
\lambda/\mu$. This system is considered to have *heavy traffic* if $\rho \uparrow 1$. In this case,
consider a rescaled queue

$$
C_\rho(t) = (1 - \rho)C\left(\frac{t}{(1 - \rho)^2}\right).
$$

The rescaled queue converges weakly in an appropriately defined topological
space to a reflected Brownian motion. That is,

$$
C_\rho \Longrightarrow B^R \qquad \text{in } D[0, T]
\tag{D.7}
$$

with the Skorokhod metric, where B^R is a reflected Brownian motion with a
drift parameter $\lambda - \mu$ and a variance parameter $\lambda + \mu$. This heavy traffic limit
can be generalized to more complex queuing systems with multiple servers,
and with general distributions for the arrival process and for the service times.
Of particular interest to LOB analysis are queues with reneging. A reneging
queue is a queue where the customers in line are assumed to be impatient, and
each individual leaves the queue according to some probability distribution.
For reneging queues, Ward and Glynn (2003, 2005) show that under some
regularity conditions, its diffusion limit is the reflected Ornstein-Uhlenbeck
process.

Bibliography

F. Abergel and A. Jedidi. Long-time behavior of a Hawkes process-based limit order book. *SIAM J. Finan. Math.*, 6(1):1026–1043, 2015.

Y. Achdou, F. J. Buera, J.-M. Lasry, P.-L. Lions, and B. Moll. Partial differential equation models in macroeconomics. *Phil. Trans. Roy. Soc. A*, 372 (2028):20130397, 2014.

H.-J. Ahn, K.-H. Bae, and K. Chan. Limit orders, depth, and volatility: Evidence from the stock exchange of Hong Kong. *J. Finance*, 56(2):767–788, 2001.

S. Ahuja, G. Papanicolaou, D. Ren, and F. Yang. Limit order trading with a mean-reverting reference price. Working paper, Department of Mathematics, Stanford University, 2016.

Y. Aït-Sahalia. Disentangling diffusion from jumps. *J. Finan. Econom.*, 74 (3):487–528, 2004.

Y. Aït-Sahalia and J. Jacod. Testing whether jumps have finite or infinite activity. *Ann. Stat.*, 39(3):1689–1719, 2011.

Y. Aït-Sahalia and J. Jacod. Analyzing the spectrum of asset returns: Jump and volatility components in high frequency data. *J. Economic Literature*, 50(4):1007–1050, 2012.

Y. Aït-Sahalia and J. Jacod. *High-Frequency Financial Econometrics*. Princeton University Press, Princeton, NJ, 2015.

Y. Aït-Sahalia, P. A. Mykland, and L. Zhang. How often to sample a continuous-time process in the presence of market microstructure noise. *Review of Financial Studies*, 18:351–416, 2005.

Y. Aït-Sahalia, J. Fan, and D. Xiu. High-frequency covariance estimates with noisy and asynchronous financial data. *J. Amer. Statist. Assoc.*, 105(492): 1504–1517, 2010.

J. Akahori, N.-L. Liu, M. E. Mancino, and Y. Yasuda. The Fourier estimation method with positive semi-definite estimators. *ArXiv e-prints*, 2014.

M. Akian, J. L. Menaldi, and A. Sulem. On an investment-consumption model with transaction costs. *SIAM J. Control Optimiz.*, 34(1):329–364, 1996.

S. S. Alexander. Price movements in speculative markets: Trends or random walks. *Industrial Management Review*, 2(2):7–26, 1961.

A. Alfonsi and P. Blanc. Dynamic optimal execution in a mixed-market-impact hawkes price model. *ArXiv e-prints*, 2014.

A. Alfonsi, A. Fruth, and A. Schied. Optimal execution strategies in limit order books with general shape functions. *Quantitative Finance*, 10(2): 143–157, 2010.

P. H. Algoet and T. M. Cover. Asymptotic optimality and asymptotic equipartition properties of log-optimum investment. *Ann. Probab.*, 16(2):876–898, 1988.

R. Almgren and N. Chriss. Value under liquidation. *Risk*, 12(12):61–63, 1999.

R. Almgren and N. Chriss. Optimal execution of portfolio transactions. *J. Risk*, 3:5–40, 2001.

R. Almgren, C. Thum, E. Hauptmann, and H. Li. Direct estimation of equity market impact. *Risk*, 3(2):5–39, 2005.

T. G. Andersen and T. Bollerslev. Intraday periodicity and volatility persistence in financial markets. *J. Empirical Finance*, 4:115–158, 1997.

C. S. Asness, T. J. Moskowitz, and L. H. Pedersen. Value and momentum everywhere. *J. Finance*, 68(3):929–985, 2013.

K. J. Åström. Theory and applications of adaptive control survey. *Automatica*, 19(5):471–486, 1983.

M. Avellaneda and S. Stoikov. High-frequency trading in a limit order book. *Quantitative Finance*, 8(3):217–224, 2008.

K. Azuma. Weighted sums of certain dependent random variables. *Tôhoku Math. J. (2)*, 19:357–367, 1967.

L. Bachelier. *Théorie de la Spéculation*. Gauthier-Villars, 1900.

L. Bachelier. *La Spéculation et le Calcul des Probabilités*. Gauthier-Villars, 1938.

K. Back and S. Baruch. Strategic liquidity provision in limit order markets. *Econometrica*, 81(1):363–392, 2013.

Z. Bai. Methodologies in spectral analysis of large dimensional random matrices, a review. *Statistica Sinica*, pages 611–662, 1999.

Z. Bai and J. W. Silverstein. No eigenvalues outside the support of the limiting spectral distribution of large-dimensional sample covariance matrices. *Ann. Probab.*, 26(1):316–345, 1998.

C. A. Ball and W. N. Torous. A simplified jump process for common stock returns. *J. Financial and Quantitative Analysis*, 18(1):53–65, 1983.

F. M. Bandi and J. R. Russell. Separating microstructure noise from volatility. *J. Finan. Econom.*, 79(3):655–692, 2006.

N. Barberis, M. Huang, and T. Santos. Prospect theory and asset prices. *Quarterly J. Economics*, 116:1–53, 2001.

M. Barigozzi and M. Hallin. Generalized dynamic factor models and volatilities: Recovering the market volatility shocks. *The Econometrics Journal*, 19(1):C33–C60, 2016.

O. E. Barndorff-Nielsen, P. R. Hansen, A. Lunde, and N. Shephard. Designing realized kernels to measure the ex-post variation of equity prices in the presence of noise. *Econometrica*, 76:1481–1536, 2008.

O. E. Barndorff-Nielsen, P. R. Hansen, A. Lunde, and N. Shephard. Multivariate realised kernels: Consistent positive semi-definite estimators of the covariation of equity prices with noise and non-synchronous trading. *J. Economet.*, 162(2):149–169, 2011.

A. G. Barto, R. S. Sutton, and C. W. Anderson. Neuronlike adaptive elements that can solve difficult learning control problems. *IEEE Trans. Systems, Man and Cybernetics*, SMC-13(5):834–846, 1983.

L. Bauwens and N. Hautsch. Modelling financial high frequency data using point processes. In *Handbook of Financial Time Series*, pages 953–979. Springer, Berlin Heidelberg, 2009.

D. S. Bayard. A forward method for optimal stochastic nonlinear and adaptive control. *IEEE Trans. Automat. Contr.*, 36(9):1046–1053, 1991.

D. Becherer, T. Bilarev, and P. Frentrup. Multiplicative limit order markets with transient impact and zero spread. *ArXiv e-prints*, 2015.

R. Bellman. *Dynamic Programming*. Princeton University Press, Princeton, NJ, 1st edition, 1957.

S. Benartzi and R. H. Thaler. Myopic loss aversion and the equity premium puzzle. *Quarterly J. Economics*, 110(1):73–92, 1995.

V. E. Benes, L. A. Shepp, and H. S. Witsenhausen. Some solvable stochastic control problems. *Stochastics*, 4(1):39–83, 1980.

A. Bensoussan and J.-L. Lions. *Impulse Control and Quasi-Variational Inequalities*. Dunod, 1982.

B. S. Bernanke and A. S. Blinder. The federal funds rate and the channels of monetary transmission. *The American Economic Review*, pages 901–921, 1992.

B. S. Bernanke, J. Boivin, and P. Eliasz. Measuring the effects of monetary policy: A factor-augmented vector autoregressive (favar) approach. *The Quarterly journal of economics*, 120(1):387–422, 2005.

D. Bertsimas and A. W. Lo. Optimal control of execution costs. *J. Financial Markets*, 1(1):1–50, 1998.

D. Bertsimas, A. W. Lo, and P. Hummel. Optimal control of execution costs for portfolios. *Computing in Science & Engineering*, 1(6):40–53, 1999.

D. Bertsimas, V. Gupta, and I. C. Paschalidis. Inverse optimization: A new perspective on the Black-Litterman model. *Operations Research*, 60(6): 1389–1403, 2012.

J. Besag. Spatial interaction and the statistical analysis of lattice systems. *J. Roy. Statist. Soc. Ser. B*, 36:192–236, 1974. With discussion by D. R. Cox, A. G. Hawkes, P. Clifford, P. Whittle, K. Ord, R. Mead, J. M. Hammersley, and M. S. Bartlett and with a reply by the author.

J. Besag. On the statistical analysis of dirty pictures. *J. Roy. Statist. Soc. Ser. B*, 48(3):259–302, 1986.

M. J. Best and R. R. Grauer. On the sensitivity of mean-variance-efficient portfolios to changes in asset means: Some analytical and computational results. *Review of Financial Studies*, 4(2):315–342, 1991.

B. Biais, P. Hillion, and C. Spatt. An empirical analysis of the limit order book and the order flow in the Paris Bourse. *J. Finance*, 50(5):1655–1689, 1995.

P. Billingsley. *Convergence of Probability Measures*. John Wiley & Sons, 1968.

F. Black and R. B. Litterman. Asset allocation: Combining investor views with market equilibrium. *J. Fixed Income*, 1(2):7–18, 1991.

J. Blanchet and X. Chen. Continuous-time Modeling of Bid-Ask Spread and Price Dynamics in Limit Order Books. *ArXiv e-prints*, 2013.

D. Bloembergen, K. Tuyls, D. Hennes, and M. Kaisers. Evolutionary dynamics of multi-agent learning: A survey. *J. Artificial Intelligence Research*, pages 659–697, 2015.

J. R. Blum. Multidimensional stochastic approximation methods. *Ann. Math. Stat.*, 25:737–744, 1954.

O. Bondarenko. Statistical arbitrage and securities prices. *Review of Financial Studies*, 16(3):875–919, 2003.

A. A. Borovkov. *Stochastic Processes in Queueing Theory*. Springer, 1976.

J.-P. Bouchaud, M. Mézard, and M. Potters. Statistical properties of stock order books: Empirical results and models. *Quantitative Finance*, 2(4): 251–256, 2002.

J.-P. Bouchaud, Y. Gefen, M. Potters, and M. Wyart. Fluctuations and response in financial markets: The subtle nature of 'random' price changes. *Quantitative Finance*, 4(2):176–190, 2004.

G. E. P. Box. Science and statistics. *J. Amer. Statist. Assoc.*, 71(356):791–799, 1976.

G. E. P. Box and N. R. Draper. *Empirical Model Building and Response Surfaces*. Wiley, Hoboken, NJ, 1987.

G. E. P. Box and G. M. Jenkins. *Time Series Analysis: Forecasting and Control*. Holden-Day, San Francisco, 1970.

G. E. P. Box and D. A. Pierce. Distribution of residual autocorrelations in autoregressive-integrated moving average time series models. *J. Amer. Statist. Assoc.*, 65(332):1509–1526, 1970.

R. C. Bradley. Basic properties of strong mixing conditions. a survey and some open questions. *Probability Surveys*, 2:107–144, 2005.

L. Breiman. Optimal gambling systems for favorable games. In *Proceedings of the Fourth Berkeley Symposium on Mathematical Statistics and Probability, Volume 1: Contributions to the Theory of Statistics*, pages 65–78, Berkeley, 1961. University of California Press.

L. Breiman. Statistical modeling: The two cultures. *Statist. Sci.*, 16(3):199–231, 2001. With invited comments and a rejoinder by the author.

M. J. Brennan. The role of learning in dynamic portfolio decisions. *European Finance Review*, 1(3):295–306, 1998.

M. Britten-Jones. The sampling error in estimates of mean-variance efficient portfolio weights. *J. Finance*, 54(2):655–671, 1999.

M. Broadie. Computing efficient frontiers using estimated parameters. *Ann. Oper. Res.*, 45(1):21–58, 1993.

W. Brock, J. Lakonishok, and B. LeBaron. Simple technical trading rules and the stochastic properties of stock returns. *J. Finance*, 47(5):1731–1764, 1992.

P. J. Brockwell and R. A. Davis. *Introduction to Time Series and Forecasting*. Springer, New York, 2nd edition, 2002.

T. Cai, A. Munk, and J. Schmidt-Hieber. Sharp minimax estimation of the variance of Brownian motion corrupted with Gaussian noise. *Statistica Sinica*, 20(3):1011–1024, 2010.

J. Y. Campbell, A. W. Lo, and A. C. MacKinlay. *The Econometrics of Financial Markets*. Princeton University Press, Princeton, NJ, 1997.

N. Canner, N. G. Mankiw, and D. N. Weil. An asset allocation puzzle. *The American Economic Review*, 87(1):181–191, 1997.

R. Carmona and F. Delarue. Probabilistic analysis of mean-field games. *SIAM J. Control Optimiz.*, 51(4):2705–2734, 2013.

R. Carmona and D. Lacker. A probabilistic weak formulation of mean field games and applications. *Ann. Appl. Probab.*, 25(3):1189–1231, 2015.

P. Carr and L. Wu. Time-changed Lévy processes and option pricing. *J. Finan. Econom.*, 71(1):113–141, 2004.

Á. Cartea, S. Jaimungal, and J. Ricci. Buy low, sell high: A high frequency trading perspective. *SIAM J. Finan. Math.*, 5(1):415–444, 2014.

G. Chamberlain and M. Rothschild. Arbitrage, factor structure, and mean-variance analysis on large asset markets. *Econometrica*, 51:1281–1304, 1983.

H. P. Chan and T. L. Lai. A general theory of particle filters in hidden Markov models and some applications. *Ann. Stat.*, 41(6):2877–2904, 2013.

H. P. Chan and T. L. Lai. MCMC with sequential state substitutions: Theory and applications. Working paper, Department of Statistics, Stanford University, 2016.

F. Chen and P. Hall. Inference for a nonstationary self-exciting point process with an application in ultra-high frequency financial data modeling. *J. Appl. Probab.*, 50(4):1006–1024, 2013.

H.-S. Chen, D. G. Simpson, and Z. Ying. Infill asymptotics for a stochastic process model with measurement error. *Statistica Sinica*, 10(1):141–156, 2000.

X. Chen and M. Dai. Characterization of optimal strategy for multiasset investment and consumption with transaction costs. *SIAM J. Finan. Math.*, 4(1):857–883, 2013.

V. K. Chopra, C. R. Hensel, and A. L. Turner. Massaging mean-variance inputs: Returns from alternative global investment strategies in the 1980s. *Management Science*, 39(7):845–855, 1993.

P. K. Clark. A subordinated stochastic process model with finite variance for speculative prices. *Econometrica*, 41(1):135–155, 1973.

R. Cont. Statistical modeling of high-frequency financial data. *Signal Processing Magazine, IEEE*, 28(5):16–25, 2011.

R. Cont and A. de Larrard. Order book dynamics in liquid markets: Limit theorems and diffusion approximations, 2012.

R. Cont and A. de Larrard. Price dynamics in a markovian limit order market. *SIAM J. Finan. Math.*, 4(1):1–25, 2013.

R. Cont and A. Kukanov. Optimal order placement in limit order markets, 2016. forthcoming.

T. M. Cover. Universal data compression and portfolio selection. In *Proceedings of the 37th Annual Symposium on Foundations of Computer Science*, pages 534–538. IEEE Computer Society, 1996.

T. M. Cover and E. Ordentlich. Universal portfolios with side information. *IEEE Trans. Information Theory*, 42(2):348–363, 1996.

A. Cowles. Can stock market forecasters forecast? *Econometrica*, 1(3):309–324, 1933.

A. Cowles. Stock market forecasting. *Econometrica*, 12(3/4):206–214, 1944.

D. R. Cox. Role of models in statistical analysis. *Statist. Sci.*, 5(2):169–174, 1990.

D. R. Cox and N. Reid. A note on pseudolikelihood constructed from marginal densities. *Biometrika*, 91(3):729–737, 2004.

J. C. Cox and C.-F. Huang. Optimal consumption and portfolio policies when asset prices follow a diffusion process. *J. Economic Theory*, 49(1):33–83, 1989.

J. Cvitanić, A. Lazrak, L. Martellini, and F. Zapatero. Dynamic portfolio choice with parameter uncertainty and the economic value of analysts' recommendations. *Review of Financial Studies*, 19(4):1113–1156, 2006.

M. Dai and Y. Zhong. Penalty methods for continuous-time portfolio selection with proportional transaction costs. *J. Computational Finance*, 13(3):1–31, 2010.

D. J. Daley and D. Vere-Jones. *An Introduction to the Theory of Point Processes Vol. I*. Springer-Verlag, New York, 2nd edition, 2003.

A. Dassios and H. Zhao. Exact simulation of Hawkes process with exponentially decaying intensity. *Electron. Commun. Probab.*, 18(62):1–13, 2013.

M. A. Davis, X. Guo, and G. L. Wu. Impulse controls for multi-dimensional jump diffusions. *SIAM J. Control Optimiz.*, 48(8):5276–5293, 2010.

M. H. Davis and A. R. Norman. Portfolio selection with transaction costs. *Mathematics of Operations Research*, 15(4):676–713, 1990.

M. H. A. Davis, V. G. Panas, and T. Zariphopoulou. European option pricing with transaction costs. *SIAM J. Control Optim.*, 31(2):470–493, 1993.

V. H. de la Peña, T. L. Lai, and Q.-M. Shao. *Self-Normalized Processes: Limit Theory and Statistical Applications*. Probability and Its Applications. Springer-Verlag, Berlin Heidelberg, 2009.

G. Demange, D. Gale, and M. Sotomayor. Multi-item auctions. *J. Political Economy*, 94(4):863–872, 1986.

M. A. Dempster, I. V. Evstigneev, and K. R. Schenk-Hoppé. Volatility-induced financial growth. *Quantitative Finance*, 7(2):151–160, 2007.

A. S. Deng, T. L. Lai, and H. Xing. *Data Science and Decision Analytics in Finance, Information Technology, and Manufacturing*. Wiley, Hoboken NJ, 2017.

J. B. Detemple. Asset pricing in a production economy with incomplete information. *J. Finance*, 41(2):383–391, 1986.

D. diBartolomeo. Smarter rebalancing: Using single period optimization in a multi-period world. Technical report, Northfield Research, 2012.

M. U. Dothan and D. Feldman. Equilibrium interest rates and multiperiod bonds in a partially observable economy. *J. Finance*, 41(2):369–382, 1986.

D. N. Dreman. *Contrarian Investment Strategy: The Psychology of Stock Market Success*. Random House, Inc., 1979.

D. N. Dreman. *Contrarian Investment Strategies: The Psychological Edge*. Simon and Schuster Co., 2012.

U. Drepper. What every programmer should know about memory. *Red Hat, Inc.*, 2007.

D. Duffie. *Dark Markets: Asset Pricing and Information Transmission in Over-the-Counter Markets*. Princeton University Press, 2012.

R. Durrett. *Probability: Theory and Examples*. Cambridge University Press, New York, NY, USA, 4th edition, 2010.

P. Dütting, T. Roughgarden, and I. Talgam-Cohen. Modularity and greed in double auctions. In *Proceedings of the Fifteenth ACM Conference on Economics and Computation*, pages 241–258, 2014.

E. Eberlein and S. Raible. Term structure models driven by general Lévy processes. *Mathematical Finance*, 9(1):31–53, 1999.

E. Eberlein, U. Keller, and K. Prause. New insights into smile, mispricing, and value at risk: The hyperbolic model. *J. Business*, 71(3):371–405, 1998.

B. Edelman, M. Ostrovsky, M. Schwarz, and others. Internet advertising and the generalized second-price auction: Selling billions of dollars worth of keywords. *American Economic Review*, 97(1):242–259, 2007.

N. El Karoui. Spectrum estimation for large dimensional covariance matrices using random matrix theory. *Ann. Stat.*, pages 2757–2790, 2008.

R. J. Elliott. *Stochastic Calculus and Applications*, volume 18. Springer-Verlag, New York, 1982.

R. F. Engle. Autoregressive conditional heteroskedasticity with estimates of the variance of U.K. inflation. *Econometrica*, 45:987–1007, 1982.

R. F. Engle. The econometrics of ultra-high frequency data. *Econometrica*, 68(1):1–22, 2000.

R. F. Engle and G. M. Gallo. A multiple indicators model for volatility using intra-daily data. *J. Econometrics*, 131(12):3–27, 2006.

R. F. Engle and J. R. Russell. Autoregressive conditional duration: A new model for irregularly spaced transaction data. *Econometrica*, 66(5):1127–1163, 1998.

R. F. Engle, J. Mezrich, and L. You. Optimal asset allocation. In *Smith-Barney Market Commentary*, 1998. January 28th issue.

R. F. Engle, R. Ferstenberg, and J. Russell. Measuring and modeling execution cost and risk. *J. Portfolio Management*, 38(2):14, 2012.

T. W. Epps. Comovements in stock prices in the very short run. *J. Amer. Statist. Assoc.*, 74(366):291–298, 1979.

E. F. Fama. Mandelbrot and the stable Paretian hypothesis. *J. Business*, 36:420–429, 1963.

E. F. Fama. Efficient capital markets: A review of theory and empirical work. *J. Finance*, 25(2):383–417, 1970.

E. F. Fama and M. E. Blume. Filter rules and stock-market trading. *J. Business*, 39(1):226–241, 1966.

E. F. Fama and K. R. French. Common risk factors in the returns on stocks and bonds. *J. Finan. Econom.*, 33(1):3–56, 1993.

E. F. Fama and K. R. French. A five-factor asset pricing model. *J. Finan. Econom.*, 116(1):1–22, 2015.

E. F. Fama and R. Roll. Parameter estimates for symmetric stable distributions. *J. Amer. Statist. Soc.*, 66(334):331–338, 1971.

J. Fan and Y. Wang. Spot volatility estimation for high-frequency data. *Statistics and its Interface*, 1(2):279–288, 2008.

J. Fan, Y. Li, and K. Yu. Vast volatility matrix estimation using high-frequency data for portfolio selection. *J. Amer. Statist. Assoc.*, 107(497): 412–428, 2012.

J. Fan, Y. Liao, and M. Mincheva. Large covariance estimation by thresholding principal orthogonal complements. *J. Roy. Statist. Soc. Ser. B*, 75(4):603–680, 2013.

J. D. Farmer, P. Patelli, and I. I. Zovko. The predictive power of zero intelligence in financial markets. *Proc. Nat. Acad. Sci. USA*, 102(6):2254–2259, 2005.

R. Fernholz and B. Shay. Stochastic portfolio theory and stock market equilibrium. *J. Finance*, 37(2):615–624, 1982.

L. Fisher. Some new stock-market indexes. *J. Business*, 39(1):191–225, 1966.

T. Fletcher, Z. Hussain, and J. Shawe-Taylor. Multiple kernel learning on the limit order book. In J. S.-T. Tom Diethe, Nello Cristianini, editor, *Journal of Machine Learning Research Workshop and Conference Proceedings*, volume 11, pages 167–74, 2010.

P. Fodra and M. Labadie. High-frequency market-making with inventory constraints and directional bets. *ArXiv e-prints*, 2012.

A. Földes and L. Rejtő. Strong uniform consistency for nonparametric survival curve estimators from randomly censored data. *Ann. Statist.*, 9(1):122–129, 1981.

M. Forni, M. Hallin, M. Lippi, and L. Reichlin. The generalized dynamic factor model: One-sided estimation and forecasting. *J. Amer. Statist. Assoc.*, 100 (471):830–840, 2005.

P. A. Forsyth, J. S. Kennedy, S. Tse, and H. Windcliff. Optimal trade execution: A mean quadratic variation approach. *J. Econom. Dynam. Contr.*, 36(12):1971–1991, 2012.

G. M. Frankfurter, H. E. Phillips, and J. P. Seagle. Performance of the Sharpe portfolio selection model: A comparison. *J. Financial and Quantitative Analysis*, 11(02):195–204, 1976.

R. S. Freeman. *Introduction to Financial Technology*. Complete Technology Guides for Financial Services. Elsevier, 2006.

D. Fudenberg, M. Mobius, and A. Szeidl. Existence of equilibrium in large double auctions. *J. Economic Theory*, 133(1):550–567, 2007.

K. Ganchev, Y. Nevmyvaka, M. Kearns, and J. W. Vaughan. Censored exploration and the dark pool problem. *Commun. ACM*, 53(5):99–107, 2010.

N. Gârleanu and L. H. Pedersen. Dynamic trading with predictable returns and transaction costs. *J. Finance*, 68(6):2309–2340, 2013.

H. Gatfaoui. Deviation from normality and Sharpe ratio behavior: A brief simulation study. *Investment Management and Financial Innovations*, 7 (4):95–107, 2010.

J. Gatheral. No-dynamic-arbitrage and market impact. *Quantitative Finance*, 10(7):749–759, 2010.

J. Gatheral and R. C. Oomen. Zero-intelligence realized variance estimation. *Finance and Stochastics*, 14(2):249–283, 2010.

J. Gatheral and A. Schied. Optimal trade execution under geometric brownian motion in the almgren and chriss framework. *Int. J. Theoret. Appl. Finance*, 14(03):353–368, 2011.

J. Gatheral and A. Schied. Dynamical models of market impact and algorithms for order execution. In J. A. L. Jean-Pierre Fouque, editor, *Handbook on Systemic Risk*, pages 579–599. Cambridge University Press, New York, 2013.

R. Ge and C. Huang. A continous approach to nonlinear integer programming. *Applied Mathematics and Computation*, 34:39–60, 1989.

I. M. Gelfand and S. V. Fomin. *Calculus of Variations*. Dover, 2000.

R. Gencay. Optimization of technical trading strategies and the profitability in security markets. *Economics Letters*, 59(2):249–254, 1998.

G. Gennotte. Optimal portfolio choice under incomplete information. *J. Finance*, 41(3):733–746, 1986.

G. Gennotte and A. Jung. Investment strategies under transaction costs: The finite horizon case. *Management Science*, 40(3):385–404, 1994.

K. Giesecke and B. Kim. Estimating tranche spreads by loss process simulation. In *Proceedings of the 39th Conference on Winter Simulation*, pages 967–975. IEEE Press, 2007.

K. Giesecke, B. Kim, and S. Zhu. Monte Carlo algorithms for default timing problems. *Management Science*, 57(12):2115–2129, 2011.

L. R. Glosten and P. R. Milgrom. Bid, ask and transaction prices in a specialist market with heterogeneously informed traders. *J. Finan. Econom.*, 14(1): 71–100, 1985.

M. Gorham and N. R. Singh. *Electronic Exchanges: The Global Transformation from Pits to Bits*. Elsevier, 2009.

M. D. Gould, M. A. Porter, S. Williams, M. McDonald, D. J. Fenn, and S. D. Howison. Limit order books. *Quantitative Finance*, 13(11):1709–1742, 2013.

B. Graham and D. L. Dodd. *Security Analysis*. McGraw-Hill Professional, New York, 1934.

R. C. Grinold. Implementation efficiency. *Finan. Analysts J.*, 61(5):52–64, 2005.

R. C. Grinold and R. N. Kahn. *Active Portfolio Management*. McGraw-Hill, New York, 2nd edition, 2000.

M. G. Gu and T. L. Lai. Functional laws of the iterated logarithm for the product-limit estimator of a distribution function under random censorship or truncation. *Ann. Probab.*, 18(1):160–189, 1990.

O. Guéant, J.-M. Lasry, and P.-L. Lions. Mean field games and applications. In *Paris-Princeton Lectures on Mathematical Finance 2010*, pages 205–266. Springer, 2011.

O. Guéant, C.-A. Lehalle, and J. Fernandez-Tapia. Optimal portfolio liquidation with limit orders. *SIAM J. Finan. Math.*, 3(1):740–764, 2012.

O. Guéant, C.-A. Lehalle, and J. Fernandez-Tapia. Dealing with the inventory risk: A solution to the market making problem. *Math. Finan. Econom.*, 7 (4):477–507, 2013.

F. Guilbaud and H. Pham. Optimal high-frequency trading with limit and market orders. *Quantitative Finance*, 13(1):79–94, 2013.

D. M. Guillaume, M. M. Dacorogna, R. R. Davé, U. A. Müller, R. B. Olsen, and O. V. Pictet. From the bird's eye to the microscope: A survey of new stylized facts of the intra-daily foreign exchange markets. *Finance and Stochastics*, 1(2):95–129, 1997.

X. Guo and M. Zervos. Optimal execution with multiplicative price impact. *SIAM J. Finan. Math.*, 6(1):281–306, 2015.

X. Guo, A. de Larrard, and Z. Ruan. Optimal placement in a limit order book: An analytical approach. Preprint, 2013.

X. Guo, Z. Ruan, and L. Zhu. Dynamics of order positions and related queues in a limit order book. Preprint, 2015.

J. Han, T. L. Lai, and V. Spivakovsky. Approximate policy optimization and adaptive control in regression models. *Computational Economics*, 27(4): 433–452, 2006.

P. R. Hansen. A test for superior predictive ability. *J. Business & Economic Statistics*, 23(4):365–380, 2005.

P. R. Hansen and A. Lunde. Realized variance and market microstructure noise. *J. Business & Economic Statistics*, 24(2):127–161, 2006.

P. R. Hansen, J. Large, and A. Lunde. Moving average-based estimators of integrated variance. *Econometric Reviews*, 27(1-3):79–111, 2008.

P. R. Hansen, Z. Huang, and H. H. Shek. Realized GARCH: A joint model for returns and realized measures of volatility. *J. Appl. Economet.*, 27(6): 877–906, 2012.

J. M. Harrison. *Brownian Motion and Stochastic Flow Systems*. Wiley Series in Probability and Mathematical Statistics. John Wiley & Sons, Inc., New York, 1985.

D. Harte. PtProcess: An R package for modelling marked point processes indexed by time. *J. Statist. Software*, 35(1):1–32, 2010.

J. Hasbrouck. Measuring the information content of stock trades. *J. Finance*, 46(1):179–207, 1991.

J. Hasbrouck. Modelling market microstructure time series. In G. S. Maddala and C. R. Rao, editors, *Handbook of Statistics, Vol. 14*, pages 647–692. North-Holland, 1996.

J. Hasbrouck. Trading fast and slow: Security market events in real time. Technical report, NYU Working Paper No. FIN-99-012, 1999.

J. Hasbrouck and G. Sofianos. The trades of market makers: An empirical analysis of NYSE specialists. *J. Finance*, 48(5):1565–1593, 1993.

T. Hastie, R. Tibshirani, and J. Friedman. *The Elements of Statistical Learning: Data Mining, Inference, and Prediction*. Springer Series in Statistics. Springer, New York, 2nd edition, 2009.

N. Hautsch. *Econometrics of Financial High-Frequency Data*. Springer-Verlag, Berlin Heidelberg, 2012.

A. G. Hawkes. Spectra of some self-exciting and mutually exciting point processes. *Biometrika*, 58(1):83–90, 1971.

T. Hayashi and N. Yoshida. On covariance estimation of non-synchronously observed diffusion processes. *Bernoulli*, 11(2):359–379, 2005.

T. Hayashi and N. Yoshida. Asymptotic normality of a covariance estimator for nonsynchronously observed diffusion processes. *Ann. Inst. Statist. Math.*, 60(2):367–406, 2008.

R. Haynes and J. S. Roberts. Automated trading in futures markets. Technical report, White Papers of the U.S. Commodity Futures Trading Commission, 2015.

D. Helbing and A. Kirman. Rethinking economics using complexity theory. *real-world economics review*, no. 64:23–51, 2013.

T. Hendershott and M. S. Seasholes. Market maker inventories and stock prices. *Amer. Econom. Review*, 97(2):210–214, 2007.

R. D. Henriksson and R. C. Merton. On market timing and investment performance. II. Statistical procedures for evaluating forecasting skills. *J. Business*, 54(4):513–533, 1981.

E. Hillebrand. Neglecting parameter changes in GARCH models. *J. Economet.*, 129(1):121–138, 2005.

T. Ho and H. R. Stoll. Optimal dealer pricing under transactions and return uncertainty. *J. Finan. Econom.*, 9(1):47–73, 1981.

M. Hoffmann, A. Munk, and J. Schmidt-Hieber. Adaptive wavelet estimation of the diffusion coefficient under additive error measurements. In *Annales de l'Institut Henri Poincaré, Probabilités et Statistiques*, volume 48, pages 1186–1216. Institut Henri Poincaré, 2012.

B. Hollifield, R. A. Miller, P. Sandås, and J. Slive. Estimating the gains from trade in limit-order markets. *J. Finance*, 61(6):2753–2804, 2006.

C. H. Hommes. Modeling the stylized facts in finance through simple nonlinear adaptive systems. *Proc. Nat. Acad. Sci. USA*, 99(Supplement 3):7221–7228, 2002.

U. Horst and D. Kreher. A weak law of large numbers for a limit order book model with fully state dependent order dynamics. *ArXiv e-prints*, 2015.

U. Horst and M. Paulsen. A law of large numbers for limit order books. *ArXiv e-prints*, 2015.

K. Huang, D. Simchi-Levi, and M. Song. Optimal market-making with risk aversion. *Operations Research*, 60(3):541–565, 2012.

M. Huang, P. E. Caines, and R. P. Malhamé. Large-population cost-coupled LQG problems with non-uniform agents: Individual-mass behavior and decentralized ε-Nash equilibria. *IEEE Trans. Automat. Contr.*, 52(9):1560–1571, 2007.

W. Huang, C.-A. Lehalle, and M. Rosenbaum. Simulating and analyzing order book data: The queue-reactive model. *J. Amer. Statist. Assoc.*, 110(509):107–122, 2015.

G. Huberman and W. Stanzl. Price manipulation and quasi-arbitrage. *Econometrica*, 72(4):1247–1275, 2004.

H. Hult and J. Kiessling. *Algorithmic trading with Markov chains*. PhD thesis, Stockholm University, Sweden, 2010.

D. L. Iglehart. Weak convergence in queueing theory. *Advan. Appl. Probab.*, 5(3):570–594, 1973a.

D. L. Iglehart. Weak convergence of compound stochastic process, I. *Stochastic Processes and their Applications*, 1(1):11–31, 1973b.

C.-K. Ing and T. L. Lai. A stepwise regression method and consistent model selection for high-dimensional sparse linear models. *Statistica Sinica*, pages 1473–1513, 2011.

J. Jacod and A. N. Shiryaev. *Limit Theorems for Stochastic Processes*, volume 288 of *Grundlehren der mathematischen Wissenschaften*. Springer-Verlag, Berlin Heidelberg, 1987.

J. Jacod, Y. Li, P. A. Mykland, M. Podolskij, and M. Vetter. Microstructure noise in the continuous case: The pre-averaging approach. *Stochastic Processes and Their Applications*, 119(7):2249–2276, 2009.

N. Jegadeesh and S. Titman. Returns to buying winners and selling losers: Implications for stock market efficiency. *J. Finance*, 48(1):65–91, 1993.

H. Jin and X. Y. Zhou. Behavioral portfolio selection in continuous time. *Mathematical Finance*, 18(3):385–426, 2008.

J. D. Jobson and B. Korkie. Estimation for Markowitz efficient portfolios. *J. Amer. Statist. Assoc.*, 75(371):544–554, 1980.

I. M. Johnstone and A. Y. Lu. On consistency and sparsity for principal components analysis in high dimensions. *J. Amer. Statist. Assoc.*, 104(486): 682–693, 2009.

P. Jorion. Bayes-Stein estimation for portfolio analysis. *J. Financial and Quantitative Analysis*, 21(03):279–292, 1986.

F. Jovanovic and P. Le Gall. Does God practice a random walk? The 'financial physics' of a nineteenth-century forerunner, Jules Regnault. *European Journal for the History of Economic Thought*, 8(3):323–362, 2001.

L. P. Kadanoff. More is the same; phase transitions and mean field theories. *J. Statist. Physics*, 137(5):777–797, 2009.

L. P. Kaelbling, M. L. Littman, and A. W. Moore. Reinforcement learning: A survey. *J. Artificial Intelligence Research*, 4:237–285, 1996.

D. Kahneman and A. Tversky. Prospect theory: An analysis of decision under risk. *Econometrica*, 47(2):263–291, 1979.

I. Karatzas. A class of singular stochastic control problems. *Advan. in Appl. Probab.*, 15(2):225–254, 1983.

I. Karatzas and S. E. Shreve. *Brownian Motion and Stochastic Calculus.* Springer-Verlag, New York, 2nd edition, 1991.

I. Karatzas, J. P. Lehoczky, and S. E. Shreve. Optimal portfolio and consumption decisions for a "small investor" on a finite horizon. *SIAM J. Control Optimiz.*, 25(6):1557–1586, 1987.

M. Kearns and Y. Nevmyvaka. Machine learning for market microstructure and high frequency trading. In D. Easley, M. de Prado, and M.O'Hara, editors, *High Frequency Trading—New Realities for Traders.* RiskBooks, 2013.

J. Kelly. A new interpretation of information rate. *Bell System Technical Journal*, 35(4):917–926, 1956.

M. Kendall. The analysis of economic time-series. Part I: Prices. *J. Roy. Statist. Soc. Ser. A*, 116(1):11–25, 1953.

J. Kiefer and J. Wolfowitz. Stochastic estimation of the maximum of a regression function. *Ann. Math. Stat.*, 23:462–466, 1952.

A. A. Kirilenko, A. S. Kyle, M. Samadi, and T. Tuzun. The Flash Crash: The impact of high-frequency trading on an electronic market, 2015.

V. Krishna. *Auction Theory.* Academic Press, New York, 2009.

M. Kritzman, S. Myrgren, and S. Page. Portfolio rebalancing: A test of the Markowitz-Van Dijk heuristic. MIT Sloan Research Paper, 2007.

Y. Kroll, H. Levy, and H. M. Markowitz. Mean-variance versus direct utility maximization. *J. Finance*, 39(1):47–61, 1984.

K. F. Kroner and J. Sultan. Time-varying distributions and dynamic hedging with foreign currency futures. *J. Financial and Quantitative Analysis*, 28 (04):535–551, 1993.

L. Kruk. Functional limit theorems for a simple auction. *Mathematics of Operations Research*, 28(4):716–751, 2003.

P. R. Kumar and P. Varaiya. *Stochastic Systems: Estimation, Identification, and Adaptive Control.* Prentice Hall, Englewood Cliffs, NJ, 1986.

H. Kushner. Necessary conditions for continuous parameter stochastic optimization problems. *SIAM J. Control*, 10(3):550–565, 1972.

A. S. Kyle. Continuous auctions and insider trading. *Econometrica*, 53(6): 1315–1335, 1985.

T. L. Lai. Sequential changepoint detection in quality control and dynamical systems. *J. Roy. Statist. Soc. Ser. B*, 57(4):613–658, 1995. With discussion and a reply by the author.

T. L. Lai. Stochastic approximation. *Ann. Stat.*, 31(2):391–406, 2003.

T. L. Lai and V. Bukkapatanam. Adaptive filtering, nonlinear state-space models, and applications to finance and econometrics. In Y. Zeng and S. Wu, editors, *State-Space Models: Applications in Economics and Finance*, Statistics and Econometrics for Finance, Vol. 1, pages 3–22. Springer, New York, 2013.

T. L. Lai and P. Gao. Approximate dynamic programming and multi-period mean-variance portfolio rebalancing with transactions costs. Working paper, Department of Statistics, Stanford University, 2016a.

T. L. Lai and P. Gao. Transaction costs and stochastic control in dynamic portfolio management. Working paper, Department of Statistics, Stanford University, 2016b.

T. L. Lai and J. Lim. Asymptotically efficient parameter estimation in hidden Markov spatio-temporal random fields. *Statist. Sinica*, 25(1):403–421, 2015.

T. L. Lai and T. W. Lim. Option hedging theory under transaction costs. *J. Econom. Dynam. Control*, 33(12):1945–1961, 2009.

T. L. Lai and K. W. Tsang. Post-selection multiple testing and a new approach to test-based variable selection. Working paper, Department of Statistics, Stanford University, 2016.

T. L. Lai and C. Z. Wei. Least squares estimates in stochastic regression models with applications to identification and control of dynamic systems. *Ann. Stat.*, 10(1):154–166, 1982.

T. L. Lai and C. Z. Wei. Extended least squares and their applications to adaptive control and prediction in linear systems. *IEEE Trans. Automat. Control*, 31(10):898–906, 1986.

T. L. Lai and S. P.-S. Wong. Valuation of American options via basis functions. *IEEE Trans. Automat. Contr.*, 49(3):374–385, 2004.

T. L. Lai and S. P.-S. Wong. Combining domain knowledge and statistical models in time series analysis. In *Time Series and Related Topics*, volume 52 of *IMS Lecture Notes Monogr. Ser.*, pages 193–209. Inst. Math. Statist., Beachwood, OH, 2006.

T. L. Lai and H. Xing. *Statistical Models and Methods for Financial Markets*. Springer Texts in Statistics. Springer, New York, 2008.

T. L. Lai and H. Xing. Stochastic change-point ARX-GARCH models and their applications to econometric time series. *Statist. Sinica*, 23(4):1573–1594, 2013.

T. L. Lai and H. Xing. *Risk Analytics and Management in Finance and Insurance*. Chapman & Hall/CRC, 2017. forthcoming.

T. L. Lai and Z. Ying. Estimating a distribution function with truncated and censored data. *Ann. Stat.*, 19(1):417–442, 1991a.

T. L. Lai and Z. Ying. Recursive identification and adaptive prediction in linear stochastic systems. *SIAM J. Control Optim.*, 29(5):1061–1090, 1991b.

T. L. Lai, M.-C. Shih, and S. P.-S. Wong. Flexible modeling via a hybrid estimation scheme in generalized mixed models for longitudinal data. *Biometrics*, 62(1):159–167, 317–318, 2006.

T. L. Lai, S. T. Gross, and D. B. Shen. Evaluating probability forecasts. *Ann. Statist.*, 39(5):2356–2382, 2011a.

T. L. Lai, H. Xing, and Z. Chen. Mean-variance portfolio optimization when means and covariances are unknown. *Ann. Appl. Statist.*, 5(2A):798–823, 2011b.

T. L. Lai, P. Gao, and L. Xu. Filtering approaches to dynamic portfolio selection in the presence of parameter uncertainty. Working paper, Department of Statistics, Stanford University, 2016a.

T. L. Lai, K. W. Tsang, and H. Yuan. Gradient boosting for high-dimensional nonlinear regression. Working paper, Department of Statistics, Stanford University, 2016b.

K. Lam and H. Yam. Cusum techniques for technical trading in financial markets. *Financial Engineering and the Japanese Markets*, 4(3):257–274, 1997.

C. G. Lamoureux and W. D. Lastrapes. Persistence in variance, structural change, and the GARCH model. *J. Business & Economic Statistics*, 8(2): 225–234, 1990.

S. Laruelle, C.-A. Lehalle, and G. Pages. Optimal split of orders across liquidity pools: A stochastic algorithm approach. *SIAM J. Finan. Math.*, 2 (1):1042–1076, 2011.

J.-M. Lasry and P.-L. Lions. Mean field games. *Japan. J. Math.*, 2(1):229–260, 2007.

H. A. Latane. Criteria for choice among risky ventures. *J. Political Economy*, 67(2):144–155, 1959.

G. Latouche. Level-independent quasi-birth-and-death processes. In J. J. Cochran, L. A. Cox, P. Keskinocak, J. P. Kharoufeh, and J. C. Smith, editors, *Wiley Encyclopedia of Operations Research and Management Science*. John Wiley & Sons, Inc., 2010.

P. Le Gall. *A History of Econometrics in France: From Nature to Models.* Routledge, 2007.

S. Le Roy. Efficient capital markets and martingales. *J. Econom. Literature*, 27:1583–1621, 1989.

O. Ledoit and M. Wolf. Improved estimation of the covariance matrix of stock returns with an application to portfolio selection. *J. Empirical Finance*, 10 (5):603–621, 2003.

O. Ledoit and M. Wolf. Honey, I shrunk the sample covariance matrix. *J. Portfolio Management*, 30(4):110–119, 2004.

O. Ledoit and M. Wolf. Nonlinear shrinkage estimation of large-dimensional covariance matrices. *Ann. Stat.*, 40(2):1024–1060, 2012.

C. M. Lee, B. Mucklow, and M. J. Ready. Spreads, depths, and the impact of earnings information: An intraday analysis. *Review of Financial Studies*, 6 (2):345–374, 1993.

S.-W. Lee and B. E. Hansen. Asymptotic theory for the GARCH(1,1) quasi-maximum likelihood estimator. *Econometric Theory*, 10(1):29–52, 1994.

B. N. Lehmann. Fads, martingales, and market efficiency. *Quarterly J. Econ.*, 105(1):1–28, 1990a.

E. L. Lehmann. Model specification: The views of Fisher and Neyman, and later developments. *Statist. Sci.*, 5(2):160–168, 1990b.

H. Leland. Optimal portfolio implementation with transactions costs and capital gains taxes. Working paper, Haas School of Business, University of California Berkeley, 2000.

D. Lepingle. La variation d'ordre p des semi-martingales. *Zeitschrift für Wahrscheinlichkeitstheorie und verwandte Gebiete*, 36(4):295–316, 1976.

H. Levy and H. M. Markowitz. Approximating expected utility by a function of mean and variance. *The American Economic Review*, 69(3):308–317, 1979.

J. Lewellen. Momentum and autocorrelation in stock returns. *Review of Financial Studies*, 15(2):533–564, 2002.

P. A. Lewis and G. S. Shedler. Simulation of nonhomogeneous Poisson processes by thinning. *Naval Research Logistics Quarterly*, 26(3):403–413, 1979.

D. Li and W.-L. Ng. Optimal dynamic portfolio selection: Multiperiod mean-variance formulation. *Mathematical Finance*, 10(3):387–406, 2000.

P.-L. Lions and J.-M. Lasry. Large investor trading impacts on volatility. In *Annales de l'institut Henri Poincaré (C) Analyse non linéaire*, volume 24, pages 311–323. Gauthier-Villars, 2007.

R. B. Litterman. The active risk puzzle. *J. Portfolio Management*, 30(5): 88–93, 2004.

R. B. Litterman. Beyond active alpha. In *CFA Institute Conference Proceedings Quarterly*, volume 25, pages 14–21. CFA Institute, 2008.

H. Liu. Optimal consumption and investment with transaction costs and multiple risky assets. *J. Finance*, 59(1):289–338, 2004.

N.-L. Liu and H.-L. Ngo. Approximation of eigenvalues of spot cross volatility matrix with a view toward principal component analysis. *ArXiv e-prints*, 2014.

G. M. Ljung and G. E. Box. On a measure of lack of fit in time series models. *Biometrika*, 65(2):297–303, 1978.

A. W. Lo. The adaptive markets hypothesis. *J. Portfolio Management*, 30(5): 15–29, 2004.

A. W. Lo and J. Hasanhodzic. *The Evolution of Technical Analysis–Financial Prediction From Babylonian Tablets to Bloomberg Terminals*. Bloomberg Press, 2010.

A. W. Lo and A. C. MacKinlay. When are contrarian profits due to stock market overreaction? *Review of Financial Studies*, 3(2):175–205, 1990.

A. W. Lo and A. C. MacKinlay. *A Non-Random Walk Down Wall Street*. Princeton University Press, 1999.

A. W. Lo and M. T. Mueller. Warning: Physics envy may be hazardous to your wealth! *J. Invest. Manage.*, 8(2):13–63, 2010.

A. W. Lo, H. Mamaysky, and J. Wang. Foundations of technical analysis: Computational algorithms, statistical inference, and empirical implementation. *J. Finance*, 55(4):1705–1770, 2000.

London Stock Exchange Group. Rebuild Order Book Service, February 2011.

F. A. Longstaff and E. S. Schwartz. Valuing American options by simulation: A simple least-squares approach. *Review of Financial Studies*, 14(1):113–147, 2001.

J. Loveless, S. Stoikov, and R. Waeber. Online algorithms in high-frequency trading. *Communications of the ACM*, 56(10):50–56, 2013.

D. G. Luenberger. *Investment Science*. Oxford University Press, Oxford, 2nd edition, 2013.

J. Ma. On the principle of smooth fit for a class of singular stochastic control problems for diffusions. *SIAM J. Control Optimiz.*, 30(4):975–999, 1992.

D. B. Madan and E. Seneta. The variance gamma (V.G.) model for share market returns. *J. Business*, 63(4):511–524, 1990.

D. B. Madan, P. P. Carr, and E. C. Chang. The variance gamma process and option pricing. *European Finance Review*, 2(1):79–105, 1998.

A. Madhavan and S. Smidt. An analysis of changes in specialist inventories and quotations. *J. Finance*, 48(5):1595–1628, 1993.

M. J. Magill and G. M. Constantinides. Portfolio selection with transactions costs. *J. Economic Theory*, 13(2):245–263, 1976.

B. G. Malkiel. *A Random Walk Down Wall Street: The Time-tested Strategy for Successful Investing*. W.W. Norton & Company, 2003.

P. Malliavin and M. E. Mancino. Fourier series method for measurement of multivariate volatilities. *Finance and Stochastics*, 6(1):49–61, 2002.

P. Malliavin and M. E. Mancino. A Fourier transform method for nonparametric estimation of multivariate volatility. *Ann. Stat.*, 37(4):1983–2010, 2009.

M. E. Mancino and M. C. Recchioni. Fourier spot volatility estimator: Asymptotic normality and efficiency with liquid and illiquid high-frequency data. *PloS One*, 10(9):e0139041, 2015.

M. E. Mancino and S. Sanfelici. Robustness of Fourier estimator of integrated volatility in the presence of microstructure noise. *Computational Statistics & Data Analysis*, 52(6):2966–2989, 2008.

M. E. Mancino and S. Sanfelici. Estimating covariance via Fourier method in the presence of asynchronous trading and microstructure noise. *J. Finan. Economet.*, 9(2):367–408, 2011a.

M. E. Mancino and S. Sanfelici. Multivariate volatility estimation with high-frequency data using Fourier method. In I. Florescu and F. Viens, editors, *Handbook of Modeling High-Frequency Data in Finance*, pages 243–294. John Wiley & Sons, Inc., New York, 2011b.

B. Mandelbrot. The variation of certain speculative prices. *J. Business*, 36(4):394–419, 1963.

B. Mandelbrot and H. M. Taylor. On the distribution of stock price differences. *Operations Research*, 15(6):1057–1062, 1967.

V. Marčenko and L. A. Pastur. Distribution of eigenvalues for some sets of random matrices. *Sbornik: Mathematics*, 1(4):457–483, 1967.

H. M. Markowitz and E. L. van Dijk. Single-period mean-variance analysis in a changing world (corrected). *Finan. Analysts J.*, 59(2):30–44, 2003.

R. Martin, S. Rachev, and F. Siboulet. Phi-alpha optimal portfolio and extreme risk management. In *The Best of Wilmott Vol 1*, pages 223–248. Wiley, Hoboken, NJ, 2005.

N. Matloff. *Parallel Computing for Data Science: With Examples in R, C++ and CUDA*. Chapman and Hall/CRC, 2015.

R. P. McAfee. A dominant strategy double auction. *J. Economic Theory*, 56 (2):434–450, 1992.

J. H. McCulloch. Simple consistent estimators of stable distribution parameters. *Communications in Statistics–Simulation and Computation*, 15(4): 1109–1136, 1986.

R. E. McCulloch and R. S. Tsay. Nonlinearity in high-frequency financial data and hierarchical models. *Studies in Nonlinear Dynamics & Econometrics*, 5(1):1–18, 2001.

R. C. Merton. Lifetime portfolio selection under uncertainty: The continuous-time case. *The Review of Economics and Statistics*, 51(3):247–257, 1969.

R. C. Merton. Optimum consumption and portfolio rules in a continuous-time model. *J. Economic Theory*, 3(4):373–413, 1971.

R. C. Merton. Option pricing when underlying stock returns are discontinuous. *J. Finan. Econom.*, 3(1-2):125–144, 1976.

R. C. Merton. *Continuous-Time Finance*. Macroeconomics and Finance Series. Basil Blackwell, 1990.

A. Meucci. *Risk and Asset Allocation*. Springer, 2005.

A. Meucci. Black–Litterman approach. In R. Cont, editor, *Encyclopedia of Quantitative Finance*. John Wiley & Sons, Ltd, 2010.

S. P. Meyn and R. L. Tweedie. *Markov Chains and Stochastic Stability*. Communications and Control Engineering Series. Springer-Verlag, 1993.

R. O. Michaud. *Efficient Asset Management*. Harvard Business School Press, 1989.

P. R. Milgrom. *Putting Auction Theory to Work*. Cambridge University Press, 2004.

M. H. Miller, J. Muthuswamy, and R. E. Whaley. Mean reversion of Standard & Poor's 500 Index basis changes: Arbitrage-induced or statistical illusion? *J. Finance*, 49(2):479–513, 1994.

C. C. Moallemi and K. Yuan. The value of queue position in a limit order book. Working paper, 2015.

C. C. Moallemi, B. Park, and B. Van Roy. The execution game. Preprint, 2009.

F. Modigliani and L. Modigliani. Risk-adjusted performance. *J. Portfolio Management*, 23(2):45–54, 1997.

J. Møller and J. G. Rasmussen. Perfect simulation of Hawkes processes. *Advan. Appl. Probab.*, 37(3):629–646, 2005.

J. Møller and J. G. Rasmussen. Approximate simulation of Hawkes processes. *Methodology and Computing in Applied Probability*, 8(1):53–64, 2006.

T. J. Moskowitz, Y. H. Ooi, and L. H. Pedersen. Time series momentum. *J. Finan. Econom.*, 104(2):228–250, 2012.

A. J. Motyer and P. G. Taylor. Decay rates for quasi-birth-and-death processes with countably many phases and tridiagonal block generators. *Adv. Appl. Probab.*, 38(2):522–544, 2006.

A. Munk, J. Schmidt-Hieber, et al. Nonparametric estimation of the volatility function in a high-frequency model corrupted by noise. *Electron. J. Stat.*, 4:781–821, 2010.

W. Murray and K.-M. Ng. An algorithm for nonlinear optimization problems withbinary variables. *Computational Optimization and Applications*, 47(2): 257–288, 2008.

W. Murray and U. V. Shanbhag. A local relaxation method for nonlinear facility location problems. *Multiscale Optimization Methods and Applications*, 82:173–204, 2006.

W. Murray and U. V. Shanbhag. A local relaxation approach for the siting of electrical substation. *Computational Optimization and Applications*, 38(3): 299–303, 2007.

W. Murray and H. Shek. A local relaxation method for the cardinality constrained portfolio optimization problem. *Computational Optimization and Applications*, 53(3):681–709, 2012.

J. Muthuswamy, S. Sarkar, A. Low, and E. Terry. Time variation in the correlation structure of exchange rates: High-frequency analyses. *J. Futures Markets*, 21(2):127–144, 2001.

R. B. Myerson and M. A. Satterthwaite. Efficient mechanisms for bilateral trading. *J. Economic Theory*, 29(2):265–281, 1983.

A. Nemirovski, A. Juditsky, G. Lan, and A. Shapiro. Robust stochastic approximation approach to stochastic programming. *SIAM J. Optimiz.*, 19 (4):1574–1609, 2009.

V. Niederhoffer and M. F. M. Osborne. Market making and reversal on the stock exchanges. *J. Amer. Statist. Assoc.*, 15(61):897–916, 1966.

N. Nisan and A. Ronen. Computationally feasible VCG mechanisms. *J. Artif. Intell. Res.*, 29:19–47, 2007.

A. A. Obizhaeva and J. Wang. Optimal trading strategy and supply/demand dynamics. *J. Financial Markets*, 16(1):1–32, 2013.

Y. Ogata. The asymptotic behaviour of maximum likelihood estimators for stationary point processes. *Ann. Inst. Statist. Math.*, 30:243–261, 1978.

Y. Ogata. On Lewis' simulation method for point processes. *IEEE Transactions on Information Theory*, 27(1):23–31, 1981.

Y. Ogata, K. Katsura, and J. Zhuang. TIMSAC84: Statistical analysis of series of events (TIMSAC84-SASE) version 2. *Computer Science Monographs*, no. 32, 2006.

B. Øksendal. *Stochastic Differential Equations: An Introduction with Applications*. Springer, New York, 6th edition, 2003.

A. Onatski. Determining the number of factors from empirical distribution of eigenvalues. *The Review of Economics and Statistics*, 92(4):1004–1016, 2010.

A. Onatski. Asymptotics of the principal components estimator of large factor models with weakly influential factors. *Journal of Econometrics*, 168(2): 244–258, 2012.

E. Ordentlich and T. M. Cover. The cost of achieving the best portfolio in hindsight. *Mathematics of Operations Research*, 23(4):960–982, 1998.

M. F. M. Osborne. Brownian motion in the stock market. *Operations Research*, 7:145–173, 1959.

T. Ozaki. Maximum likelihood estimation of hawkes' self-exciting point processes. *Ann. Inst. Statist. Math.*, 31(1):145–155, 1979.

E. Page. Continuous inspection schemes. *Biometrika*, 41(1/2):100–115, 1954.

J. Palczewski, R. Poulsen, K. R. Schenk-Hoppé, and H. Wang. Dynamic portfolio optimization with transaction costs and state-dependent drift. *European J. Oper. Res.*, 243(3):921–931, 2015.

S. Peng. A general stochastic maximum principle for optimal control problems. *SIAM J. Control Optimiz.*, 28(4):966–979, 1990.

A. F. Perold and W. F. Sharpe. Dynamic strategies for asset allocation. *Finan. Analysts J.*, 51(1):149–160, 1995.

M. H. Pesaran and A. Timmermann. A simple nonparametric test of predictive performance. *J. Business & Economic Statistics*, 10(4):461–465, 1992.

H. Pham. On some recent aspects of stochastic control and their applications. *Probab. Surveys*, 2:506–549, 2005.

S. Pliska. *Introduction to Mathematical Finance*. Blackwell Publishers, Oxford, 1997.

M. Potters and J.-P. Bouchaud. More statistical properties of order books and price impact. *Physica A: Statistical Mechanics and its Applications*, 324(1):133–140, 2003.

W. B. Powell. *Approximate Dynamic Programming: Solving the Curses of Dimensionality*. John Wiley & Sons, Inc., 2007.

W. B. Powell and I. O. Ryzhov. Optimal learning and approximate dynamic programming. In D. L. Frank L. Lewis, editor, *Reinforcement Learning and Approximate Dynamic Programming for Feedback Control*, IEEE Press Series on Computational Intelligence, pages 410–431. Wiley-IEEE Press, 2013.

D. J. Powers. Price Reporting and Dissemination; The Placement and Execution of Orders. In E. A. Gaumnitz, editor, *Futures Trading Seminar: A Commodity Marketing Forum for College Teachers of Economics*, volume III. Mimir Publishers, Inc., 1966.

S. Predoiu, G. Shaikhet, and S. Shreve. Optimal execution in a general one-sided limit-order book. *SIAM J. Finan. Math.*, 2(1):183–212, 2011.

S. J. Press. A compound events model for security prices. *J. Business*, 40(3): 317–335, 1967.

Principal Traders Group of The Futures Industry Association. Debunking the myths of high-frequency trading: Opinions published in response to Michael Lewis' Flash Boys and recent comments by NY Attorney General Eric Schneiderman, 2015.

S. Rachev and S. Mittnik. *Stable Paretian Models in Finance*. Wiley, Hoboken, NJ, 2000.

J. Regnault. *Calcul des Chances et Philosophie de la Bourse*. Mallet-Bachelier and Castel, 1863.

W. Ren. A mean-field LOB model. Ph.D. thesis, Chapter 3, Department of Mathematics, Stanford University, 2016.

D. Revuz and M. Yor. *Continuous Martingales and Brownian Motion*, volume 293 of *Grundlehren der Mathematischen Wissenschaften*. Springer-Verlag, Berlin, 3rd edition, 1999.

M. K. Richter and K.-C. Wong. Non-computability of competitive equilibrium. *Econom. Theory*, 14(1):1–27, 1999.

H. Robbins and S. Monro. A stochastic approximation method. *Ann. Math. Statist.*, pages 400–407, 1951.

R. Roll. A simple implicit measure of the effective bid-ask spread in an efficient market. *J. Finance*, 39:1127–1139, 1984.

J. P. Romano and M. Wolf. Stepwise multiple testing as formalized data snooping. *Econometrica*, 73(4):1237–1282, 2005.

S. A. Ross. The arbitrage theory of capital asset pricing. *J. Economic Theory*, 13(3):341–360, 1976.

S. M. Ross. *Simulation*. Academic Press, Inc., Orlando, FL, 5th edition, 2015.

I. Rosu. A dynamic model of the limit order book. *Review of Financial Studies*, 22:4601–4641, 2009.

A. Roth and M. A. O. Sotomayor. *Two-Sided Matching: A Study in Game-Theoretic Modeling and Analysis*. Econometric Society Monographs. Cambridge University Press, Cambridge, 1990.

T. H. Rydberg and N. Shephard. Dynamics of trade-by-trade price movements: Decomposition and models. *J. Finan. Economet.*, 1(1):2–25, 2003.

T. Sabel, J. Schmidt-Hieber, and A. Munk. Spot volatility estimation for high-frequency data: Adaptive estimation in practice. In *Modeling and Stochastic Learning for Forecasting in High Dimensions*, pages 213–241. Springer, 2015.

P. A. Samuelson. Proof that properly anticipated prices fluctuate randomly. *Industrial Management Review*, 6(2):41–49, 1965.

P. A. Samuelson. Lifetime portfolio selection by dynamic stochastic programming. *The Review of Economics and Statistics*, 51(3):239–246, 1969.

P. A. Samuelson. Mathematics of speculative price. *SIAM Review*, 15(1):1–42, 1973.

A. Schied and T. Schöneborn. Risk aversion and the dynamics of optimal liquidation strategies in illiquid markets. *Finance and Stochastics*, 13(2):181–204, 2009.

A. Schied, T. Schöneborn, and M. Tehranchi. Optimal basket liquidation for cara investors is deterministic. *Appl. Mathemat. Finance*, 17(6):471–489, 2010.

M. Scholes and J. Williams. Estimating betas from nonsynchronous data. *J. Finan. Econom.*, 5(3):309–327, 1977.

H. H. Shek. *Statistical and Algorithm Aspects of Optimal Portfolios*. PhD thesis, Stanford University Institute of Computational and Mathematical Engineering, 2010.

N. Shephard and K. Sheppard. Realising the future: Forecasting with high-frequency-based volatility (HEAVY) models. *J. Appl. Economet.*, 25(2): 197–231, 2010.

Y. Shetty and S. Jayaswal. *Practical .NET for Financial Markets*. Apress, 2006.

R. J. Shiller. *Irrational Exuberance*. Princeton University Press, 3rd edition, 2015.

D. Silver, A. Huang, C. J. Maddison, A. Guez, L. Sifre, G. van den Driessche, J. Schrittwieser, I. Antonoglou, V. Panneershelvam, M. Lanctot, S. Dieleman, D. Grewe, J. Nham, N. Kalchbrenner, I. Sutskever, T. Lillicrap, M. Leach, K. Kavukcuoglu, T. Graepel, and D. Hassabis. Mastering the game of Go with deep neural networks and tree search. *Nature*, 529(7587): 484–489, 2016.

Y. Simaan. Estimation risk in portfolio selection: The mean variance model versus the mean absolute deviation model. *Management Science*, 43(10): 1437–1446, 1997.

C. A. Sims. Interpreting the macroeconomic time series facts: The effects of monetary policy. *European Economic Review*, 36(5):975–1000, 1992.

E. Smith, J. Doyne Farmer, L. Gillemot, and S. Krishnamurthy. Statistical theory of the continuous double auction. *Quantitative Finance*, 3:481–514, 2003.

M. L. Stein. A comparison of generalized cross validation and modified maximum likelihood for estimating the parameters of a stochastic process. *Ann. Statist.*, 18(3):1139–1157, 1990.

M. L. Stein. *Interpolation of Spatial Data: Some Theory for Kriging*. Springer Series in Statistics. Springer, New York, 1999.

J. H. Stock and M. W. Watson. Forecasting using principal components from a large number of predictors. *J. Amer. Statist. Assoc.*, 97(460):1167–1179, 2002.

J. H. Stock and M. W. Watson. Dynamic factor models. *Oxford handbook of economic forecasting*, 1:35–59, 2011.

S. Stoikov and M. Saglam. Option market making under inventory risk. *Review of Derivatives Research*, 12:55–79, 2009.

R. Sullivan, A. Timmermann, and H. White. Data-snooping, technical trading rule performance, and the bootstrap. *J. Finance*, 54(5):1647–1691, 1999.

R. S. Sutton. Learning to predict by the methods of temporal differences. *Machine Learning*, 3(1):9–44, 1988.

V. N. Temlyakov. Weak greedy algorithms. *Advances in Computational Mathematics*, 12(2):213–227, 2000.

R. Tibshirani. Regression shrinkage and selection via the lasso. *J. Roy. Statist. Soc. Ser. B*, pages 267–288, 1996.

J. A. Tropp and A. C. Gilbert. Signal recovery from random measurements via orthogonal matching pursuit. *IEEE Trans. Inform. Theory*, 53(12): 4655–4666, 2007.

R. Tsay. *Analysis of Financial Time Series*. Wiley, New Jersey, 3rd edition, 2010.

J. N. Tsitsiklis and B. Van Roy. Optimal stopping of Markov processes: Hilbert space theory, approximation algorithms, and an application to pricing high-dimensional financial derivatives. *IEEE Trans. Automat. Contr.*, 44(10): 1840–1851, 1999.

G. Tsoukalas, J. Wang, and K. Giesecke. Dynamic portfolio execution, 2014.

H. R. Varian. Position auctions. *Int. J. Indust. Organiz.*, 25(6):1163–1178, 2007.

W. Vickrey. Counterspeculation, auctions, and competitive sealed tenders. *J. Finance*, 16(1):8–37, 1961.

V. Voev and A. Lunde. Integrated covariance estimation using high-frequency data in the presence of noise. *J. Finan. Economet.*, 5(1):68–104, 2007.

A. R. Ward and P. W. Glynn. A diffusion approximation for a markovian queue with reneging. *Queueing Systems*, 43(1-2):103–128, 2003.

A. R. Ward and P. W. Glynn. A diffusion approximation for a GI/GI/1 queue with balking or reneging. *Queueing Systems*, 50(4):371–400, 2005.

S. Watanabe. On discontinuous additive functionals and Lévy measures of a Markov process. *Japanese J. Math.*, 34:53–70, 1964.

C. J. Watkins and P. Dayan. Q-learning. *Machine Learning*, 8(3-4):279–292, 1992.

D. Weaver. Minimum obligations of market makers. Technical report, UK Government Foresight Project, Future of Computer Trading in Financial Markets, 2012. Economic Impact Assessment EIA7.

P. Weber and B. Rosenow. Order book approach to price impact. *Quantitative Finance*, 5(4):357–364, 2005.

P. Weiss. L'hypothèse du champ moléculaire et la propriété ferromagnétique. *J. Phys. Theor. Appl.*, 6(1):661–690, 1907.

H. White. Maximum likelihood estimation of misspecified models. *Econometrica*, 50(1):1–25, 1982.

H. White. A reality check for data snooping. *Econometrica*, 68(5):1097–1126, 2000.

W. Whitt. *Stochastic-process Limits: An Introduction to Stochastic-process Limits and Their Application to Queues*. Springer, 2002.

E. P. Wigner. Characteristic vectors of bordered matrices with infinite dimensions. *Annals of Mathematics*, pages 548–564, 1955.

J. T. Williams. Capital asset prices with heterogeneous beliefs. *J. Finan. Econom.*, 5(2):219–239, 1977.

A. S. Willsky and H. L. Jones. A generalized likelihood ratio approach to the detection and estimation of jumps in linear systems. *IEEE Trans. Automat. Contr.*, 21(1):108–112, 1976.

M. Wooldridge. *An Introduction to Multiagent Systems*. John Wiley & Sons, 2009.

Y. Xia. Learning about predictability: The effects of parameter uncertainty on dynamic asset allocation. *J. Finance*, 56(1):205–246, 2001.

D. Xiu. Quasi-maximum likelihood estimation of volatility with high-frequency data. *J. Economet.*, 159:235–250, 2010.

V. I. Zakamouline. A unified approach to portfolio optimization with linear transaction costs. *Mathematical Methods of Operations Research*, 62(2): 319–343, 2005.

L. Zhang. Estimating covariation: Epps effect, microstructure noise. *J. Econometrics*, 160(1):33–47, 2011.

L. Zhang, P. A. Mykland, and Y. Aït-Sahalia. A tale of two time scales: Determining integrated volatility with noisy high frequency data. *J. Amer. Statist. Assoc.*, 100(472):1394–1411, 2005.

B. Zheng, E. Moulines, and F. Abergel. Price jump prediction in a limit order book. *J. Math. Finance*, 3(2):242–255, 2013.

B. Zhou. High-frequency data and volatility in foreign exchange rates. *J. Business & Econom. Stat.*, 14(1):45–52, 1996.

X. Y. Zhou and D. Li. Continuous-time mean-variance portfolio selection: A stochastic LQ framework. *Appl. Math. Optimiz.*, 42(1):19–33, 2000.

H. Zou and T. Hastie. Regularization and variable selection via the elastic net. *J. Roy. Statist. Soc. Ser. B*, 67(2):301–320, 2005.

Index

Printed in the United States
by Baker & Taylor Publisher Services